D1074326

Geometry and Symmetry

HOLDEN–DAY SERIES IN MATHEMATICS

Earl A. Coddington and Andrew M. Gleason, Editors

Geometry and Symmetry

Paul B. Yale
Pomona College

Holden–Day
San Francisco, Cambridge, London, Amsterdam

Library of Congress Catalog Card Number: 67–28042
Printed in the United States of America

To my wife

Preface

This book is an introduction to the geometry of euclidean, affine, and projective spaces with special emphasis on the important groups of symmetries of these spaces. The two major objectives of the text are to introduce the main ideas of affine and projective spaces and to develop facility in handling transformations and groups of transformations. Since there are many good texts on affine and projective planes, I have concentrated on the n-dimensional ($n \geq 3$) cases.

There is at the present time a resurgence of interest in "bringing the undergraduate geometry course up to date." This book is my nomination for one semester of that course. I believe that it is ideal preparation for a modern approach to differential geometry, algebraic geometry, the classical groups, or (if augmented by a course in topology) algebraic topology. In all of these areas facility in juggling transformations and a thorough knowledge of the classical geometric spaces are essential. With the exception of Chapter 3, my guiding principle has been to restrict my attention to those topics and techniques in geometry that seem to me to be of maximal use in all areas of mathematics. I do not claim to cover all such topics (convexity is the most notable exception), but I do claim to avoid long excursions into those areas of geometry (e.g., the axiomatics of projective planes) which, while fascinating, have few direct applications.

Chapters 1 and 2 form the introductory portion of the book. In these two chapters groups of transformations are presented in two intuitive settings: combinatorics and euclidean space. In both chapters the geometric significance of cosets and conjugacy is stressed. In Chapter 1 this view of cosets leads to a natural interpretation of Lagrange's theorem and provides the basis for the proof of the Polya-Burnside theorem. In Chapter 2 a careful study of the geometric implications of conjugacy within the similarities group prepares the way for the high point of the chapter, a proof that the similarities group is isomorphic in the natural way to its own group of automorphisms.

The introduction to crystallography in Chapter 3 is a natural sequel to Chapter 2. I have tried to present an accurate treatment of enough basic material to enable the reader to proceed on his own in the extensive technical literature in this field. This chapter is a digression from the main stream of the book and may be omitted or simply used as a source of examples for Chapters 1 and 2.

To complete the prerequisites for the study of affine and projective spaces, Chapter 4 is devoted to a quick review of fields and vector spaces. Once again (see the discussion of similar matrices) the connection between "conjugates in a group" and "the same type of geometric transformation" is stressed.

Chapters 5 and 6 contain parallel presentations of affine and projective spaces. The interconnections among vector spaces, affine spaces, and projective spaces are emphasized. At the end of Chapter 5 there are several interesting results about volume, lattices, and collineations in affine spaces over the field of real numbers.

The prerequisites for a course using this text depend on how rapidly one presents the material in the early parts of Chapters 1 and 4. At least a cursory knowledge of groups, fields, and vector spaces is required. In view of the present fluid state of the mathematics curriculum, it seemed wise to include, at least in condensed form, essentially all the material needed in these three areas of algebra. By assigning some sections as outside reading and covering others slowly and thoroughly, the instructor may tailor the "review" material on groups, fields, and vector spaces to the particular background of his class.

I am very grateful to Pat Kelly and Donna Beck for the typing they did so graciously and competently. I am also grateful to Andrew Gleason, one of the editors of the Holden-Day Series in Mathematics, for his painstaking work reading the manuscript and suggesting numerous improvements in the exposition. I found his use of a tape recording to communicate these suggestions very efficient and feel that this technique should be used more often.

I greatly appreciate any notes readers send to me suggesting improvements, calling my attention to better examples, pointing out errors, or merely commenting on the exposition.

Paul B. Yale

Contents

Chapter 1: Algebraic and Combinatoric Preliminaries

1.1. Basic notions about sets and groups 1
1.2. The algebra of permutations and examples of groups . . 8
1.3. Homomorphisms and permutation representations of groups 16
1.4. Automorphisms of a group 24
1.5. Cosets and orbits, Lagrange's theorem 29
1.6. The Polya–Burnside theorem 37
Bibliography and suggestions for further reading 43

Chapter 2: Isometries and Similarities: An Intuitive Approach

2.1. Isometries and similarities 46
2.2. Involutions in **S** 52
2.3. The classification of isometries 57
2.4. Geometric implications of conjugacy in **S** 65
2.5. An exercise: Isometries in the plane 67
2.6. Dilatations and spiral similarities 68
2.7. Automorphisms of **E** *and* **S** 72
2.8. Homomorphisms of **E** *and* **S** 76
Bibliography and suggestions for further reading 83
Review problems for Chapters 1 and 2 84

Chapter 3: An Introduction to Crystallography

3.1. Discrete groups of isometries 86
3.2. Finite groups of isometries 89
3.3. Lattices and lattice groups 99
3.4. Crystallographic point groups 103
3.5. The seven crystal systems 108
3.6. Crystallographic space groups 113
3.7. Generalizations 117
Bibliography and suggestions for further reading 118

Chapter 4: Fields and Vector Spaces: A Quick Review

4.1. Fields 120
4.2. Vector spaces, subspaces, echelon form 125
4.3. Linear transformations 132
4.4 Coordinate mappings, matrices for linear transformations . 138
4.5. Similar matrices and commutative diagrams 142
4.6. Applications of matrix similarity 149
4.7. Symmetries of $V_n(\mathbf{F})$, $GL_n(\mathbf{F})$ 154
Bibliography and suggestions for further reading 157

Chapter 5: Affine Spaces

5.1. Axioms for affine spaces 159
5.2. Affine subspaces 161
5.3. Parallel and skew subspaces 167
5.4. Affine coordinates 169
5.5. Affine symmetries I: Dilatations 171
5.6. Affine symmetries II: Affine transformations 174
Term-paper topics 180
5.7. The analytic representation of affine transformations . 180
5.8. Affine symmetries III: Collineations 184
5.9. Volume in real affine spaces 188
5.10. Lattices in real affine spaces 191
5.11. Collineations in real affine spaces 196
Bibliography and suggestions for further reading 199

Chapter 6: Projective Spaces

6.1. Extended affine spaces and collapsed vector spaces . . 201
6.2. Projective subspaces 206

6.3. Projective planes 210
6.4. Homogeneous coordinates 214
6.5. Projective symmetries I: Perspectivities 224
6.6. Projective symmetries II: Projective transformations . . 234
6.7. Projective symmetries III: Collineations 244
6.8. Dual spaces and the principle of duality 252
6.9. Correlations and semi-bilinear forms 257
6.10. Quadrics and polarities 262
6.11. Real projective spaces 271
6.12. Projective spaces over noncommutative fields . . . 275
Bibliography and suggestions for further reading 278
Term-paper topics 280
Index 281
Index of notation 286

Chapter 1

Algebraic and Combinatoric Preliminaries

This chapter is devoted entirely to the algebraic and combinatoric tools that we use in our study of geometry and symmetry. These tools are centered around the concepts of set, function, and group. A thorough understanding of these concepts and facility in using the associated notation are essential in order to read the rest of the book.

1.1. Basic notions about sets and groups

To avoid ambiguity we state our notational conventions for sets even though they conform to the standard conventions. As usual $x \in X$ means that x is an element of the set X; $\{x_1, x_2, \ldots, x_n\}$ denotes the set whose elements are $x_1, x_2, \ldots, x_n$; and $\{x|$ (some conditions on x)$\}$ denotes the set of all objects satisfying the given conditions. It is well known that one can get into trouble with this last notation, e.g., $X = \{x|x$ is a set and $x \notin x\}$ is at the root of Russell's paradox. (Does $X \in X$?) We avoid trouble by being reasonably cautious about the collections of objects that we agree to call sets. In fact, we usually start with a well-known set X and form a subset of X by imposing requirements for membership. In this situation we often write $\{x \in X|$ (conditions on x)$\}$ in place of $\{x|x \in X$ and (conditions on x)$\}$.

The standard symbols, $\cup$ and $\cap$, are used for set unions and intersections. Just as long sums and products are abbreviated using the Σ and Π notations, so we abbreviate long unions or intersections; namely, $\bigcup_{A \in \mathfrak{F}} A$ and $\bigcap_{A \in \mathfrak{F}} A$ denote the union and intersection, respectively, of all sets A in the family of sets, $\mathfrak{F}$. If $\mathfrak{F}$ is empty, then $\bigcup_{A \in \mathfrak{F}} A$ is also the empty set, but for practical reasons we avoid using $\bigcap_{A \in \mathfrak{F}} A$ if $\mathfrak{F}$ is empty.

For subset relations we use $X \subset Y$ or $Y \supset X$ to mean "X is a *proper* subset of Y" and use $X \subseteq Y$ or $Y \supseteq X$ to mean "X is a subset of Y which *may* equal Y." The empty set, which we denote by $\varnothing$, is a subset of any set X, i.e., $\varnothing \subseteq X$. Note that the two statements, "$\varnothing \neq X$" and "$\varnothing \subset X$," are equivalent.

For any finite set X, $\#(X)$ denotes the number of elements in X. Thus $\#(\varnothing) = 0$, $\#\{a\} = 1$, etc. The extension of this notation to infinite sets is used in the sense of *cardinality*, i.e., $\#(X) = \#(Y)$ means that there is a one-to-one correspondence between the elements of the two sets.

We often consider ordered n-tuples of elements chosen from a set X. The ordered n-tuple with ith component x_i is denoted by $(x_1, x_2, \ldots, x_n)$. Thus the ordered pair (a, b) has first component a and second component b. Two other common types of brackets used for ordered n-tuples, $\langle\,\rangle$ and $[\,]$, are reserved for special uses (vectors and projective coordinates, respectively).

A *function*, f, from X into Y is defined formally as a set of ordered pairs with first component in X and second component in Y such that: (1) no two ordered pairs in f have the same first component but different second components, i.e., f is single-valued; and (2) each element of X appears once as the first component of an ordered pair in f. Of more importance in reading this book is the intuitive idea that a function f from X into Y is determined whenever we have an unambiguous rule assigning to each element $x \in X$ a definite element $y \in Y$. Often, notably in calculus courses, one denotes the object y assigned to x by $f(x)$; however, since we want products of functions (defined below) to read in their natural order, we *never* use this notation. In order to emphasize its importance we state our convention as a formal definition.

Definition. If f is a function from X into Y and if $(x, y) \in f$, then we say that y is the *image* of x under f and write $(x)f = y$. In special situations we may replace this by $x^f = y$ or $x_f = y$ in order to avoid too many parentheses. If Z is a subset of X, then $(Z)f$ (respectively, Z^f or Z_f), called the *image* of Z under f, is the set of all images of elements of Z under f, i.e., $(Z)f = \{(z)f \mid z \in Z\}$.

As usual, if f is a function from X into Y, then we call X the *domain* and Y the *codomain* of f. The *range* of a function is the image of its domain. The words *mapping* and *transformation* are used as synonyms for function.

Three special types of functions occur often enough to merit special definitions. First, we distinguish carefully between the words "into" and "onto" and say f is a function from X *onto* Y only if the image of X under f is Y itself, not some proper subset of Y. Second, we say that a function is *one to one*, or 1–1, if $(x)f = (w)f$ implies $x = w$. Third, we say that a function, f, is a *one-to-one correspondence* between X and Y if f is a function from X onto Y that is also one to one.

Definition. Let f be a function from X into Y and g a function from Y into Z. The *product*, fg, of f and g is the function from X into Z such that for each $x \in X$, $(x)fg = ((x)f)g$.

Strictly speaking, we should prove that there is exactly one function fg determined by the rule above. This is not difficult, but if the reader has never seen a formal proof he should try to prove it himself, i.e., find a suitable set of ordered pairs and show that it is unique.

Note carefully that products of functions are always read from left to right in this book. In other books, especially in some that are referred to later on, this convention is reversed. The student should *always* check a reference to see if the author is using the $f(x)$ or the $(x)f$ convention for functional notation.

An interesting function (one for each set X) is the identity function, $\mathbf{1}_X$. Formally, $\mathbf{1}_X = \{(x, x) | x \in X\}$. Intuitively, $\mathbf{1}_X$ is the function from X onto X which assigns each $x \in X$ to itself, i.e., leaves each $x \in X$ unchanged. If the context determines the set X, we often use $\mathbf{1}$ in place of $\mathbf{1}_X$.

The identity map and the concept of the product of two functions can be combined to give an interesting characterization of "1–1" or "onto." Intuitively, a function is one to one if and only if it can be untangled, i.e., if and only if it has a post-inverse. Similarly, a function maps X onto Y if and only if each element in Y is the image of at least one object in X, i.e., if and only if the function has a pre-inverse. These intuitive ideas are formalized in the following theorem.

Theorem 1.1. Let f be a function from X into Y, then:
(1) f maps X onto Y if and only if there is a function g from Y into X such that $gf = \mathbf{1}_Y$.
(2) f is one to one if and only if there is a function h from Y into X such that $fh = \mathbf{1}_X$.

Proof. Assume the function g exists. Then for each $y \in Y$, $y = (y)\mathbf{1}_Y = (y)gf = ((y)g)f$. Hence y is the image of $(y)g$ under f. Since $(y)g \in X$ for each $y \in Y$, this shows that f maps X onto Y. Conversely, assume f maps X onto Y. Then for each $y \in Y$ choose *one* $x \in X$ such that $(x)f = y$. (If Y is infinite, we appeal to the axiom of choice in order to accomplish this infinite set of choices.) Define g by $(y)g = x$, and we have our desired function g from Y into X.

The proof of (2) is similar (but simpler!) so we leave it as an exercise (exercise 1.1, problem 3).

In the case of a 1–1 correspondence, f, between X and Y both parts of the theorem apply, and we have both a pre-inverse, g, and a post-inverse, h. Clearly, $g = h$, so we call this function the *inverse* of f and denote it by f^{-1}.

This completes our listing of elementary notions about sets and functions. We turn now to the basic type of algebraic system that is essential in any study of symmetry.

Definition. A *group*, **G**, is a non-empty set G and a "product" operation defined on G such that:

(1) For any pair of elements a, b in G, the product ab is uniquely determined by a and b and is in G.

(2) There is an element, e, in G, called the *identity*, such that $eb = be = b$ for any b in G.

(3) For any a in G, there is an element x in G, called the *inverse* of a, such that $ax = xa = e$. The inverse of a is denoted by a^{-1}.

(4) For all $a, b, c \in G$, $a(bc) = (ab)c$, i.e., products are *associative*.

In general, we try to use a boldface capital letter for the name of an algebraic system and the corresponding lightface capital for the name of the associated set. If the reader first sees H, and then **H** appears without explanation, he should look for a product in H that is determined by the context.

Strictly speaking, we should first show that the identity is unique before calling it *the* identity, and similarly we should show that *the* inverse of a is unique.

Note that we do *not* assume that the product is *commutative*, i.e., we do not assume that $ab = ba$ for all a, b in a group. A group in which this property does hold is called an *abelian* group. If **G** is an abelian group, we *may* write $a + b$ and $-a$ in place of ab and a^{-1}. In other words, **we reserve additive notation for abelian groups.**

At this point, we should present some examples of groups; however, the reader will often need to refer to the basic definitions and theorems about groups so it helps if the reference material is not cluttered up with extraneous material. For this reason we defer all examples until the next section and restrict the remainder of this section to the bare statements of major definitions and their immediate consequences.

Definition. Assume **G** and **H** are groups. **H** is a *subgroup* of **G** if and only if $H \subseteq G$ and the product in **H** is inherited from the product in **G**, i.e., if $a, b \in H$, then ab in **H** is the same as ab in **G**.

In practice we use one of the following methods to recognize a subgroup.

Theorem 1.2. Let **G** be a group with identity e. Let H be a subset of G and **H** the algebraic system consisting of the set H and the product inherited from **G**. **H** is a group (and hence a subgroup of **G**) if and only if *either* of the following statements holds.

(1) $e \in H$, H is closed with respect to products ($a, b \in H$ implies $ab \in$

H), and H is closed with respect to taking inverses ($a \in H$ implies $a^{-1} \in H$).

(2) H is non-empty, and $a, b \in H$ implies $ab^{-1} \in H$.

In order to verify that either of these conditions is sufficient the reader should reread the definition of a group and check that if either condition holds, then all four properties are satisfied by **H.** It is easy to prove that either condition is necessary *after* one verifies that if **H** is a subgroup of **G,** then the identity in **H** is the identity in **G,** and inverses in **H** coincide with inverses taken in **G.** Probably the easiest way to verify that identities in **G** and **H** are equal is to exploit the following result.

Proposition 1.3. Let **K** be a group. If a, b, and $c \in K$, then $ab = ac$ if and only if $b = c$. In particular, the identity is the only solution in **K** of $xx = x$.

Proof. Assume $b = c$. Then by property 1 for groups, $ab = ac$. Conversely, if we assume $ab = ac$, then $a^{-1}(ab) = a^{-1}(ac)$. But $a^{-1}(ab) = (a^{-1}a)b = eb = b$ and similarly, $a^{-1}(ac) = c$. Thus, $b = c$. For any x in K, $x = xe$ if e is the identity. Hence if $xx = x$, then $xx = xe$, and hence $x = e$ by the first part of the theorem.

The technique used in the proof generalizes as follows.

Theorem 1.4. We may pre-multiply (respectively post-multiply) both members of an equation involving group elements by any element of the group without changing the validity of the equation. For example, the equations $f = g$, $fh = gh$, and $hf = hg$ are all equivalent if f, g, h belong to the same group. Similarly, we may pre-divide (respectively post-divide) both members by any element in the group. For example, $fg = h$, $g = f^{-1}h$, and $f = hg^{-1}$ are all equivalent.

The proof is omitted. Note carefully that unless the product operation in **G** is commutative, we must be careful to *pre*-multiply on *both* sides or else *post*-multiply on *both* sides of an equation.

Returning now to subgroups, we discuss one of the basic situations in which subgroups arise.

Definition. Let **G** be a group and X a non-empty subset of G. If **K** is the smallest subgroup of **G** containing X (existence and uniqueness of **K** to be proved by the reader in exercise 1.1, problem 8), then **K** is called the subgroup *generated* by X. The elements of X are called *generators* of **K.** In the special case in which X contains just one element, x, we denote the subgroup by (x) and call it a *cyclic* subgroup.

If **G** is a group with identity e and $g \in G$, it is customary to define $g^0 = e$, $g^n = gg \cdots g$ (n times), and $g^{-n} = (g^{-1})(g^{-1}) \cdots (g^{-1})$ (n times) for any positive integer n. With these conventions the usual laws of exponents apply. Using this notation, we can describe (g) explicitly.

Proposition 1.5. The subgroup (g) is the set of all integral (positive, zero, and negative!) powers of g.

Proof. The identity, g^0, is in this set. If g^j and g^k are in the set, $g^j g^k = g^{j+k}$ is, too. Finally, if g^j is in the set, so is $(g^j)^{-1} = g^{-j}$. Hence the set of integral powers of g is a subgroup by part (1) of theorem 1.2. Since (g) is the *smallest* subgroup containing g, (g) must be a subset of the set of integral powers of g.

To reverse the subset relation we note that since (g) is closed with respect to products, an easy proof by induction shows that $g^j \in (g)$ for all positive integers j. Since $e \in (g)$ and (g) is closed with respect to taking inverses, $e = g^0$ and $g^{-j} = (g^j)^{-1}$ are in (g). Thus the set of integral powers of g is contained in (g). This completes the proof that this set equals (g).

A similar proof, but with a slightly more complicated inductive step, shows that the subgroup generated by X is just the set of all finite products of integral powers of the various elements in X. *Caution!* Since multiplication is not commutative, there may be no way to simplify a product such as $a^2ba^{-10}b^3c^5b^{-17}$; moreover, it is not always true that g^j and g^k are distinct if $j \neq k$. The conditions under which $g^j = g^k$ are easily stated in terms of the order of g, which we now define.

Definition. Let $\mathbf{G}$ be a group with identity e, and let $g \in G$. The *order* of g is the smallest positive integer n such that $g^n = e$, if there is at least one such integer. If there is no such integer, then the order of g is infinite or, more precisely, the cardinality of the set of integers.

Theorem 1.6. Let $\mathbf{G}$ be a group and $g \in G$. The order of g equals $\#(g)$. If the order of g is infinite, then $g^j \neq g^k$ if $j \neq k$, but if the order of $g = n$, then $g^j = g^k$ if and only if n divides $j - k$.

Proof. Let $N = \{p | p \text{ is an integer and } g^p = e\}$. Since $g^j = g^k$ if and only if $g^{j-k} = e$, it follows that if $N = \{0\}$, then the correspondence $j \leftrightarrow g^j$ is a one-to-one correspondence between the set of integers and (g). Thus, if $N = \{0\}$, order $g = \#(g)$, and $g^j \neq g^k$ for $j \neq k$.

If, on the contrary, $N \neq \{0\}$, then (since $g^j = e$ implies $g^{|j|} = e$) there are positive integers in N, and hence the order of g is n, the smallest positive integer in N. Using the division algorithm for integers and the laws of exponents in $\mathbf{G}$, it is easy to verify that N is the set of all multiples of n, and therefore that $g^j = g^k$ if and only if n divides $j - k$. This in turn implies that the correspondence $j \leftrightarrow g^j$ is a one-to-one correspondence between the set

$\{0, 1, \ldots, n-1\}$ and (g). Hence, in this case also, the order of $g = \#(g)$, and the proof is complete.

The word "order" is used in one other sense in relation to groups; namely, if **G** is a group, we often call $\#(G)$ the *order* of **G**. Fortunately, in the only case in which confusion could arise, these two senses are consistent since (theorem 1.6) the order of g equals the order of (g).

Exercise 1.1

1. Find all ways of defining a product on the set $\{x, y, z, e\}$ such that the group axioms are satisfied, e plays the role of the identity, and $xx = e$. Repeat, assuming $xx = y$ instead of e. After completing this you should have some understanding of the statement that there are essentially only two four-element groups. It may help you to first try the analogous problem with one-, two-, or three-element sets.

2. Let **G** be a group and $\mathbf{H}_1$ and $\mathbf{H}_2$ two subgroups of **G** such that neither subgroup is contained in the other. Show that $H_1 \cup H_2$ cannot be a subgroup of **G**. The subgroup *generated* by $H_1 \cup H_2$ is usually called the *join* of $\mathbf{H}_1$ and $\mathbf{H}_2$ and is denoted by $\mathbf{H}_1 \vee \mathbf{H}_2$.

3. Prove part (2) by theorem 1.1.

4. Let X and Y be finite sets, say, $\#X = p$ and $\#Y = q$. Count the number of one-to-one functions from X into Y. *Caution!* Consider the case, $p > q$, first.

5. Explore the difficulties in counting the number of functions from X *onto* Y, X, and Y as in the preceding problem.

6. Let Y^X be the set of *all* functions from X into Y. Show that if X and Y are finite, then $\#(Y^X) = (\#Y)^{\#X}$. If $Y = \{0,1\}$, determine a natural one-to-one correspondence between Y^X and the set of all subsets of X.

7. Let $\mathfrak{F}$ be a non-empty family of subgroups of a group **G**. Show that $\mathbf{K} = \bigcap_{H \in \mathfrak{F}} \mathbf{H}$ is also a subgroup of **G**. Moreover show that **K** is the only subgroup, **L**, of **G** that satisfies *both* of the properties:

1. $\mathbf{L} \subseteq \mathbf{H}$ for all $\mathbf{H} \in \mathfrak{F}$.
2. If **M** is a subgroup of **G** such that $\mathbf{M} \subseteq \mathbf{H}$ for all $\mathbf{H} \in \mathfrak{F}$ then $\mathbf{M} \subseteq \mathbf{L}$.

8. Let **G** be a group and X a non-empty subset of G. Let $\mathfrak{F}$ be the family of all subgroups of **G** containing X.

1. Show that $\mathfrak{F}$ is non-empty.
2. Let $\mathbf{K} = \bigcap_{H \in \mathfrak{F}} \mathbf{H}$ as in problem 7. Show that **K** is the subgroup of **G** generated by X, i.e., show that $X \subseteq K$ and that if **H** is any subgroup of **G** such that $X \subseteq H$, then $K \subseteq H$.

9. Assume that **G** is a *finite* group and that H is a subset of G that is non-empty and closed under multiplication. Show that **H** is a subgroup of **G**. Give an example showing that this may not be true if **G** is not finite.

10. Let G be the set of all linear functions on the set of real numbers R, i.e., $f \in G$ if and only if $f:R \to R$ and there are real numbers a and b such that $a \neq 0$ and $(x)f = ax + b$ for all $x \in R$.

(1) Using as the product operation ordinary composition of functions as defined in this section, show that **G** is a group.

(2) If p and q are distinct real numbers, show that there is *exactly one* linear function, f, such that $(0)f = p$ and $(1)f = q$.

(3) Use parts (1) and (2) and theorem 1.4 to show that if p, q, r, and s are real numbers with $p \neq q$ and $r \neq s$, then there is exactly one linear function sending p to r and q to s.

(4) Show that f is a linear function if and only if f is a one-to-one function from R onto R such that: For all p, q, r, s in R, if $r \neq s$, then

$$\frac{p - q}{r - s} = \frac{(p)f - (q)f}{(r)f - (s)f}.$$

1.2. The algebra of permutations and examples of groups

Among the various examples of groups, the permutation groups play a central role. Since they are important in the study of symmetry and also easily accessible, we present them among our first examples of groups.

Definition. Let X be a non-empty set. A *permutation* of X is a one-to-one transformation of X onto itself. The set of all permutations of X is denoted S_X or, in the special case in which $X = \{1, 2, 3, \ldots, n\}$, S_n. The special permutation $\mathbf{1}_X$ is called the *trivial permutation* of X and all other permutations are referred to as *nontrivial.*

In section 1.1 we defined a natural product for functions. Since permutations are special functions, we have a natural product defined on the set S_X. The set S_X, together with this product, forms an algebraic system $\mathbf{S}_X$. For example, if $X = \{1, 2\}$, then $S_2 = \{\mathbf{1}, \phi\}$ where $(1)\phi = 2$ and $(2)\phi = 1$. The multiplication table for $\mathbf{S}_2$ is

	1	ϕ
1	1	ϕ
ϕ	ϕ	**1**.

It is easy to verify that $\mathbf{S}_2$ is a group. The following theorem generalizes this to an arbitrary set X.

Theorem 1.7. The algebraic system $\mathbf{S}_X$ is a group. This group is abelian if and only if X has one or two elements. If X is a finite set and $\#(X) = n$, then $\#(S_X) = n!$.

Proof. $\mathbf{1}_X$ is a permutation of X; hence $S_X \neq \varnothing$. Let f and g be permutations of X; then fg is a function from X into X. Given $x \in X$, since both f and g are onto, we can first find a $y \in X$ such that $(y)g = x$ and then a $z \in X$ such that $(z)f = y$. But then $(z)fg = x$, so fg is onto. Similarly, since f and g are both one to one, fg is one to one. (Verify!) Thus for any pair of elements of S_X, their product is in S_X. Clearly, the product, fg, is uniquely determined by the functions f and g.

Since $(x)\mathbf{1}_X f = (x)f = (x)f\mathbf{1}_X$ for any $x \in X$ and any function f from X into X, it follows that $\mathbf{1}_X f = f\mathbf{1}_X = f$ for any permutation, f, of X. Thus $\mathbf{1}_X$ is the identity in $\mathbf{S}_X$.

If f is a permutation of X, then by theorem 1.1 there is a function, f^{-1}, from X into X such that $ff^{-1} = f^{-1}f = \mathbf{1}_X$. Again using theorem 1.1, we see that since $ff^{-1} = \mathbf{1}_X$, f^{-1} is onto, and since $f^{-1}f = \mathbf{1}_X$, f^{-1} is one to one, i.e., f^{-1} is in S_X, and thus each f in S_X has an inverse in S_X.

As long as f, g, and h are functions such that either $f(gh)$ or $(fg)h$ is defined, it follows that the other product is also defined and $f(gh) = (fg)h$ (exercise 1.2, problem 2), i.e., products of functions are associative. Since our permutations in S_X are special functions, it follows that the product in $\mathbf{S}_X$ is associative. This completes the proof that $\mathbf{S}_X$ is a group.

If X has only one element, then S_X has only one element, $\mathbf{1}_X$, and thus in this case $\mathbf{S}_X$ is abelian. For any set with two elements, the only two permutations are $\mathbf{1}$, which leaves the two elements fixed, and the permutation which interchanges the two elements. Clearly, the group $\mathbf{S}_X$ is abelian in this case. However, if X has more than two elements, then we can select three elements a, b, and c and construct permutations, f and g, of X such that $fg \neq gf$. One suitable pair is defined by $(a)f = b$, $(b)f = a$, and $(x)f = x$ for any $x \in X$ different from a and b; and $(b)g = c$, $(c)g = b$, and $(x)g = x$ for any $x \in X$ different from b and c. Since $(a)fg = c$, but $(a)gf = b$, $fg \neq gf$. Thus $\mathbf{S}_X$ is not abelian if X has more than two members.

Finally we count the number of permutations of X if X is finite, say, $\#(X) = n$. To do this we order the elements in X, call them $x_1, x_2, \ldots, x_n$. If we are to construct a one-to-one function f from X into X, then we have n choices for $(x_1)f$, $(n - 1)$ choices for $(x_2)f$ [since we must avoid $(x_1)f$], $\ldots$, $(n - j)$ choices for $(x_{j+1})f, \ldots$, and 1 choice for $(x_n)f$. Thus there are $n(n - 1)(n - 2)\ldots(1) = n!$ possible one-to-one functions from X into X. Since X is finite, and one cannot distribute n objects one to a slot into n distinct slots without filling up each slot, each one-to-one function is also onto, hence there are $n!$ permutations of X. This completes the proof of the theorem.

Definition. For any non-empty set, X, the group $\mathbf{S}_X$ is called the *full symmetric group* on X. $\mathbf{S}_n$ is called the *symmetric group of degree n*.

There is a standard notation, called the *cycle notation,* for those permutations of a set X which leave all but a finite number of elements of X fixed. If $a_1, a_2, \ldots, a_n$ are n *distinct* objects in X, then denote by $(a_1, a_2, \ldots, a_n)$ that permutation of X sending a_1 to a_2, a_2 to $a_3, \ldots, a_n$ to a_1 while leaving all other elements of X invariant. For example, the two permutations used in the proof above are $f = (a, b) = (b, a)$ and $g = (b, c) = (c, b)$. Note that if a, b, and c are distinct then $(a, b)(b, c) = (a, c, b) = (c, b, a) = (b, a, c)$, whereas $(b, c)(a, b) = (a, b, c) = (b, c, a) = (c, a, b)$. The permutation $(a_1, a_2, \ldots, a_n)$ is called a *cycle* of *length n*. A cycle of length 2 is often called a *transposition*. Note that a transposition is its own inverse in $\mathbf{S}_X$, and, in general, if f is a cycle of length n then $f^n = \mathbf{1}$ in $\mathbf{S}_X$. As indicated above for cycles of length 2 or 3, each cycle of length n has n different names in the cycle notation, one name using each of the n elements as the leading element.

If g is a nontrivial permutation of X which leaves fixed all but a finite number of elements of X, then we can represent g as a product of disjoint cycles. For example, $g = (1, 2, 3)(2, 3, 4)$ (viewed as a permutation of $\{1, 2, 3, 4\}$) is a product of the overlapping cycles $(1, 2, 3)$ and $(2, 3, 4)$ but is also the product of the disjoint cycles $(1, 3)$ and $(2, 4)$, i.e., $(1, 2, 3)(2, 3, 4) = (1, 3)(2, 4)$. In general a representation for g as a product of disjoint cycles can be obtained by choosing any element x which g does not leave fixed and constructing the cycle $(x, (x)g, (x)g^2, \ldots, (x)g^n)$ where $(x)g^{n+1} = x$, then choosing any element y left over such that $(y)g \neq y$ and constructing $(y, (y)g, \ldots, (y)g^k)$ where $(y)g^{k+1} = y$, etc. The product $(x, (x)g, \ldots, (x)g^n)(y, \ldots, (y)g^k)(z, \ldots) \ldots$ is the desired representation.

To illustrate the cycle notation and the concepts defined in section 1.1 we now examine the group $\mathbf{S}_4$ and some of its subgroups.

Example 1. The group $\mathbf{S}_4$. This group consists of the 24 permutations of $\{1, 2, 3, 4\}$. Since the set itself is finite, each nontrivial permutation moves only a finite number of objects and therefore can be represented in the cycle notation. There are only five types of products of disjoint cycles: $\mathbf{1}$, (x, y), (x, y, z), $(x, y)(z, w)$, and (x, y, z, w). The subgroups generated by these types of permutations contain one, two, three, two, or four elements, respectively. For example, the subgroup generated by $(1, 3, 2, 4)$ contains $(1, 3, 2, 4)$, $(1, 3, 2, 4)^2 = (1, 2)(3, 4)$, $(1, 3, 2, 4)^3 = (1, 4, 2, 3)$, and $(1, 3, 2, 4)^4 = \mathbf{1}$.

In addition to the subgroups generated by a single permutation, there are subgroups of order 4, 6, 8, and 12 which require more than one generator, e.g., the minimum number of generators for the subgroup $\{\mathbf{1}, (1, 2), (3, 4), (1, 2)(3, 4)\}$ is two (exercise 1.2, problem 7). The (unique!) subgroup of $\mathbf{S}_4$ of order 12 also requires at least two generators, e.g., $(1, 2)(3, 4)$ and $(2, 3, 4)$ (exercise 1.2, problem 8).

To illustrate some of the shortcomings of the cycle notation for permutations we consider permutations of an infinite set X.

Example 2. Let Z be the set of integers. There are many permutations of Z that move infinitely many elements and hence cannot be represented in the cycle notation. In fact, most permutations of Z are extremely awkward (and often impossible) to describe in any reasonable way. *Any* finite product of cycles of finite length involving only integers can be viewed as defining a permutation of Z. An example of a permutation that is easy to describe and yet is not a (finite) product of cycles is the permutation ϕ defined by: $(x)\phi = 1 - x$. Clearly, there is no integer for which $(x)\phi = x$, so ϕ moves all of the integers. Note that the order of ϕ is 2. An interesting subgroup of $\mathbf{S}_Z$ is the subgroup, $\mathbf{G}$, generated by ϕ_1 and ϕ_2, where $(x)\phi_1 = x + 1$ and $(x)\phi_2 = -x$. An easy computation shows that $(x)\phi_1^n = x + n$ for any integer n (+ or −), and that $\phi_2^{-1} = \phi_2$. The reader should check that $(x)\phi_1^{-n}\phi_2\phi_1^n = 2n - x$. Later we shall have the necessary tools to show that ϕ_1^n, $\phi_1^{-n}\phi_2\phi_1^n$, and $\phi_1^{-n}\phi\phi_1^n$ make up all of the elements of $\mathbf{G}$.

Permutations are relevant to the study of symmetry since any symmetry operation is a permutation of a set. Of course, a symmetry operation is usually not an arbitrary permutation but rather one that leaves the "structure" under consideration invariant. For example, the symmetries of a cube can be viewed as those permutations of the eight vertices that keep adjacent vertices adjacent. Similarly the symmetries of euclidean space are those permutations of the points in space that preserve distance. The simplest kind of "structure" one could have for a set X is a distinguished subset Y. There are two obvious ways that this "structure" can be preserved. The weakest requirement is that the subset Y be mapped onto itself. A more restrictive requirement is to insist that each element in Y remain fixed. The following definitions and theorems establish basic tools for these two situations.

Definition. Let Y be a non-empty subset of X. We say that a permutation, ϕ, of X leaves Y *globally invariant* or *globally fixed* if $(Y)\phi = Y$, and that it leaves Y *pointwise invariant* or *pointwise fixed* if $(y)\phi = y$ for every $y \in Y$.

A permutation leaving Y pointwise fixed also leaves it globally fixed, and if Y has only one element, the converse is also true. If $\{a, b\} \subseteq Y$, $a \neq b$, then (a, b) is an example of a permutation leaving Y globally but not pointwise fixed.

Definition. Let Y be a non-empty subset of X and $\mathbf{G}$ a subgroup of $\mathbf{S}_X$. The set of all permutations in G leaving Y globally invariant is denoted $[\mathbf{G}, Y; gi]$. Similarly, $[\mathbf{G}, Y; pi]$ denotes the set of permutations in G that leave Y pointwise invariant.

Theorem 1.8. With the product inherited from $\mathbf{G}$, $[\mathbf{G}, Y; gi]$ and $[\mathbf{G}, Y; pi]$ are subgroups of $\mathbf{G}$.

Proof. The trivial permutation of X, $\mathbf{1}_X$, must belong to G since $\mathbf{G}$ is a subgroup of $\mathbf{S}_X$ (proposition 1.3), and thus $\mathbf{1}_X$ is in both $[\mathbf{G}, Y; gi]$ and $[\mathbf{G}, Y; pi]$. If $(y)g = (y)f = y$, then $(y)gf^{-1} = y$, and thus, by theorem 1.2, $[\mathbf{G}, Y; pi]$ is a subgroup of $\mathbf{G}$. If $(Y)g = Y$, then since g is one to one and maps Y *onto* Y, it follows that $(Y)g^{-1} = Y$ (exercise 1.2, problem 3). Obviously if f and g both leave Y globally invariant, then fg does also. Thus $[\mathbf{G}, Y; gi]$ is also a subgroup of $\mathbf{G}$.

Example 3. Let X be a finite set with n elements, and let Y be a subset with p elements, $0 < p < n$. In order to count the number of permutations of X leaving Y pointwise invariant we observe that there is a natural one-to-one correspondence between $[\mathbf{S}_X, Y; pi]$ and the permutations of $X - Y = \{x \in X | x \notin Y\}$. To find the permutation of $X - Y$ corresponding to $f \in [\mathbf{S}_X, Y; pi]$ simply ignore the fact that f is defined anywhere except in $X - Y$; in other words, restrict f to $X - Y$. Since $\mathbf{S}_{X-Y}$ contains $(n - p)!$ elements, so does $[\mathbf{S}_X, Y; pi]$. For each permutation of X that leaves Y pointwise fixed, there are $p!$ permutations of X leaving Y globally fixed; thus $\#[\mathbf{S}_X, Y; gi] = p!(n - p)!$.

We shall use the following example extensively, so it is important to spend an hour or two becoming familiar with the symmetries of a cube and their products. A cube with its faces labeled, e.g., a die, is helpful when computing products of rotational symmetries.

Example 4. The vertex symmetries of a cube. Number the eight vertices of a cube as indicated in Figure 1. Let $\mathbf{G}$ be the group consisting of those per-

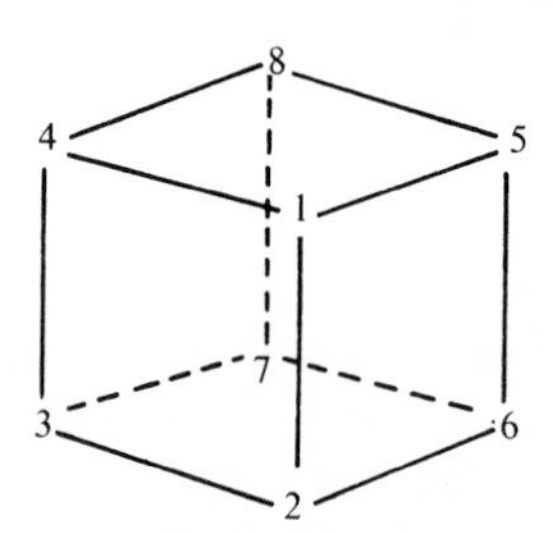

FIG. 1. Numbering the vertices of a cube.

mutations of $\{1, 2, \ldots, 8\}$ which preserve the "adjacent vertex" relationship. There are 48 such permutations (24 corresponding to rigid motions of the cube and 24 requiring that the cube be turned inside out) which split rather naturally into ten different types of "geometrically equivalent" permutations. A brief description of each of the ten types, an example of each, and the number of symmetries of each type are listed in Table 1.

TABLE 1. The Ten Types of Symmetries of a Cube

Type	*Example*	*Number of this type*
The identity	**1**	1
Inversion in the centroid	(17)(28)(35)(46)	1
90° rotation—facial axis	(1234)(5678)	6
90° rotatory inversion	(1836)(2547)	6
180° rotation—facial axis	(13)(24)(57)(68)	3
Reflection in an axial plane	(15)(26)(37)(48)	3
180° rotation—semi-diagonal axis	(14)(67)(28)(35)	6
Reflection in a diagonal plane	(16)(47)	6
120° rotation—diagonal axis	(245)(386)	8
120° rotatory inversion	(17)(265843)	8

Many interesting subgroups of **G** are of the form [**G**, Y; *pi*] or [**G**, Y; *gi*] for suitable subsets Y. For example, the subgroup of 8 symmetries leaving the top face fixed is [**G**, Y; *gi*] for $Y = \{1, 4, 5, 8\}$. The subgroup [**G**, $\{1, 2\}$; *gi*] contains 4 permutations only, two of which are in [**G**, $\{1, 2\}$; *pi*].

In example 4 above we introduced the vague notion of "geometrically equivalent" symmetries of the cube. Note that even though $180° = 180°$ and $90° \neq 270°$, we considered a 180° rotation about a facial axis as not equivalent to a 180° rotation about a semi-diagonal axis, and yet we considered all 90° and 270° rotations as equivalent. In order to explain this quixotic behavior and to make the "equivalence" idea precise we turn now to the study of conjugate elements in a group.

Definition. If f and g are elements of a group **G**, then the product $g^{-1}fg$ is called the *conjugate* of f by g. The set of all conjugates of f, $\{g^{-1}fg | g \in G\}$ is called the *conjugacy class* of f in **G**.

Roughly speaking, if f and g are permutations, then the conjugate of f by g behaves as if it were f shifted by g. For instance, if f is the 90° rotation looking down on the top face of a cube and g is any one of the four rigid motions of the cube carrying the top face to the bottom face, then $g^{-1}fg$ is the 90° rotation looking up at the bottom face, i.e., the 270° rotation from the top. In general, for permutations, we can assert the following:

Proposition 1.9. A permutation, f, sends x to y if and only if $g^{-1}fg$ sends $(x)g$ to $(y)g$. If Y is the set of points left fixed by f, then $(Y)g$ is the set of points left fixed by $g^{-1}fg$.

Proof. The conjugate of f by g, $g^{-1}fg$, sends $(x)g$ to $(x)fg$. However, $(x)fg = (y)g$ if and only if $(x)f = y$. In particular, f sends x to x if and only if

$g^{-1}fg$ sends $(x)g$ to $(x)g$, i.e., $(x)g$ is left fixed by $g^{-1}fg$ if and only if x is left fixed by f.

Proposition 1.10. If f, g, and h are elements of a group **G**, then:
(1) The conjugate of fg by h is the product of the conjugates of f and g by h, i.e., $h^{-1}(fg)h = (h^{-1}fh)(h^{-1}gh)$.
(2) The conjugate of f^{-1} by h is the inverse of the conjugate of f by h, i.e., $h^{-1}f^{-1}h = (h^{-1}fh)^{-1}$.
(3) The conjugate of f by h is f if and only if f and h *commute*, i.e., $h^{-1}fh = f$ if and only if $hf = fh$.
(4) The conjugate of (the conjugate of f by g) by h is the conjugate of f by gh, i.e., $h^{-1}(g^{-1}fg)h = (gh)^{-1}f(gh)$.
(5) If h is the conjugate of f by g, then f is the conjugate of h by g^{-1}.

The proofs of (1), (3), and (4) are left as exercises. The easiest way to prove (2) is to apply (1) and theorem 1.4 as follows: By part (1), $h^{-1}(ff^{-1})h = (h^{-1}fh)(h^{-1}f^{-1}h)$, but (if e is the identity in **G**) $e = h^{-1}eh = h^{-1}(ff^{-1})h$ and by definition $(h^{-1}fh)(h^{-1}fh)^{-1} = e$. Combining these three equations we have $(h^{-1}fh)(h^{-1}fh)^{-1} = (h^{-1}fh)(h^{-1}f^{-1}h)$, and this, by theorem 1.4, is equivalent to $(h^{-1}fh)^{-1} = (h^{-1}f^{-1}h)$.

Parts (4) and (5) imply that the relation "g is the conjugate of f by at least one element in **G**" is transitive and symmetric for elements in **G**. Since $f = e^{-1}fe$, the relation is also reflexive and is therefore an equivalence relation. The equivalence classes are, of course, just the conjugacy classes. In particular, note that conjugacy classes never overlap unless they are equal.

Using induction and (1) and (2), it follows that $h^{-1}(f^n)h = (h^{-1}fh)^n$ for any integer n, and thus the order of f is the same as the order of any of its conjugates. *Caution!* Elements in a group need not be conjugate just because they have the same order.

Example 5. Conjugacy classes in $\mathbf{S}_n$. If $a_1, a_2, \ldots, a_k$ are distinct elements of a set X and g is a permutation of X, then the conjugate of $(a_1, a_2, \ldots, a_k)$ by g is the k-cycle $((a_1)g, (a_2)g, \ldots, (a_k)g)$ by proposition 1.9. By proposition 1.10, part (1), we may compute the conjugate of a product of cycles by computing the conjugate of each cycle. Thus the conjugate of a product, f, of cycles by g is the product of cycles obtained by replacing each symbol in f by its image under g. For example, if $f = (1, 3, 5)(2, 4)$ and $g = (1, 2)(3, 6, 7)$, then $g^{-1}fg = (2, 6, 5)(1, 4)$. With this technique for computing conjugates in $\mathbf{S}_n$, it is clear that two permutations in $\mathbf{S}_n$ are conjugate if and only if they have the same *type of cycle decomposition*, i.e., if and only if they can both be factored into k disjoint cycles of length $n_1, n_2, \ldots, n_k$ with $1 < n_1 \leq n_2 \leq \cdots \leq n_k$ and $\Sigma n_i = n$. There are, for instance, five conjugacy classes in $\mathbf{S}_4$ corresponding to the five "types" of permutations mentioned in example 1.

Example 6. Conjugacy classes in the group of vertex symmetries of a cube. Let **G** be the group of example 4. The identity and the inversion of the cube in its centroid each commute with every element of **G**. Hence the conjugacy class of each contains just one element. Any permutations conjugate in **G** must also be conjugate in $\mathbf{S}_8$; by examining the types of cycle representations given in example 4 and recalling the results of example 5, we see that there are *at least* five more conjugacy classes, and in particular no reflection in a diagonal plane can be conjugate to any other type of symmetry of order 2. We claim that there are in fact ten conjugacy classes in **G** corresponding exactly to the ten types of symmetries listed in Table 1. In Chapter 2 we develop the tools to verify this assertion efficiently.

There is an obvious way to generalize conjugacy from elements of **G** to subsets of **G**. This is especially important for those subsets that are subgroups of **G**.

Definition. If $\mathbf{H}_1$ and $\mathbf{H}_2$ are both subgroups of **G** then we say they are *conjugate subgroups* if and only if there is some element $g \in \mathbf{G}$ such that $g^{-1}\mathbf{H}_1 g = \mathbf{H}_2$, i.e., such that $\{g^{-1}hg | h \in \mathbf{H}_1\} = \mathbf{H}_2$.

At the end of section 1.5 we shall introduce one of the major themes of this book, the geometric significance of conjugate subgroups. We shall show that if **G** is the group of all symmetries of a certain mathematical structure then conjugate subgroups within **G** have essentially the same algebraic structure and have the same significance for the mathematical structure preserved by **G**. But before doing this we must define carefully what it means for two groups to have "essentially the same algebraic structure" and show that "conjugating by g" is a mapping that preserves algebraic structure. We accomplish these tasks in the next two sections.

Exercise 1.2

1. Prove that **G** is an abelian group if and only if **G** is a group in which each conjugacy class contains exactly one element.

2. Let X_i, $i = 1, 2, 3, 4$, be sets and f_j, $j = 1, 2, 3$, be functions such that f_j maps X_j into X_{j+1}. Show that $f_1(f_2f_3)$ and $(f_1f_2)f_3$ are both functions mapping X_1 into X_4 and that $f_1(f_2f_3) = (f_1f_2)f_3$. Thus we see that the composition of functions is always associative.

3. Let Z be the set of integers and P the set of positive integers. Let ϕ be the permutation of Z defined by $(x)\phi = x + 1$. Show that ϕ maps P into but not onto P and that ϕ^{-1} does not map P into P. Keeping this example in mind, prove that $g \in [\mathbf{G}, Y; gi]$ implies $g^{-1} \in [\mathbf{G}, Y; gi]$. Point out where the assumption that g maps Y onto Y enters your proof.

4. Let **G** be the group of vertex symmetries of a cube as in example 4 of this section. With the vertices numbered as in Figure 1, let $D = \{1, 7\}$, and find $[\mathbf{G}, D; pi]$ and $[\mathbf{G}, D; gi]$. Conjugate each element of $[\mathbf{G}, D; gi]$ by the 90° rotation whose axis hits face $\{1, 2, 3, 4\}$ and which sends vertex 1 to 2. Show that the resulting set of symmetries is a subgroup of the form $[\mathbf{G}, E; gi]$ for a suitable set, E, of vertices. How are D and E related?

5. Use the results in example 5 of this section to simplify

$$(3, 2, 1)(4, 5)(3, 6, 2)(1, 2, 3).$$

6. Give an example of a subgroup **H** of $\mathbf{S}_4$ and two permutations f and g in H such that f and g are conjugate in $\mathbf{S}_4$ but not in **H**.

7. Show that the subgroup $\{\mathbf{1}, (1, 2), (3, 4), (1, 2)(3, 4)\}$ of $\mathbf{S}_4$ requires at least two generators. Generalize this result and show that $\mathbf{S}_{2n}$ has an abelian subgroup in which each element satisfies $x^2 = e$ (not quite the same as saying *every* element has order two) and which requires n generators.

8. Find a subgroup of order 8 and a subgroup of order 12 in $\mathbf{S}_4$. Determine the minimum number of generators for each subgroup.

9. Let Y be a non-empty subset of X, and let **G** be a subgroup of $\mathbf{S}_X$. Let $H = \{g \in G | (Y)g \subseteq Y\}$. Use problem 3 above to show that **H** need not be a subgroup of **G**. Show that $H = [\mathbf{G}, Y; gi]$ if either Y or $X - Y$ is a finite set.

10. Let X be an infinite set, and let F be the set of all permutations of X leaving all but a finite number of elements of X fixed. Note that different permutations in F are allowed to "move" different finite subsets of X. Show that **F** is the subgroup of $\mathbf{S}_X$ generated by the transpositions.

1.3. Homomorphisms and permutation representations of groups

In order to have a relatively simple but nontrivial example with which to illustrate the concepts of this section we introduce the group of symmetries of a square.

Example 1. Let the sides, vertices, and axes of symmetry of a square be labeled as in Figure 2, and let ρ be the 90° (clockwise) rotation of the square. The set of eight symmetries of the square, i.e., $e =$ the identity, ρ, ρ^2, ρ^3, and the four reflections h, v, m, n with axes A_4, A_2, A_1, A_3, respectively, then form a group **G**, called *the group of the square*, with the natural product xy meaning first perform x then y. We have chosen the names of the reflections h, v, m, n (for horizontal, vertical, main diagonal, and minor diagonal) to remind the reader that these reflections are defined with respect to the *original positions* of their respective axes and thus $nh = \rho$, not ρ^{-1}. The eight symmetries in this group separate into five conjugacy classes: $\{e\}$, $\{\rho^2\}$, $\{\rho, \rho^3\}$, $\{h,v\}$, and $\{m,n\}$. By considering the subgroups generated by various small sub-

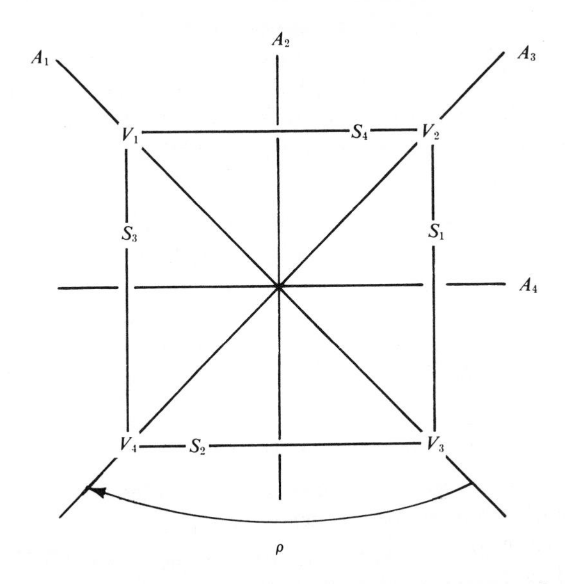

FIG. 2. Numbering the vertices, sides, and axes of symmetry of a square.

sets of **G**, it is not hard to find all of the subgroups of **G** (exercise 1.3, problem 1).

There are three sets with four objects in each that are naturally associated with the square, namely, the vertices $\{V_1, V_2, V_3, V_4\}$, the sides $\{S_1, S_2, S_3, S_4\}$, and the axes of symmetry $\{A_1, A_2, A_3, A_4\}$. Each symmetry of the square induces a permutation of each of these sets as shown in Table 2.

TABLE 2. Permutations Induced by the Symmetries of a Square

Symmetry	*Vertices*	*Sides*	*Axes*
e	**1**	**1**	**1**
ρ	$(V_1V_2V_3V_4)$	$(S_1S_2S_3S_4)$	$(A_1A_3)(A_2A_4)$
ρ^2	$(V_1V_3)(V_2V_4)$	$(S_1S_3)(S_2S_4)$	**1**
ρ^3	$(V_4V_3V_2V_1)$	$(S_4S_3S_2S_1)$	$(A_1A_3)(A_2A_4)$
h	$(V_1V_4)(V_2V_3)$	(S_2S_4)	(A_1A_3)
v	$(V_1V_2)(V_3V_4)$	(S_1S_3)	(A_1A_3)
m	(V_2V_4)	$(S_1S_2)(S_3S_4)$	(A_2A_4)
n	(V_1V_3)	$(S_1S_4)(S_2S_3)$	(A_2A_4)

The three correspondences between symmetries and permutations of each of the three sets all share the following important property: if x and

y are symmetries corresponding to the permutations f and g, then xy corresponds to fg. The third correspondence (between symmetries and permutations of the axes) is different from the other two in that it is not one to one. In this correspondence there are two symmetries for *each* permutation. Note also that the order of a permutation always divides the order of the corresponding symmetry or symmetries.

With this example in mind, we turn to the formal development of concepts designed to study correspondences between elements in two groups.

Definition. Let **G** and **H** be groups. A *homomorphism* of **G** into **H** is a function $\phi: G \to H$ which preserves the group operation, i.e., for all $g_1, g_2 \in G$, $(g_1g_2)\phi = (g_1)\phi(g_2)\phi$. The *kernel* of ϕ is the set of all $g \in G$ such that $(g)\phi$ is the identity element in **H**.

Example 1 (continued). Using the table of induced permutations, Table 2, for the eight symmetries of the square and recalling our observation that if the symmetries x and y correspond to the permutations f and g, then xy corresponds to fg, we can define homomorphisms of the group of the square into the full symmetry groups on $\{V_1V_2V_3V_4\}$, $\{S_1S_2S_3S_4\}$, and $\{A_1A_2A_3A_4\}$ simply by viewing each column of the table as a table of function values. If we focus on the numbers of the vertices, sides, or axes then we obtain three more homomorphisms, ϕ_V, ϕ_S, ϕ_A, of **G** into $\mathbf{S}_4$. Table 3 shows the values for these three homomorphisms.

TABLE 3. Three Homomorphisms of the Group of the Square

g	$(g)\phi_V$	$(g)\phi_S$	$(g)\phi_A$
e	**1**	**1**	**1**
ρ	(1234)	(1234)	(13)(24)
ρ^2	(13)(24)	(13)(24)	**1**
ρ^3	(4321)	(4321)	(13)(24)
h	(14)(23)	(24)	(13)
v	(12)(34)	(13)	(13)
m	(24)	(12)(34)	(24)
n	(13)	(14)(23)	(24)

The kernels of ϕ_V and ϕ_S are both $\{e\}$, but the kernel of ϕ_A is $\{e, \rho^2\}$.

Since a homomorphism is a function that preserves the group operation, we should expect that it also preserves everything that is defined in terms of the group operation, e.g., inverses, conjugates, subgroups, etc. The only note of caution arises from the fact that a homomorphism *may* collapse several elements in G to a single element in H, and thus subgroups, conjugacy classes, and the order of an element in G may be "collapsed" in somewhat the same way. Theorem 1.11 states precisely the way in which some concepts are preserved, and theorem 1.12 states the manner in which the size of the kernel and the amount of "collapsing" are related.

Theorem 1.11. Let **G** and **H** be groups with identities e and e', respectively. Let ϕ be a homomorphism of **G** into **H.** The basic properties of ϕ, in addition to preserving products, are as follows:

(1) ϕ preserves the identity, i.e., $(e)\phi = e'$.
(2) ϕ preserves inverses, i.e., $(g^{-1})\phi = ((g)\phi)^{-1}$ for all $g \in G$.
(3) ϕ preserves conjugacy, i.e., if f, g, and h are all in G and f is the conjugate of g by h, then $(f)\phi$ is the conjugate of $(g)\phi$ by $(h)\phi$ in **H.**
(4) ϕ preserves subgroups, i.e., if **F** is a subgroup of **G,** then $(F)\phi = \{(f)\phi | f \in F\}$ is a subgroup of **H.**
(5) ϕ either preserves or collapses order, i.e., if $g \in G$ and the order of g is finite, then the order of $(g)\phi$ divides the order of g. If the order of g is infinite, then, depending on ϕ, the order of $(g)\phi$ may be finite or infinite.

Proof. Let $g \in G$; then $(g)\phi = (ge)\phi = (g)\phi(e)\phi$. By cancellation in **H** (theorem 1.4), this implies $e' = (e)\phi$, thus proving (1). Using this result it follows that $e' = (g^{-1}g)\phi = (g^{-1})\phi(g)\phi$. But we also have $e' = ((g)\phi)^{-1}(g)\phi$, so again by cancellation in **H,** $(g^{-1})\phi = ((g)\phi)^{-1}$, thus proving (2). Part (3) follows immediately from (1) and (2).

If **F** is a subgroup of **G,** then $F \neq \varnothing$, and hence $(F)\phi \neq \varnothing$. Moreover, if $x, y \in (F)\phi$, then we can choose $f, g \in F$ such that $x = (f)\phi$ and $y = (g)\phi$. Thus $xy^{-1} = (fg^{-1})\phi$ is also in $(F)\phi$, and hence $(F)\phi$ is a subgroup (theorem 1.2), thus proving (4).

An easy induction proof shows that for each $g \in G$ and each integer n, $(g^n)\phi = ((g)\phi)^n$. Thus if $g^n = e$, it follows that $((g)\phi)^n = e'$. Hence the order of $(g)\phi$ divides the order of g if the order of g is finite. The natural homomorphisms of the additive groups of integers onto themselves and onto cyclic groups of order $n = 2, 3, 4, \ldots$, demonstrate that anything can happen if the order of g is infinite. This completes the proof.

Theorem 1.12. Let **G, H,** e, e', and ϕ be as above, and let K be the kernel of ϕ. Then **K** is a subgroup of **G** with the additional property that whenever $g \in K$, the conjugacy class of g in **G** is entirely contained in K. If $h \in H$, then either $h \notin (G)\phi$ or else $\#\{f \in G | (f)\phi = h\} = \#K$; in other words, the number of elements in G sent to a particular $h \in H$ is either zero or else the same as the number of elements in G sent to the identity in **H.**

Proof. By theorem 1.11, $e \in K$ and K is closed with respect to products and inverses. Hence **K** is a subgroup of **G.** If $k \in K$ and g is the conjugate of k by h, then $(g)\phi = (h^{-1})\phi(k)\phi(h)\phi = ((h)\phi)^{-1}e'(h)\phi = e'$, so g is also in K.

For any $f, g \in G$, $(f)\phi = (g)\phi$ if and only if $fg^{-1} \in K$. Using this the reader can verify that the correspondence $k \leftrightarrow kg$ is a one-to-one correspondence between K and $\{f \in G | (f)\phi = (g)\phi\}$. Thus, $\#K = \#\{f \in G | (f)\phi = (g)\phi\}$ for each $g \in G$, and the proof is complete.

Subgroups of a group **G** with the special property that they never contain an element of **G** without containing its entire conjugacy class are called *normal* or *invariant* subgroups of **G.** Thus, we could have rephrased part of theorem 1.12 to say that the kernel of a homomorphism is always a normal subgroup. The converse is also true: any normal subgroup of **G** can be realized as the kernel of at least one homomorphism of **G** into some group. The reader should return to the group of the square and determine which subgroups are normal and try to realize each of these normal subgroups as the kernel of a homomorphism similar to those presented in example 1.

As a special case of theorem 1.12, we see that no collapsing takes place if and only if the kernel contains just one element, e. Thus, a homomorphism is one to one everywhere if and only if it is one to one at e. This special type of homomorphism merits special attention.

Definition. Let ϕ be a homomorphism of **G** into **H.** If ϕ is one to one, then ϕ is called an *isomorphism* of **G** *into* **H.** If, in addition, $(G)\phi = H$, then ϕ is called an isomorphism of **G** *onto* **H** or an isomorphism *between* **G** and **H,** and in this case (only!) **G** and **H** are called *isomorphic groups.*

The relation "**G** and **H** are isomorphic" is clearly an equivalence relation in any set of groups. Many results state that a group of a certain type is *unique up to isomorphism,* meaning that any two groups of this type must be isomorphic. This is often all that can be reasonably expected since isomorphic groups are indistinguishable insofar as their group structure is concerned. For example, all groups with two elements are isomorphic, although the two elements in one group may not be the same as the two elements in another.

In the study of symmetry and groups of symmetries most homomorphisms and isomorphisms arise naturally in the process of trying to find effective computational tools to keep track of products of symmetries. The two major tools for this problem are permutation representations, especially effective for small finite groups, and matrix representations, effective for both finite and infinite groups. Both techniques have their limitations although permutation representations are in a certain sense (*cf.* theorem 1.13) universal.

Definition. A *representation* of a group **G** is a homomorphism of **G** into a group **H.** If ϕ is an isomorphism of **G** into **H,** then we say that the representation is *faithful.* If $\mathbf{H} = \mathbf{S}_X$ for some non-empty set X, we say that ϕ is a *permutation representation* of **G** and say that **G** can be *viewed* (via ϕ) *as operating on* X. Similarly, if **H** is a group of matrices (product the ordinary product of matrices), then we say that ϕ is a *matrix representation* of **G**. The *degree* of a permutation representation refers to the number of elements being permuted, i.e., if $(\mathbf{G})\phi \subseteq \mathbf{S}_X$ then the degree of ϕ is $\#(X)$. If ϕ is a matrix representation then the *degree* of ϕ refers to the size of the matrices, i.e., if $(\mathbf{G})\phi$ is contained in the group of non-singular n by n matrices then the *degree* of ϕ is n.

We shall ignore matrix representations of groups until Chapter 4. In example 1 of this section, we demonstrated three permutation representations in $\mathbf{S}_4$ of the group of the square. The first two representations, ϕ_V and ϕ_S, are faithful, and the third, ϕ_A, collapses the group onto a group of order 4. Note that it is often advantageous to have a representation that is *not* faithful in order to "chop the group down to a manageable size." Moreover, several different representations often give much more information than just one representation. We return to the group of symmetries of the cube in order to illustrate these points.

Example 2. Let **G** be the group of all 48 symmetries of the cube, classified according to type in example 4 of section 1.2. In that example we gave a faithful representation of **G** in $\mathbf{S}_8$ by numbering the eight vertices and letting each symmetry in **G** correspond to the induced permutation of the vertex numbers. In a similar way we obtain a faithful representation of this group in $\mathbf{S}_6$ by numbering the six faces and viewing the symmetries as permuting the faces. We number the faces as they are numbered on a die: opposite numbers add up to 7, and faces 1, 2, and 3 cycle counterclockwise around a vertex.

Two other representations yield some interesting normal subgroups of **G.** Let A be the set of the three facial axes numbered so that axis j hits the center of face j, for $j = 1, 2, 3$. The symmetries of the cube permute these three axes so we have a representation of **G** in $\mathbf{S}_3$. An easy computation shows that this representation maps **G** *onto* $\mathbf{S}_3$, and hence by theorem 1.12 the kernel K, of this representation is a normal subgroup of **G,** and each of the six elements in $\mathbf{S}_3$ corresponds to $\#K$ elements in G. It follows that $\#K = 8$. K consists of the identity, the inversion in the centroid, the three reflections in facial planes, and the three 180° rotations about facial axes. $\mathbf{S}_3$ contains a normal subgroup of order 3, $\{\mathbf{1}, (1, 2, 3), (3, 2, 1)\}$, and hence (exercise 1.3, problem 5) the set of 24 symmetries represented by these three permutations is a normal subgroup of **G**. Note that this subgroup contains both rigid motions *and* symmetries requiring that the cube be turned inside out. The 24 rigid motions of the cube also form a normal subgroup, $\mathbf{G}^+$. Thus we have found two distinct normal subgroups of order 24.

Let D be the set of four diagonals of the cube. The symmetries of the cube permute these four diagonals; moreover, the only symmetries leaving all four diagonals fixed are the identity and inversion in the center. Thus there is a representation of **G** in $\mathbf{S}_4$ in which each of the 24 permutations in $\mathbf{S}_4$ corresponds to two of the 48 symmetries of the cube. Note that of these two symmetries one is a rigid motion and the other is not.

The permutation representation of **G** in $\mathbf{S}_6$ obtained by viewing the symmetries as permuting the numbers of the six faces gives additional information concerning conjugacy classes in **G.** Recall that in example 6 of section 1.2 we showed that the ten types of symmetries of the cube split into at least six conjugacy classes since they are represented by nonconjugate permutations of the eight vertices. We also showed that inversion in the centroid is in a conjugacy class by itself since it commutes with all of the symmetries. Let us add to this information by considering, for the six types not split into distinct classes in $\mathbf{S}_8$, their representatives in $\mathbf{S}_6$ obtained by viewing symmetries as permuting the six faces numbered as on a die. See Table 4.

TABLE 4

Representation in $\mathbf{S}_8$	*Type of symmetry*	*Cycle decomposition in* $\mathbf{S}_6$	
Represented by conjugate elements in $\mathbf{S}_8$	90° rotation	$(abcd)$	nonconjugate
	90° rotatory inversion	$(abcd)(ef)$	
Represented by conjugate elements in $\mathbf{S}_8$	180° rotation—facial axis	$(ab)(cd)$	nonconjugate
	reflection in axial plane	(ab)	
	180° rotation—semi-diagonal axis	$(ab)(cd)(ef)$	conjugate
	inversion in center	$(ab)(cd)(ef)$	

Thus these types split into at least five conjugacy classes. Combining this information with that of example 6 of section 1.2, we see that there are at least ten conjugacy classes in **G.** In Chapter 2 we show that there are at most ten.

If we write out the complete multiplication table of a finite group, then, in view of the cancellation law in groups, each column is a permutation of the index column. This suggests that we can look at the table and view each column as a table of values for a permutation of the group elements. This can be generalized to infinite groups also and leads to the (*right*) *regular representation* of **G.** See exercise 1.3, problem 7 concerning what happens if we do this with rows instead of columns.

Theorem 1.13 (Cayley). Any group **G** has a faithful representation as permutations of its own elements. More precisely, for each $g \in G$ define

$g^\rho: G \to G$ by $(x)g^\rho = xg$ for all $x \in G$. The mapping ρ is a faithful representation of **G** in $\mathbf{S}_G$.

Proof. The individual functions, g^ρ, are one to one by cancellation in **G** and map G onto G since $g^{-1} \in G$. Thus, $\rho: \mathbf{G} \to \mathbf{S}_G$. The mapping ρ is faithful since $g^\rho = h^\rho$ implies $g = eg = (e)g^\rho = (e)h^\rho = eh = h$, and it is a homomorphism since $(x)(gh)^\rho = xgh = ((x)g^\rho)h = (x)g^\rho h^\rho$ for all $x \in G$. Thus, ρ is a faithful representation.

Definition. The representation ρ as defined in the theorem above is called the *right regular representation* of **G.**

Exercise 1.3

1. Find and classify according to conjugacy all subgroups of the group of the square. Find at least one pair of isomorphic but not conjugate subgroups.

2. Find a homomorphism of $\mathbf{S}_3$ onto $\mathbf{S}_2$ and a subset of the kernel that is not a subgroup of $\mathbf{S}_3$. Discuss the converse of part 4 of theorem 1.11 in the light of your example. Compare with problem 5 below.

3. Let X be a non-empty set, **G** a group, and G^X the set of all functions from X into G. Define a multiplication in G^X by the following: $f \cdot g$ is that function such that $(x)f \cdot g = (x)f(x)g$ for all $x \in X$. Show that $\mathbf{G}^X$ is a group and that for any $x \in X$ the mapping sending $f \in G^X$ to $(x)f$ is a homomorphism of $\mathbf{G}^X$ onto **G.** When is $\mathbf{G}^X$ abelian?

4. Let **G** be a group and ϕ a permutation representation of **G** in $\mathbf{S}_X$. Let Y be a non-empty subset of X and extend our notation [**G**, Y; *pi*] and [**G**, Y; *gi*] in the obvious way. Show that both the kernel of ϕ and [**G**, Y; *pi*] are *normal* subgroups of [**G**, Y; *gi*].

5. Let ϕ be a homomorphism of the group **G** onto the group **H,** and let F be a subset of H. Show that **F** is a subgroup of **H** if and only if $\mathbf{E} = \{g \in G | (g)\phi \in F\}$ is a subgroup of **G.** Show that **E** is normal in **G** if and only if **F** is normal in **H.**

6. Consider the polynomial $p_n = \prod_{i<j} (x_i - x_j)$ in the n variables $x_1, x_2, \ldots, x_n$. For each permutation θ in $\mathbf{S}_n$, define $(p_n)\theta = \prod_{i<j} (x_{(i)\theta} - x_{(j)\theta})$. Since θ is one to one and onto, it is clear that $(p_n)\theta$ is either $+p_n$ or $-p_n$. Accepting this, find a natural permutation representation of $\mathbf{S}_n$ in $\mathbf{S}_2$. The kernel of this representation if called $\mathbf{A}_n$, the *alternating group of degree n*. Show that a cycle is in $\mathbf{A}_n$ if and only if its length is odd, i.e., no transpositions, four cycles, . . . , are in $\mathbf{A}_n$, but all three cycles, five cycles, . . . , are in $\mathbf{A}_n$.

7. For each g in the group **G,** define $g^\lambda: G \to G$ by $(x)g^\lambda = gx$. Show that the function λ which maps **G** into $\mathbf{S}_G$ is a faithful *anti*-representation of **G,** i.e., it is one to one and $(gh)^\lambda = h^\lambda g^\lambda$. If we view the *rows* of **G**'s multiplication table as tables of function values, then λ is the resulting anti-representation.

8. Let λ be the function defined in the previous problem, and let $\mu: G \to G$ be defined by $(x)\mu = x^{-1}$. Show that the composite mapping $\mu\lambda$ is a faithful representation of **G** in $\mathbf{S}_G$ and is different from the right regular representation if **G** is not abelian.

9. Let **G** be a subgroup of $\mathbf{S}_X$ and f a one-to-one correspondence mapping X onto Y. Show that $\mathbf{H} = \{f^{-1}gf | g \in G\}$ is a subgroup of $\mathbf{S}_Y$ and that **G** and **H** are isomorphic groups. This type of isomorphism is important to us later in the book when we consider coordinate mappings and matrices. In that context X is the abstract "space," Y the space of coordinates, f the coordinate mapping, **G** the abstract group, and **H** the coordinate form of **G**, usually a group of matrices.

10. Let **G** be the group of all 48 symmetries of a cube. In example 2 of this section, we introduced two subgroups of order 24 in **G**: $\mathbf{G}^+$ = the group of rigid motions, and **H** = the group of 24 symmetries that permute the three facial axes cyclically, i.e., represented by **1**, (1, 2, 3), or (1, 3, 2) in the permutation representation of **G** in $\mathbf{S}_3$. Determine the orders of the elements in $\mathbf{G}^+$ and **H**, and show that these two normal subgroups of **G** are not isomorphic.

11. Let $\mathbf{G}^+$ be the group of rigid motions of a cube and **T** be the group of those symmetries of a cube leaving one of the inscribed tetrahedra globally invariant. By finding natural permutation representations of $\mathbf{G}^+$ and **T**, show that they are both isomorphic to $\mathbf{S}_4$. Are they conjugate subgroups in the group of symmetries of the cube?

12. Let Z be the set of integers and E be the set of even integers. Use problem 9 to show that $\mathbf{S}_Z$ and $\mathbf{S}_E$ are isomorphic. Show that $\mathbf{S}_E$ and $[\mathbf{S}_Z, E; pi]$ are isomorphic. Thus we see that an infinite group may be isomorphic to a proper subgroup of itself.

1.4. Automorphisms of a group

Our guiding principle in this book is that in order to understand a particular geometry one should study its group of symmetries and various subgroups of that group. This approach can be followed for any mathematical system, in particular for groups. In this section we study the group of symmetries (automorphisms) of a group, and investigate the subgroup consisting of those automorphisms induced by elements within the group.

Definition. An *automorphism* of a group **G** is a permutation of G that preserves the algebraic structure (the product in **G**!!) of **G**. In other words, an *automorphism* of **G** is an isomorphism of **G** onto itself. The set, A(**G**), of *all* automorphisms of **G** is of course a subgroup of $\mathbf{S}_G$ since it is the set of all permutations of G preserving a certain structure on G, in this case algebraic instead of geometric. Thus we call **A(G)** the *automorphism group* of **G**.

Caution! Given a group **G** and a function $f: G \to G$ the following properties of f are independent, i.e., any one can fail even though both of the others hold for f:

1. f is one to one.
2. f maps G onto G.
3. $(xy)f = (x)f(y)f$ for all $x, y \in G$.

Therefore, one needs to verify all three properties in order to be certain that f is an automorphism of **G**. Of course if **G** is finite then properties 1 and 2 are equivalent.

Since an automorphism of **G** is a homomorphism of **G** onto **G** in which no "collapsing" takes place, we see from theorems 1.11 and 1.12 that an automorphism preserves such things as the identity element, inverses, conjugacy, subgroups, and the order of elements in **G**.

Example 1. Let **G** be the group of the square. (We use the notation established in example 1 of section 1.3.) Let **H** be the subgroup $\{e, h, v, \rho^2\}$. Within **H** the elements h, v, and ρ^2 are algebraically indistinguishable. (Each is of order 2, and the product of any two is the third.) Hence if we let ϕ be any of the six permutations of H for which $(e)\phi = e$, then ϕ is an automorphism of **H**.

Within **G**, however, ρ^2 *is* distinguishable from h and v. (ρ^2 is a perfect square, but h and v are not.) Thus of the six automorphisms of **H** only two [$\mathbf{1}_H$ and (h, v)] could possibly be extended to an automorphism of **G**. Since $\{h, v\}$ and $\{m, n\}$ are two conjugacy classes in **G**, any automorphism of **G** must either interchange these sets or leave each globally invariant. Since h and n generate **G**, their images completely determine an automorphism. And so the following eight permutations of G (given in cycle notation) are the only possible automorphisms of **G**: $\mathbf{1}_G$, (h, n, v, m), $(h,v)(m, n)$, (m, v, n, h), $(h, v)(\rho, \rho^3)$, $(m, n)(\rho, \rho^3)$, $(h, m)(n, v)(\rho, \rho^3)$, and $(h, n)(v, m)(\rho, \rho^3)$. We show later (after theorem 1.15) that these are all, in fact, automorphisms of **G**.

In the example above it is clear that the automorphism group of **H** is isomorphic to $\mathbf{S}_3$. The astute reader will also have noticed that the automorphism group of the group of the square is isomorphic to the group of the square. We return to this observation in a moment, but first let us consider a special type of automorphism, one that is induced by an element in the group and hence called an inner automorphism.

Definition. Let **G** be a group. For each $g \in G$ define a function g^i: $G \to G$ by $(x)g^i = g^{-1}xg$ for each $x \in G$, i.e., g^i is simply conjugating by g. The function i sending each g to g^i maps G into the set of functions from G into G. Here, as for the right regular representation ρ, we are using g^i in place of $(g)i$ in order to avoid long strings of parentheses. In anticipation of the following result, we call g^i the *inner automorphism* induced by g. All automorphisms of **G** that are not inner automorphisms are called *outer automorphisms*.

Theorem 1.14. The inner automorphism induced by g is, in fact, an automorphism of **G**. Furthermore, it leaves each conjugacy class in **G** globally invariant.

Proof. By part (5) of proposition 1.10 $(g^{-1})^i$ is the (function) inverse of g^i. Hence g^i is both one to one and onto. Part (1) of the same proposition asserts that g^i preserves multiplication. Hence g^i is an automorphism of **G**. By definition, x, gxg^{-1}, and $g^{-1}xg$ are in the same conjugacy class. Hence g^i maps a conjugacy class into itself ($(x)g^i = g^{-1}xg$) and also maps it onto itself ($x = (gxg^{-1})g^i$), i.e., g^i leaves the conjugacy class globally invariant.

Theorem 1.15. If **G** is a group, then:

(1) The mapping i sending each $g \in G$ to its induced inner automorphism, g^i, is a homomorphism of **G** into **A(G)**.

(2) The set of all inner automorphisms, $\mathbf{G}^i$, is a normal subgroup of **A(G)**.

(3) The kernel of i is the *center* of **G**, i.e., it consists of those elements in G that commute with every element in G.

Proof. The mapping i sending each $g \in G$ to g^i is a mapping of G into **A(G)** by theorem 1.14. It preserves products by part (4) of proposition 1.10. Hence i is a homomorphism of **G** into **A(G)**. Part (3) of proposition 1.10, when translated into the language of inner automorphisms, asserts that $(f)h^i = f$ if and only if f and h commute. Thus, $h^i = \mathbf{1}_G$ if and only if h commutes with every $f \in G$, i.e., the kernel of i consists of those elements in G that commute with every element in G.

Finally we must show that $\mathbf{G}^i$ is a normal subgroup of **A(G)**. Let g^i be in $\mathbf{G}^i$, and let ϕ be any automorphism of **G**. Then for every $f \in G$, if $h = (g)\phi$, $(f)\phi^{-1}g^i\phi = (f)h^i$. (Verify!) Thus, the conjugate of g^i by any automorphism is an inner automorphism, so $\mathbf{G}^i$ is normal in **A(G)**.

Example 1 (continued). Returning to the automorphisms of **G**, the group of the square, and of one of its subgroups, $\mathbf{H} = \{e, \rho^2, h, v\}$, we consider the inner automorphisms of these two groups. Since **H** is abelian, every element commutes with every other element, and so **H** has only one inner automorphism, $\mathbf{1}_H$, and five outer automorphisms. A few elementary computations show that the center of **G** is $\{e, \rho^2\}$ and thus the eight elements of **G** pair off as follows to induce four inner automorphisms of **G**: $e^i = (\rho^2)^i = \mathbf{1}_G$, $\rho^i = (\rho^3)^i = (h, v)(m, n)$, $h^i = v^i = (m, n)(\rho, \rho^3)$, and $m^i = n^i = (h, v)(\rho, \rho^3)$. This proves that four of the eight permutations we listed as possible automorphisms of **G** are, in fact, automorphisms of **G**. What about the remaining four? Recall the permutation representations ϕ_V and ϕ_S of **G** in $\mathbf{S}_4$ that were introduced in example 1 of section 1.3. These are both faithful; hence they are isomorphisms of **G** into $\mathbf{S}_4$. They map G onto the same subgroup of $\mathbf{S}_4$; hence $\phi_V\phi_S^{-1}$ is an automorphism of **G**, but $\phi_V\phi_S^{-1} = (h, n, v, m)$. Since **A(G)** is a group, this new automorphism multiplied by each of the four inner automorphisms yields four outer automorphisms of **G**, i.e., all eight permutations listed earlier are actually automorphisms. Since these were the only possibilities, we have found all of the automorphisms of the group of the square. This method of proving that certain permutations are in fact automorphisms may seem pretentious. In fact a brute force computation using the multiplication table for **G** is just about as efficient *for this small group.* For a larger group, e.g., the symmetry group of a cube, this or an analogous method is essential.

The simplest possible situation as far as automorphisms are concerned is one in which distinct elements in the group induce distinct inner automorphisms, and there are *no* outer automorphisms. Such a group is called a *complete group.* In this case the homomorphism i is an isomorphism of **G** onto **A(G)**, i.e., **G** and **A(G)** are isomorphic in a natural way. Note that the group of the square is an example of a group that is *not* complete and yet *is* isomorphic to its group of automorphisms. Almost all of the full symmetric groups are complete, namely, $\mathbf{S}_X$ is complete if $\#X \neq 2$ or 6. We shall not prove this here, but instead refer the reader to [9],* pages 92–95 or [15] for the case in which X is a finite set, and to [14] if X is infinite. For an explicit construction of an outer automorphism of $\mathbf{S}_6$, see [10]. Note that $\mathbf{S}_2$ has no outer automorphisms but fails to be complete since it is abelian. It is not difficult to show (exercise 1.4, problem 3) that every abelian group with at least one element whose order is not 2 has an outer automorphism. It is true but much more difficult to show that an abelian group in which every element ($\neq e$) is of order 2 has a nontrivial automorphism if it has more than two elements (see [7], pp. 85–86, for a proof if **G** is finite).

Recall that in section 1.3 we defined a normal or invariant subgroup of **G** to be a subgroup **H** such that $h \in H$ implies $g^{-1}hg \in H$ for every $g \in G$. Translating this into a statement about inner automorphisms, we see that **H** is normal in **G** if $(H)g^i \subseteq H$ for every $g^i \in \mathbf{G}^i$. There *are* groups **G** with subgroups **H** for which $(H)g^i \subseteq H$ and yet $(H)g^i \neq H$ for *certain* $g^i \in \mathbf{G}^i$ (exercise 1.4, problem 7). However, if $(H)g^i \subseteq H$ for *every* $g^i \in \mathbf{G}^i$, then in particular $(H)(g^{-1})^i \subseteq H$, and since $(g^{-1})^i = (g^i)^{-1}$, this implies that $(H)g^i = H$. Thus, a subgroup is normal if and only if it is left globally invariant by each inner automorphism. This is the reason normal subgroups are often called invariant subgroups. The obvious generalization—to subsets and to arbitrary automorphisms—leads to the following definition.

---

* Wherever bracketed references appear in text, they refer to the reference list at the end of that chapter in which they are mentioned. For instance, [9] cited here is found at the end of Chapter 1.

Definition. Let **G** be a group and X a subset of G. X is *normal* or *invariant* in **G** if X is left globally invariant by each inner automorphism of **G**. Similarly, X is *characteristic* in **G** if it is left globally invariant by *every* automorphism of **G**.

Any subset of G that can be defined solely in terms of the product in **G** is clearly a characteristic subset of **G**. Two important examples are the set of elements commuting with every element in G and the set of *commutators* in $G = \{a^{-1}b^{-1}ab | a, b \in G\}$. Some other examples of characteristic subsets: the set of elements of order 2, the perfect squares, the union of all conjugacy classes with at most 3 elements, etc. Clearly the subgroup generated by a characteristic (respectively normal) subset is a characteristic (respectively normal) subgroup. For the first two examples above, this leads to two important subgroups.

Definition. Let **G** be a group. Then $\mathbf{Z}_G = \{g \in G | xg = gx \text{ for all } x \in G\}$ is called the *center* of **G**. Since we have shown that it is the kernel of i, it is a normal subgroup of **G**. The set of commutators is not always a subgroup nevertheless, **G**$'$, the subgroup generated by the commutators, is called the *commutator subgroup* of **G**.

The following theorem follows directly from the definition and the remarks preceding the definition.

Theorem 1.16. The center, $\mathbf{Z}_G$, and the commutator subgroup, $\mathbf{G}'$, are both characteristic subgroups of **G**. The three statements, **G** is abelian, $\mathbf{Z}_G = \mathbf{G}$, and $\mathbf{G}' = \{e\}$, are equivalent.

We might paraphrase the last part of the theorem by saying that $\mathbf{Z}_G$, which is always abelian, measures how much of the group is abelian, while $\mathbf{G}'$ measures how much the group misses being abelian. We consider the center of each of the various groups of symmetries throughout the book, but aside from one or two exercises at the end of this section we very seldom refer to commutators or the commutator subgroup. In another area of symmetry, Galois theory (the study of symmetries between roots of polynomials), the commutator subgroup plays a very important role (see [8], pp. 210–214).

Exercise 1.4

1. Prove that $\mathbf{S}_4$ is a complete group. *Hint.* Use example 5, section 1.2 to classify the elements into conjugacy classes and to count the number of elements commuting with a particular element.

2. Find all inner and all outer automorphisms of the symmetry group of a regular hexagon. For uniformity in notation, let ρ be the clockwise rotation through $\pi/3$

radians, θ_i the reflection in the axis hitting vertices V_i and V_{i+3}, $i = 1, 2, 3$, and ϕ_j the reflection whose axis hits the midpoint of V_jV_{j+1}, $j = 1, 2, 3$.

3. Show that every non-abelian group has a nontrivial automorphism and that every abelian group has an outer automorphism provided it has at least one element x such that $x^3 \neq x$.

4. Let ϕ be an isomorphism of **G** onto **H** and θ an automorphism of **G**. Show that $\theta^{-1}\phi$ is an isomorphism of **G** onto **H** and that $\phi^{-1}\theta\phi$ is an automorphism of **H**. When we study coordinate mappings, we consider this again. In that context, $\theta^{-1}\phi$ is a new coordinate mapping, and $\phi^{-1}\theta\phi$ is the "coordinate form" of θ.

5. Let **G**$'$ be the commutator subgroup of **G**.

(1) Assume that ϕ is a homomorphism of **G** onto **H**. Show that **H** is abelian if and only if **G**$'$ is contained in the kernel of ϕ.

(2) Assume that **K** is a subgroup of **G** and that $G' \subseteq K$. Show that **K** is normal in **G**.

6. Find the commutator subgroup of the group of all 48 symmetries of the cube. *Hint*. Problem 5, part (1) and the various permutation representations of the cube group will help.

7. Let Z be the set of integers and P the set of positive integers. Let g be the permutation of Z defined by $(n)g = n - 1$. Show that the inner automorphism, g^i, of $\mathbf{S}_Z$ induced by g carries the subgroup $[\mathbf{S}_Z, P; pi]$ into but not onto itself.

8. Prove that the symmetry groups of a cube and an octahedron are isomorphic. Do the same for a dodecahedron and icosahedron. *Hint*. Explore the geometric relationships between the various solids, and look for faithful representations of the symmetry groups exploiting these relationships.

1.5. Cosets and orbits, Lagrange's theorem

To motivate the theoretical material in this section we first consider two examples.

Example 1. Let **G** be the group of the 48 symmetries of a cube. (See example 4 in section 1.2 and example 2 in section 1.3.) There are several permutation representations of **G** that arise naturally by viewing the symmetries as permuting the vertices, faces, edges, etc. Consider, for example, the symmetries leaving one edge (globally) invariant. There are four such symmetries: **1**, one reflection in a diagonal plane, one reflection in an axial plane, and one 180° rotation about a semi-diagonal axis. This edge can be carried to any of the 12 edges by at least one symmetry; moreover, the number of symmetries carrying the edge to a specific edge is always 4. For example, to carry it to one of the adjacent edges we can use either a 90° or 120° rotation, a 120° rotatory inversion or a reflection in a diagonal plane. Let us tabulate the number of places an object in the cube can be carried and the number of symmetries leaving it (globally) fixed. See Table 5.

TABLE 5

Type of object	*No. of equivalent objects*	*No. of symmetries leaving it fixed*
Vertex	8	6
Face	6	8
Diagonal line	4	12
Inscribed tetrahedron	2	24
Semi-diagonal line	6	8
Centroid	1	48

It is not difficult to recognize the pattern in Table 5. The number of equivalent objects times the number of symmetries leaving one of these objects fixed is always 48.

Example 2. Let X be the set of all 2 by 2 matrices with 0 or 1 as entries. There are 16 such matrices. Let $\mathbf{G}$ be the group of the square as in example 1 of section 1.3. The symmetries in $\mathbf{G}$ can be viewed as permuting these 16 matrices, e.g., the 90° clockwise rotation, ρ, carries

$$\begin{bmatrix} 1 & 0 \\ 0 & 0 \end{bmatrix} \text{ to } \begin{bmatrix} 0 & 1 \\ 0 & 0 \end{bmatrix} \text{ and leaves } \begin{bmatrix} 1 & 1 \\ 1 & 1 \end{bmatrix} \text{ fixed .}$$

Thus we have a natural permutation representation of $\mathbf{G}$ in $\mathbf{S}_X$. Again we tabulate, for six inequivalent matrices, the number of equivalent matrices and the number of symmetries leaving the matrix invariant as shown in Table 6. Once again the product is always the number of elements in $\mathbf{G}$. In this example let us also compute the number of matrices left fixed by each symmetry, as in Table 7.

Two observations are in order. We were able to tabulate results according to conjugacy classes because the number of objects left fixed is constant within a conjugacy class (proposition 1.9). Note also that the average number of matrices left fixed, $(16 + 2 + 2 + 4 + 6 + 6 + 6 + 6)/8 = 6$, is just the number of inequivalent matrices.

An extremely important point in the examples above is that the relationships hold whether or not the permutation representation is faithful. Thus in all the theorems that follow we are careful not to assume that just one symmetry (group element) induces a given permutation. In the first theorem, the most important one, $\mathbf{G}$ need not be a finite group. We use g^p in place of $(g)p$ in order to avoid extra parentheses.

Theorem 1.17. Let $\mathbf{G}$ be a group and $p:\mathbf{G} \to \mathbf{S}_X$ be a permutation representation of $\mathbf{G}$. For each $x \in X$, let $[\mathbf{G}, x] = \{g \in G | (x)g^p = x\}$. $[\mathbf{G}, x]$ is a

TABLE 6

Matrix	No. of equivalent matrices	No. of symmetries leaving the matrix invariant
$\begin{bmatrix} 0 & 0 \\ 0 & 0 \end{bmatrix}$	1	8
$\begin{bmatrix} 1 & 1 \\ 1 & 1 \end{bmatrix}$	1	8
$\begin{bmatrix} 1 & 0 \\ 0 & 1 \end{bmatrix}$	2	4
$\begin{bmatrix} 1 & 0 \\ 0 & 0 \end{bmatrix}$	4	2
$\begin{bmatrix} 0 & 1 \\ 1 & 1 \end{bmatrix}$	4	2
$\begin{bmatrix} 1 & 1 \\ 0 & 0 \end{bmatrix}$	4	2

TABLE 7

Type of symmetry, i.e., conjugacy class	Symmetries of this type	No. of matrices left fixed by each symmetry
Identity	e	16
90° rotation	ρ, ρ^3	2
180° rotation	ρ^2	4
Reflection, horizontal or vertical	h, v	6
Reflection, diagonal	m, n	6

subgroup of **G**, and for each $f \in G$ the set of products $\{gf|g \in [\mathbf{G}, x]\} = [\mathbf{G}, x]f$ is the same set as $\{h \in G|(x)h^p = (x)f^p\}$. In other words, $[\mathbf{G}, x]f$ is the set of all group elements sending x to the same place f does.

Before proving the theorem, we introduce a new term and paraphrase the theorem.

Definition. If **H** is a subgroup of **G** and $g \in G$, then $Hg = \{hg|h \in H\}$ is called a *right coset* of **H**.

In terms of right cosets, the theorem reads roughly as follows: *If* **H** *is the subgroup of* **G** *leaving x fixed, then each right coset of* **H** *consists of all ele-*

ments in **G** *sending* x *to a specific place.* This is the geometric significance of right cosets just as "same type of transformation" is the geometrical significance of conjugacy classes. In view of their naturalness from a geometric standpoint, it is no wonder that these two concepts are so useful in group theory. If we had used the left-hand functional notation, $g(x)$, in place of $(x)g$, then left cosets would replace right cosets; fgf^{-1} would replace $f^{-1}gf$ as the conjugate of g by f, and the left instead of the right regular representation of a group would be the natural one.

Proof of theorem 1.17. We leave to the reader the proof that $[\mathbf{G}, x]$ is a subgroup of **G**. Reread theorems 1.8 and 1.11 before attempting it, and remember that p is a homomorphism! To complete the proof we assume that $gf \in [\mathbf{G}, x]f$. Then $(x)(gf)^p = (x)g^pf^p = (x)f^p$, so $gf \in \{h \in G | (x)h^p = (x)f^p\}$. Conversely, if $(x)h^p = (x)f^p$ and $h \in G$, then $(x)h^p(f^p)^{-1} = x$. Since $(hf^{-1})^p = h^p(f^p)^{-1}$, this implies that $hf^{-1} \in [\mathbf{G}, x]$. Thus, $h = (hf^{-1})f \in [\mathbf{G}, x]f$, and the proof is complete.

In view of the cancellation law in **G**, it is clear that $\#H = \#Hf$ for each $f \in G$. Hence if **G** is a finite group, we have the following:

Corollary 1.18. Let **G** be a group and p a permutation representation of **G** in $\mathbf{S}_X$. For each $x \in X$, $\#G = \#[\mathbf{G}, x]\ \#\{(x)g^p | g \in G\}$. In other words, the number of group elements leaving x fixed times the number of places x is sent is $\#G$.

Proof. Each place x is sent corresponds to a unique coset, but each coset has the same number of elements as $[\mathbf{G}, x]$. Since the cosets are disjoint and the union of *all* cosets is clearly G, $\#G =$ the product of $\#[\mathbf{G}, x]$ times the number of places x is sent.

Theorem 1.17 and its corollary are powerful tools for discovering symmetries or predicting the number of symmetries with certain properties. When writing example 1, we found the four symmetries of a cube carrying one edge to an adjacent edge by multiplying each of the four symmetries leaving the edge fixed by the reflection in the appropriate diagonal plane. For another example, let **G** be the group of six symmetries of a cube leaving one diagonal D pointwise invariant. If E is an edge of the cube with one end of E on D, then **G** sends E to three places; hence there must be two symmetries in **G** leaving E fixed. If, however, $E \cap D = \varnothing$, then **G** sends E to six places. Hence **1** must be the only symmetry in **G** leaving E fixed.

Example 3. The symmetry group of a dodecahedron. A regular dodecahedron has twelve faces, each of which is a regular pentagon. There are five rotations and five reflections leaving the face globally fixed. Hence there are $5 \cdot 12 = 60$ rigid motions that are symmetries of a dodecahedron and 120 symmetries if we allow the solid to be turned inside out. If the reader has a model of a dodecahedron (they are easy to make), he may find it challenging to find the five conjugacy classes of rigid motions and the five conjugacy classes of symmetries turning the dodecahedron inside out.

In addition to applications to symmetry groups, theorem 1.17 and its corollary have important applications in the study of groups themselves. We now present two of the more elementary (and important!) applications.

Definition. For each g in a group $\mathbf{G}$, the set $N_g = \{h \in G | gh = hg\}$ is called the *normalizer* of g in $\mathbf{G}$.

Theorem 1.19. For each $g \in G$, $\mathbf{N}_g$ is a subgroup of $\mathbf{G}$. If $\mathbf{G}$ is finite, then $\#G/\#N_g =$ the number of conjugates of g. Thus, both $\#N_g$ and the number of elements in a conjugacy class are always divisors of $\#(G)$.

Proof. The mapping i sending $h \in G$ to its induced inner automorphism is a permutation representation of $\mathbf{G}$ in $\mathbf{S}_G$. Since h^i sends g to the conjugate of g by h and leaves g fixed if and only if $hg = gh$, then $\{(g)h^i | h \in G\}$ is the conjugacy class of g and $[\mathbf{G}, g] = N_g$. Thus, by theorem 1.17, $\mathbf{N}_g$ is a subgroup, and, if $\mathbf{G}$ is finite, $\#G$ is the product of $\#N_g$ and the number of conjugates of g.

Corollary 1.20 (the class equation). Let $\mathbf{G}$ be a finite group, then

$$\#G = \sum_{a \in A} \frac{\#G}{\#N_a} = \#Z_G + \sum_{b \in B} \frac{\#G}{\#N_b},$$

where A is a subset of G with exactly one element in each conjugacy class, Z_G is the center of $\mathbf{G}$, and B is a subset of G with exactly one element in each conjugacy class containing more than one element.

Proof. Since Z_G is the union of the one-element conjugacy classes, it suffices to prove the first equation. Since $\#G/\#N_a =$ the number of conjugates of a, the first equation follows from the fact that the conjugacy classes partition G.

The class equation has many important applications in group theory, e.g., in the proof that if p is prime and divides $\#G$, then there are elements of order p in G, and in the development of the Sylow theory. Consult any standard text on group theory for examples of these applications.

Our second application is the famous theorem of Lagrange. We approach this theorem as a special case of corollary 1.18.

Theorem 1.21 (Lagrange). If $\mathbf{G}$ is a finite group and $\mathbf{H}$ is a subgroup of $\mathbf{G}$, then $\#H$ divides $\#G$.

Proof. **G** can be viewed as permuting (via the right regular representation) the subsets of G. In this representation a subset X is sent by g to the set $Xg = \{xg|x \in X\}$. For a sub*group* **H**, $Hg = H$ if and only if $g \in H$. Thus, $[\mathbf{G}, H] = H$, and the result follows from corollary 1.18.

Definition. If **H** is a subgroup of **G** and C is the set of cosets of **H** in **G**, then $\#C$ is called the *index* of **H** in **G** and is denoted by $[\mathbf{G}:\mathbf{H}]$. Note that if any two of the numbers $\#G$, $\#H$, $[\mathbf{G}:\mathbf{H}]$ are finite, then so is the third and $\#G = [\mathbf{G}:\mathbf{H}](\#H)$. There *are* infinite groups **G** and **H** for which $[\mathbf{G}:\mathbf{H}]$ is finite.

In the last part of this section, we develop a symbolism to help us talk about "the places **G** sends x," and "equivalent objects."

Definition. Let **G** be a group and p a permutation representation of **G** in $\mathbf{S}_X$. For each x and y in X, we say x and y are *equivalent under* **G** and write $x \overset{\mathbf{G}}{\sim} y$ if and only if there is a $g \in G$ such that $(x)g^p = y$. The set of all y such that $x \overset{\mathbf{G}}{\sim} y$ is called the **G**-*orbit* of x or simply the *orbit* of x. The set of all **G**-orbits is denoted $X/\mathbf{G}$.

Strictly speaking, our notation is poor in that the particular permutation representation is not identified. It is so seldom necessary to contend with more than one permutation representation of **G** in a given set that it is not worth the effort to have "p" tagging along in the notation. There are two other terms often used by other authors for **G**-orbits, *transitive sets* or *transitive constituents.* These terms are popular among algebraists since a group is said to be *transitive* on a set if for each pair of elements in the set there is at least one group element sending one to the other. Geometers, differential geometers, and topologists seem to prefer the term "orbit" since they often consider continuous groups.

Theorem 1.22. If **G** is a group which can be viewed as permuting the elements in X, then the relation "$x \overset{\mathbf{G}}{\sim} y$" is an equivalence relation in X, i.e.,

(1) $x \overset{\mathbf{G}}{\sim} x$ for all $x \in X$ (reflexive)

(2) for all $x, y \in X$, if $x \overset{\mathbf{G}}{\sim} y$, then $y \overset{\mathbf{G}}{\sim} x$ (symmetric)

(3) for all $x, y, z \in X$, if $x \overset{\mathbf{G}}{\sim} y$ and $y \overset{\mathbf{G}}{\sim} z$, then $x \overset{\mathbf{G}}{\sim} z$ (transitive).

The various **G**-orbits in X are the equivalence classes of this relation, and as such they partition the set X into disjoint subsets.

The proof is left to the reader with the observation that the three properties of equivalence correspond exactly to the requirements that **G** have an identity, be closed under "taking inverses," and be closed under "taking products."

Our final result in this section formalizes the intuitive idea that objects equivalent under a certain group ought to be left fixed by the same type of subgroup.

Theorem 1.23. If $x \overset{\mathbf{G}}{\sim} y$, then $[\mathbf{G}, x]$ and $[\mathbf{G}, y]$ are conjugate (and hence isomorphic) subgroups of **G**. Moreover, if $(x)g^p = y$ then $[\mathbf{G}, x]g^i = [\mathbf{G}, y]$.

Proof. Since $x \overset{\mathbf{G}}{\sim} y$, there is a $g \in G$ such that $(x)g^p = y$. An easy computation shows that $h \in [\mathbf{G}, x]$ if and only if $g^{-1}hg \in [\mathbf{G}, y]$. Thus, g^i is the required inner automorphism mapping $[\mathbf{G}, x]$ isomorphically onto $[\mathbf{G}, y]$.

We can rephrase the theorem as follows: The elements in **G** leaving two **G**-equivalent objects fixed form conjugate subgroups of **G.** Thus, the subgroups of the symmetries of the cube leaving the various vertices fixed are all conjugate subgroups. One must be a little cautious about the converse to this theorem. By proposition 1.9 it is true that if $\mathbf{H}_1$ and $\mathbf{H}_2$ are conjugate subgroups, then the sets $F_j = \{x \in X|(x)h = x \text{ for all } h \in H_j\}$, $j = 1, 2$ are **G**-equivalent sets. However, it can happen that $[\mathbf{G}, x]$ and $[\mathbf{G}, y]$ are conjugate subgroups, but x and y are not equivalent under **G.** For example, the subgroup of symmetries of a cube leaving a vertex fixed is the same as the subgroup leaving fixed a point two-thirds of the way out on the diagonal from the center to that vertex. There is clearly no symmetry of the cube sending the vertex to this point.

Exercise 1.5

1. Consider the semi-regular solid (Figure 3) obtained by joining two regular tetrahedra at one edge. Illustrate theorem 1.17 and corollary 1.18 by computing $[\mathbf{G}, x]$ and the **G**-orbit of x if **G** is the group of eight symmetries of this solid and the following:

(1) x is a vertex
(2) x is an edge
(3) x is a face.

Note that the results depend on which vertex, edge, or face you choose!

2. Let **H** be a subgroup of **G** such that $[\mathbf{G}:\mathbf{H}] = 2$. Show that **H** is a normal subgroup of **G.** Use the group of the square to show that **H** need not be a characteristic subgroup if $[\mathbf{G}:\mathbf{H}] = 2$.

3. Show that **H** is a normal subgroup of **G** if and only if it is a subgroup for which every right coset, Hg, is also the left coset, gH.

4. Let the group of the square operate on the set of 3 by 3 zero–one matrices in a manner analogous to that in example 2 of this section. Give a representative of each **G**-orbit, the subgroup leaving that matrix fixed, and the right coset carrying the matrix to a second representative of that **G**-orbit.

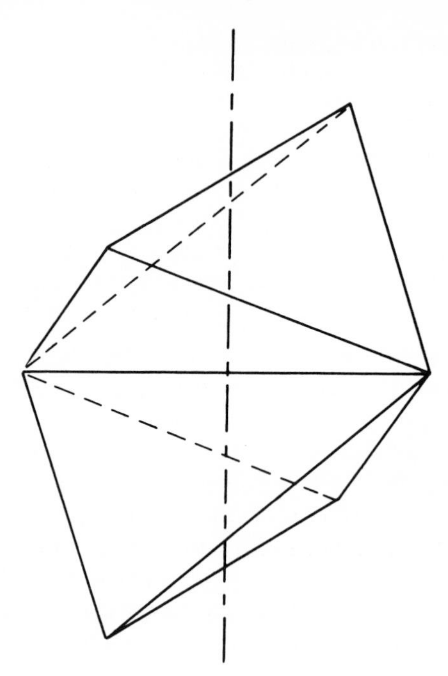

FIG. 3. Two regular tetrahedra joined at one edge, showing the vertical axis of symmetry.

5. Find all subgroups of $\mathbf{S}_4$ and group them into sets of conjugate subgroups. Viewing $\mathbf{S}_4$ as operating on the set of subgroups via conjugation, illustrate corollary 1.18 and theorem 1.23.

6. Use the class equation and Lagrange's theorem to show that the center of a group of order p^n, p prime, has more than one element. Derive as a corollary that if $\#G = p^2$, p prime, then $\mathbf{G}$ is abelian.

7. $\mathbf{G}$ operates on G via the mapping i sending each g in $\mathbf{G}$ to its induced inner automorphism. Thus, for each subset, S, of G, we have two natural subgroups of $\mathbf{G}$ to consider:

$\mathbf{N}_S = [\mathbf{G}, S; gi] =$ the normalizer of S, and
$\mathbf{Z}_S = [\mathbf{G}, S; pi] =$ the centralizer of S.

Show that:

(1) $\mathbf{N}_S = \{g \in G | gS = Sg\}$.
(2) $\mathbf{Z}_S = \{g \in G | gs = sg \text{ for all } s \in S\}$.
(3) $\mathbf{Z}_S$ is a normal subgroup of $\mathbf{N}_S$.
(4) If $\mathbf{H}$ is any subgroup of $\mathbf{G}$ such that $\mathbf{Z}_S$ is a normal subgroup of $\mathbf{H}$, then $H \subseteq N_S$.

1.6. The Polya–Burnside Theorem

Recall our observation in example 2 of the last section that the number of inequivalent matrices was just the average number of matrices left fixed by the elements in **G**. As another simple example, consider the six symmetries of a cube leaving one vertex fixed. They form a group **G** which can be viewed as permuting several objects connected with the cube. A few simple computations show that the average number of objects left fixed is always the number of inequivalent objects, e.g., 4 for the vertices, 2 for the faces, 2 for the diagonals, and 2 for the semi-diagonals. Note once again that several symmetries may induce the same permutation without harm.

The theorem asserting the universal validity for finite groups of the observation above appears in Burnside's classic on group theory (see bibliography at the end of chapter 1, [2], section 145, theorem VII). In a fundamental paper on combinatorics, Polya exploited this result and developed new techniques for counting "distinguishable" cases when symmetry was involved. For a delightful introduction to applications of the Polya–Burnside theorem, see chapter 5 of reference [5]. For a careful description of the combinatorial techniques Polya combined with this theorem on groups, see [4]. This theory of counting plays a very important role in the present art of combinatorial mathematics.

Theorem 1.24 (Polya–Burnside). Let **G** be a finite group which operates (via the permutation representation p) on the finite set X. For each $g \in G$, let $g^{\#} = \#\{x \in X | (x)g^p = x\}$. Then

$$(\#G)(\#X/G) = \sum_{g \in G} g^{\#}.$$

Proof. Consider a rectangular table whose index column down the left side consists of the elements of X and whose index row along the top consists of the elements of G. Since X and G are both finite, this table is finite. For each $g \in G$ and each $x \in X$, place a check in the table at row x, column g if and only if $(x)g^p = x$. (See Figure 4 for a sample table.) We count the number of checks in two ways. The number of checks in column g is $g^{\#}$; hence the total number of checks is $\sum_{g \in G} g^{\#}$. Now choose an $x \in X$ and consider the rows indexed by those y's in X such that $x \overset{\mathbf{G}}{\sim} y$. Each of these rows has $\#[\mathbf{G}, y]$ checks, and, by theorem 1.23, $\#[\mathbf{G}, y] = \#[\mathbf{G}, x]$. Moreover, by corollary 1.18, the number of y rows is $\#G/\#[\mathbf{G}, x]$. Hence the total number of checks in this group of rows is $\#G$. (See Figure 4.) Thus, we get $\#G$ checks for each set of rows corresponding to a **G**-orbit, and, therefore, the total number of checks is $(\#G)(\#X/G)$. Comparing the two counts, we see that the theorem is proved.

Orbit	Matrix	e	ρ^2	h	v	m	n	ρ	ρ^3	
1st orbit	$\begin{bmatrix}1 & 1\\1 & 1\end{bmatrix}$	√	√	√	√	√	√	√	√	8 checks
2nd orbit	$\begin{bmatrix}0 & 0\\0 & 0\end{bmatrix}$	√	√	√	√	√	√	√	√	8 checks
3rd orbit	$\begin{bmatrix}1 & 0\\0 & 1\end{bmatrix}$	√	√			√	√			8 checks
	$\begin{bmatrix}0 & 1\\1 & 0\end{bmatrix}$	√	√			√	√			
4th orbit	$\begin{bmatrix}1 & 1\\0 & 0\end{bmatrix}$	√			√					8 checks
	$\begin{bmatrix}0 & 0\\1 & 1\end{bmatrix}$	√			√					
	$\begin{bmatrix}1 & 0\\1 & 0\end{bmatrix}$	√		√						
	$\begin{bmatrix}0 & 1\\0 & 1\end{bmatrix}$	√		√						
5th orbit	$\begin{bmatrix}1 & 0\\0 & 0\end{bmatrix}$	√				√				8 checks
	$\begin{bmatrix}0 & 1\\0 & 0\end{bmatrix}$	√					√			
	$\begin{bmatrix}0 & 0\\0 & 1\end{bmatrix}$	√				√				
	$\begin{bmatrix}0 & 0\\1 & 0\end{bmatrix}$	√					√			
6th orbit	$\begin{bmatrix}0 & 1\\1 & 1\end{bmatrix}$	√				√				8 checks
	$\begin{bmatrix}1 & 0\\1 & 1\end{bmatrix}$	√					√			
	$\begin{bmatrix}1 & 1\\1 & 0\end{bmatrix}$	√				√				
	$\begin{bmatrix}1 & 1\\0 & 1\end{bmatrix}$	√					√			

FIGURE 4. Table of fixed values for the group of the square permuting the 2 by 2 zero–one matrices.

Three observations can help simplify applications of the Polya–Burnside theorem. *First:* The number $g^\#$ is constant for all g's in a particular conjugacy class (proposition 1.9). *Second:* If objects are to be formed by a sequence of n choices with c_i possibilities for choice number i, then there are $c_1 c_2 \cdots c_n$ possible objects (the choice principle). *Third:* Assume that we are trying to count the number of inequivalent patterns formed by placing n types of objects at certain positions on a regular figure. Assume g is a symmetry of that regular figure and that p is the number of (g)-orbits among the positions. If we have enough objects of each type so that we do not run out of objects of a certain type while filling up a (g)-orbit, then the choice principle implies: (1) There are n^p patterns left fixed by g if we may use the same type in more than one orbit. (2) There are $n(n-1)(n-2)\cdots(n-p+1)$ patterns left fixed by g if repetition is not allowed. Example 2, section 1.5 (the 2 by 2 zero–one matrices) is a good example where repetitions *are* allowed.

Example 1. The number of conjugacy classes in a group, **G**. The function i sending each $g \in G$ to its induced inner automorphism is a permutation representation (often unfaithful) of **G** in $\mathbf{S}_G$. In this representation the **G**-orbits are just the conjugacy classes and $g^\# = \#\{h \in G | gh^i = g\} = \#\{h \in G | gh = hg\} = \#N_g$, so the number of conjugacy classes is just

$$\frac{1}{\#G}\sum_{g\in G}\#N_g = \sum_{g\in G}\frac{1}{[\mathbf{G}:\mathbf{N}_g]}.$$

Example 2. Consider the problem of arranging n beads on a circular wire. Two arrangements are equivalent if and only if sliding the beads around the wire or moving the wire ring will carry one arrangement into the other in such a way that adjacent beads remain adjacent. If we distribute the beads evenly around the ring, it is clear that the allowable permutations represent the symmetries of a regular n-gon. There are $2n$ such symmetries, n rotations (including **1**), and n reflections. The reflections fall into one or two conjugacy classes according as n is odd or even. The identity is, of course, in a conjugacy class by itself. The nontrivial rotations fall into $(n-1)/2$ or $n/2$ conjugacy classes according as n is odd or even.

If the n beads are all distinguishable, there are $n!$ arrangements to permute. Of the $2n$ symmetries, only **1** leaves any of these invariant. Hence the average number of fixed arrangements is $n!/2n = (n-1)!/2$. By the Polya–Burnside theorem, this is the number of *distinct* arrangements on the wire.

If some of the beads are indistinguishable, then the problem is more complicated (and more interesting!). For example, let us count the distinguishable arrangements of five black and five white beads on a circular wire. We first count the number of ways of placing the ten beads at the vertices of a decagon. There are 10! orderings of the beads, but since the ordering of the black beads (5!) and of the white beads (5! again) is immaterial, we have $10!/(5!)^2 = 252$ arrangements to permute just as we have 16 zero–one matrices in example 2 in section 1.5. The symmetries of a decagon split into eight conjugacy classes (we number the vertices 0 through 9 with V_0 at the top):

(1) **1**, the identity
(2) θ_j, the reflection whose axis joins vertices V_j and V_{j+5}, $j = 0, 1, 2, 3, 4$
(3) ϕ_j, the reflection whose axis contains the midpoint of the edge joining vertices V_j and V_{j+1}, $j = 0, 1, 2, 3, 4$
(4) ρ and $\rho^{-1} = \rho^9$, ρ a rotation through $\pi/5$ radians
(5) ρ^2 and $\rho^{-2} = \rho^8$, ρ as above
(6) ρ^3 and ρ^7
(7) ρ^4 and ρ^6
(8) ρ^5.

Counting (by conjugacy class) the arrangements left invariant by the symmetries, we find that:

(1) **1** leaves all 252 invariant.
(2) Although there are 6 (θ_0)-orbits among the ten positions, we run out of beads of one color if we try to assign colors to orbits arbitrarily. Since we have an odd number of beads of each color, the beads at the ends of the axis must be of opposite colors. We "choose" the arrangements left fixed by first assigning the two pairs of black and two pairs of white beads to the four orbits off the axis, then choose which end of the axis gets the white bead. There are $4!/(2!)^2$ ways for the first choice and 2 for the second. Hence twelve arrangements are left fixed by θ_j, $j = 0, 1, 2, 3, 4$.
(3) and (8) Neither ρ^2 nor any of the ϕ_j leave an arrangement fixed since a fixed arrangement would require an even number of beads of each color.
(4) and (6) The rotations ρ, ρ^3, ρ^7, ρ^9 all have order 10; hence there is only one orbit among the ten positions. We cannot fill this orbit with beads of one color, and so no arrangements are left fixed.
(5) and (7) Each of ρ^2, ρ^4, ρ^6, ρ^8 is a power of any one of the others so they all leave the same arrangements invariant. There are two (ρ^2)-orbits, each with five positions in it, so ρ^2 leaves two arrangements fixed.

An easy computation shows that the average number of fixed arrangements is 16, so there are 16 inequivalent arrangements. These are shown in Figure 5.

Example 3. Let us consider the number of different ways to place red and white balls at the eight vertices of a cube. Assume that we have at least eight balls of each color. There are two natural ways to consider arrangements

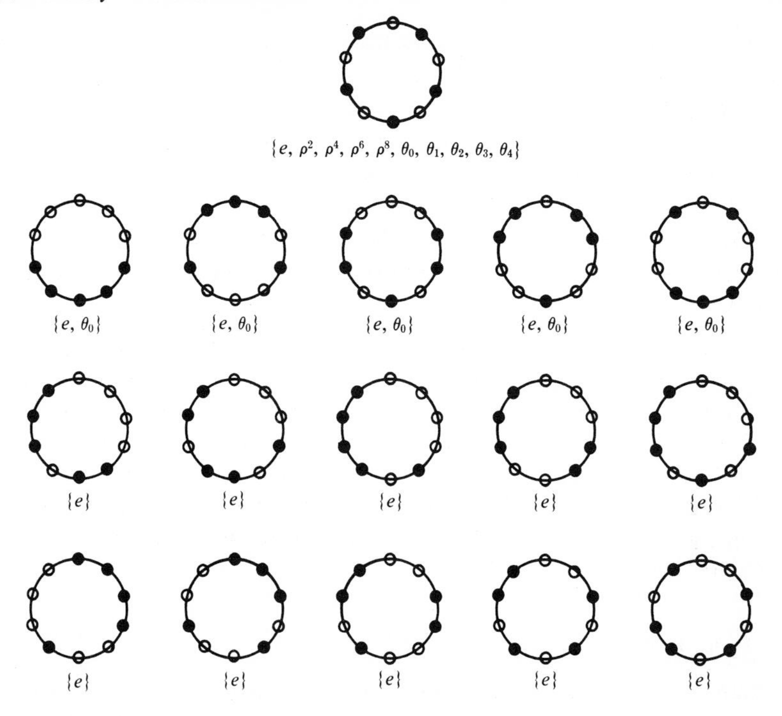

FIG. 5. The 16 distinct arrangements of 5 black and 5 white beads on a ring, showing $[\mathbf{G}, x]$ for each arrangement.

equivalent: relative to the group of 24 rigid motions, or else relative to the group of all 48 symmetries. We tabulate, by conjugacy class, the results of our computations, as shown in Table 8.

TABLE 8. Vertex Orbits for Symmetries of the Cube

Type of symmetry, i.e., conjugacy class	*No. of symmetries of this type*	*No. of (g)-orbits among the vertices*
1	1	8
90° rotation	6	2
120° rotation	8	4
180° rotation, facial axis	3	4
180° rotation, semi-diagonal axis	6	4
Inversion	1	4
90° rotatory inversion	6	2
120° rotatory inversion	8	2
Reflection, axial plane	3	4
Reflection, diagonal plane	6	6

The two orbits for a 90° rotatory reflection consist of the vertices of the two inscribed tetrahedra. Since there are 2^p arrangements fixed by g if there are p (g)-orbits, we find that the average number left fixed by the 24 rigid motions is 23, whereas the average number left fixed by all 48 symmetries is 22. Thus there is only one pair of "enantiomorphic" arrangements, i.e., one pair of oppositely oriented arrangements—one left-handed, the other right-handed. We leave it to the reader to find them. See pages 28–30 of [16] for an interesting discussion of some scientific ramifications of levo and dextro forms of crystals.

With this example we conclude our excursion into combinatorics. For those who desire to go further, another good example showing how the Polya–Burnside theorem can be applied is a paper, [6], on the number of different polygons (not necessarily convex) with n sides. For an introduction to "generating functions," one of the tools Polya combined with the Polya–Burnside theorem to prove his fundamental theorem on combinatorics, we recommend Polya's expository paper, [12], and the more technical article by De Bruijn, [4].

Exercise 1.6

1. Count the number of distinct, completed tic-tac-toe configurations if the game is played on an opaque board, i.e., the board can only be rotated, not flipped. What if the board is transparent so that it can be reflected in an axis of symmetry as well as rotated?

2. Assume that **G** operates transitively on the set X, i.e., that X is the only **G**-orbit in X. Choose $x \in X$ and let $\mathbf{H} = [\mathbf{G}, x]$. Show that the number of **H**-orbits in X is the average of the squares of the numbers of elements left fixed by elements of **G** ([2], section 145, theorem VII).

3. Count the number of inequivalent arrangements (see example 2) of:
 (1) six red and four black beads on a ring
 (2) five red, three black, and two yellow beads on a ring.

4. Count the number of different ways of placing three red and three black balls at the vertices of a regular octahedron:
 (1) if we are only allowed to rotate the octahedron
 (2) if we may also turn the octahedron inside out.

5. In Table 6, section 1.5, we gave a representative of each of the six inequivalent classes of 2 by 2 zero–one matrices. How many classes are there if, instead of the complete symmetry group of the square, we only allow rotations? What if, in addition to rotations, we can replace all 0's by 1's and vice versa? Count the inequivalent 3 by 3 zero–one matrices in these two situations.

6. See chapter V in [5], pages 49–52 in [13], or section 5.4 in [4] for more exercises on the Polya–Burnside theorem.

Bibliography and suggestions for further reading

[1] Birkhoff, Garrett, and Saunders MacLane, *A Survey of Modern Algebra*, 3rd ed., Macmillan, New York, 1965.
[2] Burnside, William, *Theory of Groups of Finite Order*, 2nd ed., Cambridge Univ. Press, Cambridge, 1911, or Dover, New York, 1955.
[3] Coxeter, H. S. M., and W. O. J. Moser, *Generators and Relations for Discrete Groups, Ergebnisse der Mathematik und ihrer Grenzgebiete No. 14*, Springer, Berlin, 1957.
[4] De Bruijn, N. G., "Polya's Theory of Counting," in *Applied Combinatorial Mathematics* (Edwin F. Beckenbach, ed.), Univ. of Calif. Engineering and Physical Sciences Extension Series, Wiley, New York, 1964.
[5] Golomb, Solomon W., *Polyominoes*, Scribner's, New York, 1965.
[6] Golomb, Solomon W., and L. R. Welch, "On the Enumeration of Polygons," *Am. Math. Monthly* **67,** 349–353 (1960).
[7] Hall, Marshall, *The Theory of Groups*, Macmillan, New York, 1959.
[8] Herstein, I. N., *Topics in Algebra*, Blaisdell, Boston, 1964.
[9] Kurosh, A. G., *The Theory of Groups*, Vol. 1 (transl. by K. A. Hirsch), Chelsea, New York, 1955.
[10] Miller, Donald W., "On a Theorem of Holder," *Bull. Am. Math. Soc.* **65,** 252–254 (1958).
[11] Parker, E. T., "A Memorable Teacher," *Am. Math. Monthly* **72,** 1127–1128 (1965).
[12] Polya, G., "On Picture-Writing," *Am. Math. Monthly* **63,** 689–697 (1956).
[13] Rotman, Joseph J., *The Theory of Groups, An Introduction*, Allyn and Bacon, Boston, 1965.
[14] Schreier, J., and S. Ulam, "Über die Automorphismen der Permutationsgruppe der natürlichen Zahlenfolge," *Fund. Math.* **28,** 258–260 (1937).
[15] Segal, Irving E., "The Automorphisms of the Symmetric Group," *Bull. Am. Math. Soc.* **46,** 565 (1940).
[16] Weyl, Hermann, *Symmetry*, Princeton Univ. Press, Princeton, 1952.
[17] Wielandt, Helmut, *Finite Permutation Groups* (transl. by R. Bercov), Academic Press, New York, 1964.

For more information on sets, functions, and elementary abstract algebra, consult [1] or [8]. These books, and most others on abstract algebra, contain the details that we have omitted in some proofs in the early part of this chapter.

There are many excellent books on group theory. Certainly no bibliography should omit the classic by Burnside, [2]. The research monograph by Wielandt, [17], contains (in highly condensed form) everything that we have mentioned about permutation groups in this chapter. A wide variety of examples of finite groups can be gleaned from the tables in [3].

If one combines [10], [14], and [15], then he will have the complete picture concerning automorphism of $\mathbf{S}_X$, regardless of whether or not X is finite. The proof in [9], pp. 92–95, that $\mathbf{S}_X$ is complete when X is finite is essentially the same as the proof in [15].

Our references to combinatorial mathematics avoid the standard works on the subject. If one consults [4] and the other articles bound in the same volume, he will have a more than adequate supply of references. We have found chapter V in [5] a readable and delightful introduction to combinatorics.

Everyone who studies this book should read the short book by Weyl, [16]. This is a slightly modified version of four lectures given at Princeton in February, 1951, just before Weyl retired from the Institute for Advanced Study.

Finally, we have included a testimonial, [11], to the importance of considering good examples in a mathematics course. Every teacher, or potential teacher, of mathematics should be dedicated to the quest for illuminating examples.

Chapter 2

Isometries and Similarities: An Intuitive Approach

Our main objective in this chapter is to help the reader develop facility in handling transformations, products of transformations, and groups of transformations. To do this we consider a familiar geometry, ordinary euclidean space, and study the structure of two important groups associated with this geometry: the euclidean group, **E**, and the similarities group, **S**. Keep in mind that we are not attempting to present a careful development of euclidean geometry. In particular, we use without proof many of the elementary properties of euclidean space, especially those referring to congruence, parallelism, and perpendicularity. The intuitive material presented in this chapter provides important background material for the introduction to crystallography in Chapter 3 and the formal study of affine and projective geometry in Chapters 5 and 6.

Before defining the euclidean and similarities groups, we must establish a few notational conventions for points, lines, and planes in euclidean space. We view euclidean space as a set of points with certain distinguished subsets called lines and certain other subsets called planes. This viewpoint has certain implications concerning notation, e.g., if A and B are planes, then $A \cap B$, their intersection, is either A itself (if $A = B$), or a line, or the empty set (if $A \parallel B$ and $A \neq B$). Similarly, we abbreviate "the point P or the line L is in the plane A" by "$P \in A$ or $L \subseteq A$."

We wish to reserve lowercase letters for names of similarities, so we choose names for planes from the set $\{A, B, C, D\}$, for lines from $\{K, L, M, N\}$, and for points from $\{W, X, Y, Z\}$, using subscripts if too many objects need names simultaneously.

The concept of independent points is not always covered in elementary geometry courses. Still it will prove to be very convenient for us. It is defined as follows.

Definition. Two, three, or four points in euclidean space are *independent* if they are distinct, noncollinear, or noncoplanar, respectively.

Given two points, W and X, the distance between them is denoted $|W, X|$ and the closed line segment consisting of W, X, and all points between W and X is denoted by $[W, X]$. If W, X, and Y are independent points, then $W + X$ denotes the line joining W and X, and $W + X + Y$ denotes the plane containing W, X, and Y. As usual, $\perp$ and $/\!/$ are used for perpendicular and parallel. We consider each line and each plane to be parallel to itself and insist that perpendicular lines intersect.

2.1. *Isometries and similarities*

The structure of euclidean space is completely determined by the various properties of the distance between two points. Thus the two natural types of symmetries of euclidean space are those permutations that leave distances invariant, i.e., map a segment to a congruent segment, and those that preserve equal distances, i.e., map pairs of congruent segments to pairs of congruent segments. In this section we investigate the basic properties of the symmetries that preserve equal distances. In particular we show that a permutation of euclidean space preserves equal distances if and only if it stretches (or contracts) *all* distances by a constant factor, i.e., if and only if it corresponds to our intuitive concept of a similarity transformation.

Definition. A permutation, ϕ, of euclidean space is called a *similarity* if for all points W, X, Y, and Z, $|W, X| = |Y, Z|$ if and only if $|(W)\phi, (X)\phi| = |(Y)\phi, (Z)\phi|$. In other words, a similarity is a permutation that *preserves equal distances.* If ϕ *preserves distances*, i.e., if $|W, X| = |(W)\phi, (X)\phi|$ for all W and X, then we call ϕ an *isometry* [from the Greek *isos* (equal) and *metron* (measure)].

It is easy to verify (do so!) that the similarities form a subgroup of the full permutation group of euclidean space; moreover, the isometries form a subgroup of the group of similarities. These are the two groups we shall study so we give them names.

Definition. The *similarities group*, **S**, is the group of similarities of euclidean space with product the standard product for permutations. The *euclidean group*, **E**, is that subgroup of **S** consisting of all isometries.

Theorem 2.1. Similarities preserve planes, lines, perpendicularity, parallelism, midpoints, inequalities between distances, and line segments. More precisely, if ϕ is a similarity, then the following are true:

(1) If A is a plane, so is $(A)\phi$.
(2) If K is a line, so is $(K)\phi$.
(3) If F and G are lines or planes and $F \perp G$, then $(F)\phi \perp (G)\phi$.
(4) If F and G are lines or planes and $F /\!/ G$, then $(F)\phi /\!/ (G)\phi$.
(5) If W is the midpoint of $[X, Y]$, then $(W)\phi$ is the midpoint of $[(X)\phi, (Y)\phi]$.
(6) If $|W, X| < |Y, Z|$, then $|(W)\phi, (X)\phi| < |(Y)\phi, (Z)\phi|$.
(7) $[X, Y]\phi = [(X)\phi, (Y)\phi]$.

Proof. We attack the seven assertions in order.

1. If A is a plane, then we can choose independent points Y and Z such that A is the perpendicular bisector of Y and Z. Let B be the perpendicular bisector of $(Y)\phi$ and $(Z)\phi$. Since $X \in A$ if and only if $|X, Y| = |X, Z|$ and $|X, Y| = |X, Z|$ implies $|(X)\phi, (Y)\phi| = |(X)\phi, (Z)\phi|$, then it follows that $(A)\phi \subseteq B$. Since ϕ^{-1} is also a similarity (note how the "if and only if" clause of the definition is essential here), we also have $(B)\phi^{-1} \subseteq A$. Thus, $(A)\phi = B$, showing that ϕ preserves planes.

2. If K is a line, choose two planes, A and B, such that $K = A \cap B$. Then $(K)\phi = (A)\phi \cap (B)\phi$ is the non-empty intersection of two distinct planes; hence it is a line.

3. If either F or G, say, F, is a plane and $F \perp G$, then we can choose points Y and Z in G such that F is the perpendicular bisector of Y and Z. But then, as in the proof of part 1, $(F)\phi$ is the perpendicular bisector of $(Y)\phi$ and $(Z)\phi$. Since these two points are in $(G)\phi$, it follows that $(F)\phi \perp (G)\phi$. The case in which both F and G are lines is left to the reader (exercise 2.1, problem 1).

4. If F and G are both planes, then $F /\!/ G$ if and only if there is a line K such that $F \perp K$ and $K \perp G$. By 3 above, ϕ preserves perpendicularity; hence $(F)\phi /\!/ (G)\phi$. Similar proofs hold for the cases in which either F or G or both are lines (exercise 2.1, problem 1).

5. The midpoint of $[X, Y]$ is $A \cap K$ where $K = X + Y$ and A is the perpendicular bisector of X and Y. As in part 1, $(A)\phi$ is the perpendicular bisector of $(X)\phi$ and $(Y)\phi$, and by part 2, $(K)\phi = (X)\phi + (Y)\phi$. Hence $(W)\phi = (A)\phi \cap (K)\phi$ is the midpoint of $[(X)\phi, (Y)\phi]$.

6. Given points W, X, Y, Z such that $|W, X| < |Y, Z|$, choose a point Z_1 such that $|W, Z_1| = |Y, Z|$. The distance between two points is less than s *if and only if* one of the points is the midpoint of some chord of the sphere of radius s centered at the second point. Thus, $|W, X| < |Y, Z| = |W, Z_1|$ implies

that we can choose Z_2, Z_3 such that $|W, Z_1| = |W, Z_2| = |W, Z_3|$, and such that X is the midpoint of $[Z_2, Z_3]$. By definition of a similarity, $|(W)\phi, (Z_1)\phi| = |(W)\phi, (Z_2)\phi| = |(W)\phi, (Z_3)\phi|$, and, by part 5, $(X)\phi$ is the midpoint of the chord $[(Z_2)\phi, (Z_3)\phi]$. Thus, $|(W)\phi, (X)\phi| < |(W)\phi, (Z_1)\phi| = |(Y)\phi, (Z)\phi|$.

7. Let X and Y be distinct points, and let Z be the midpoint of $[X, Y]$. A point, W, is between X and Y if and only if W is on $X + Y$ and $|Z, W| < |Z, X|$. However, $(W)\phi$ is on $(X)\phi + (Y)\phi$ (part 2), and $|(Z)\phi, (W)\phi| < |(Z)\phi, (X)\phi|$ (part 6). Hence $(W)\phi$ is between $(X)\phi$ and $(Y)\phi$. Clearly, the endpoints cause no trouble, so $[X, Y]\phi \subseteq [(X)\phi, (Y)\phi]$. Since ϕ^{-1} is also a similarity, $[X, Y]\phi \supseteq [(X)\phi, (Y)\phi]$, and the proof is complete.

Theorem 2.2. A permutation, ϕ, of euclidean space is a similarity if and only if there is a positive number α such that for all points X and Y, $|(X)\phi, (Y)\phi| = \alpha|X, Y|$.

Proof. If ϕ is a permutation such that the constant α exists, then clearly ϕ is a similarity. Now, to prove the converse, assume that ϕ is a similarity. Choose two independent points W and Z, and let $\alpha = |(W)\phi, (Z)\phi|/|W, Z|$, i.e., α is the factor by which the fixed segment $[W, Z]$ is "stretched." We must show that every segment, $[X, Y]$, is stretched by the same amount. This becomes obvious if we can show that ϕ is a "homogeneous" mapping, i.e., if we can show that $|X, Y| = \beta|W, Z|$ implies $|(X)\phi, (Y)\phi| = \beta|(W)\phi, (Z)\phi|$. To do this we use a standard device, covering the cases in which β is integral, rational, or real, in that order.

Case 1. β is a nonnegative integer. If $\beta = 0$ or 1, then

$$|(X)\phi, (Y)\phi| = \beta|(W)\phi, (Z)\phi|$$

whenever $|X, Y| = \beta|W, Z|$, since ϕ is a function ($\beta = 0$) and is a similarity ($\beta = 1$). Now assume this is true for all pairs of points for which $\beta = k$, and assume that X, Y is a new pair for which

$$|X, Y| = (k + 1)|W, Z| .$$

Choose a point Y_1 such that $Y_1 \in [X, Y]$ and $|XY_1| = k|W, Z|$. Since Y_1 is between X and Y, this implies $|Y_1, Y| = |W, Z|$. By our induction hypothesis,

$$|(X)\phi, (Y_1)\phi| = k|(W)\phi, (Z)\phi|$$

and by definition of ϕ,

$$|(Y_1)\phi, (Y)\phi| = |(W)\phi, (Z)\phi| .$$

By part (7) of theorem 2.1, $(Y_1)\phi$ is between $(X)\phi$ and $(Y)\phi$, and thus,

$$|(X)\phi, (Y)\phi| = |(X)\phi, (Y_1)\phi| + |(Y_1)\phi, (Y)\phi| = (k + 1)|(W)\phi, (Z)\phi| .$$

This completes the induction step, and hence the homogeneity property holds for all nonnegative integers β.

Case 2. β is a positive rational, say, $\beta = m/n$. In this case, if X and Y are two points such that $|X, Y| = \beta|W, Z|$, then $n|X, Y| = m|W, Z|$. Choose a point Y_1 such that

$$n|X, Y| = |X, Y_1| = m|W, Z| .$$

Two applications of case 1 yield

$$n|(X)\phi, (Y)\phi| = |(X)\phi, (Y_1)\phi| = m|(W)\phi, (Z)\phi| ,$$

and thus the homogeneity property holds whenever β is rational.

Case 3. β is a positive real number. Assume that $|X, Y| = \beta|W, Z|$, but that

$$|(X)\phi, (Y)\phi| = \gamma|(W)\phi, (Z)\phi|$$

with $\gamma \neq \beta$. Choose a rational number λ between β and γ and a point Y_1 such that $|X, Y_1| = \lambda|W, Z|$. When we apply ϕ, we find the inequality between $|X, Y_1|$ and $|X, Y|$ reversed, e.g., if $\beta < \lambda < \gamma$, then

$$|X, Y_1| = \lambda|W, Z| > \beta|W, Z| = |X, Y| ,$$

but (by case 2)

$$|(X)\phi, (Y_1)\phi| = \lambda|(W)\phi, (Z)\phi| < \gamma|(W)\phi, (Z)\phi| = |(X)\phi, (Y)\phi| .$$

This contradicts part (6) of theorem 2.1, so we cannot have $\beta \neq \gamma$. Thus, the homogeneity property holds if β is any nonnegative real number.

Definition. If ϕ is a similarity, then the positive constant α such that $|(X)\phi, (Y)\phi| = \alpha|X, Y|$ for all points X and Y is called the *stretching factor* of ϕ.

The theorem above is often quoted but seldom proved in detail. In fact, many authors choose the easy course of *defining* similarities to be those permutations stretching distances by a fixed factor. This is often a wise choice since it is a bit tedious to prove that the stretching factor exists. However as soon as it is known to exist, many proofs concerning similarities can be simplified. From now on we exploit the stretching factor and do not return to the basic definition of a similarity until section 2.7, where we investigate the automorphisms of **S**.

Corollary 2.3. If ϕ is a similarity, then ϕ preserves angles, i.e., angle XYZ is congruent to angle $(X)\phi(Y)\phi(Z)\phi$.

Proof. Since ϕ has a stretching factor, the corresponding sides of triangles X, Y, Z and $(X)\phi, (Y)\phi, (Z)\phi$ are proportional. Hence the triangles are similar, and angle XYZ is congruent to angle $(X)\phi(Y)\phi(Z)\phi$.

Corollary 2.4. A similarity, ϕ, is a continuous transformation, i.e., if X_n, $n = 1, 2, \ldots$ is a sequence of points converging to the point X, then the sequence $(X_n)\phi$ converges to $(X)\phi$.

Proof. Let α be the stretching factor of ϕ. Given $\epsilon > 0$, choose an integer N such that $n > N$ implies $|X_n, X| < \epsilon/\alpha$. Then for $n > N$, we also have $|(X_n)\phi, (X)\phi| < \epsilon$. Hence for each $\epsilon > 0$, we can find a suitable N and thus $\lim_{n\to\infty} (X_n)\phi = (X)\phi$.

Theorem 2.5. The mapping which assigns to each similarity its stretching factor is a homomorphism of **S** into the multiplicative group of positive real numbers. The kernel of this homomorphism is **E**; hence **E** is a normal subgroup of **S**.

The proof is left to the reader (exercise 2.1, problem 4).

In the following theorem we show that a similarity has a unique fixed point if it is not an isometry. The technique used is a specialization of the general theorem that any mapping of a complete metric space into itself must have a unique fixed point if it "contracts" distances. This theorem, known as the "principle of contractive mappings," has many important applications in abstract analysis. See pp. 43–51 in [7] for examples of such applications and a proof of the general theorem.

Theorem 2.6. If ϕ is a similarity but not an isometry, then ϕ has exactly one fixed point.

Proof. Let α be the stretching factor of ϕ. We are assuming that $\alpha \neq 1$, so it is clear that ϕ cannot have more than one fixed point since the distance between two fixed points is not changed by applying ϕ. ϕ and ϕ^{-1} have the same fixed points and reciprocal stretching factors so we can assume without loss of generality that $\alpha < 1$. This implies that α is a contraction mapping, i.e., that there is a constant $\alpha < 1$ such that $|(X)\phi, (Y)\phi| \leq \alpha|X, Y|$ for all points X and Y.

Choose a point X and let $X_0 = X$, $X_1 = (X_0)\phi, \ldots, X_n = (X_{n-1})\phi, \ldots$. If $X_n = X_{n-1}$ for some n, then we have our desired fixed point. If not, then we have an infinite sequence of points which (we shall show) has a limit point, and this limit point is our desired fixed point. For any n and m such that $n < m$,

$$|X_n, X_m| = \alpha|X_{n-1}, X_{m-1}| = \alpha^2|X_{n-2}, X_{m-2}| = \cdots = \alpha^n|X_0, X_{m-n}| .$$

Thus for any n and any k,

$$\begin{aligned} |X_n, X_{n+k}| &= \alpha^n|X_0, X_k| \leq \alpha^n\{|X_0, X_1| + |X_1, X_2| + \cdots + |X_{k-1}, X_k|\} \\ &\leq \alpha^n|X_0, X_1|\{1 + \alpha + \alpha^2 + \cdots + \alpha^{k-1}\} \\ &\leq \alpha^n \frac{|X_0, X_1|}{1 - \alpha} . \end{aligned}$$

Since $\alpha < 1$, $|X_n X_{n+k}| \rightarrow 0$ as $n \longrightarrow \infty$ (independent of k). Hence the sequence $X_0, X_1, X_2, \ldots$ is a Cauchy sequence and therefore has a limit point X. ϕ is a continuous transformation, and so

$$(X)\phi = \lim_{n \to \infty} (X_n)\phi = \lim_{n \to \infty} X_{n+1} = X\,.$$

Thus X is the desired fixed point, and the proof is complete.

Once we are sure that the fixed point exists it is relatively easy to find it as one of the two points of intersection of three spheres (see p. 103 of [5]). But without knowing that the fixed point exists, it seems to be difficult to prove that the spheres actually intersect.

To conclude this section we present a theorem which asserts that euclidean space is at least as "rigid" as one would intuitively expect. In the next section we show that euclidean space is as "movable" as one would expect.

Theorem 2.7. If ϕ is a similarity, then it must be an isometry if it has more than one fixed point. Furthermore, if ϕ has two, three, or four independent fixed points, then it must leave pointwise fixed the associated line, plane, or entire space. In particular, the only similarity with four independent fixed points is the identity.

Proof. Since distances between fixed points are not changed, only isometries can have more than one fixed point.

If ϕ is an isometry leaving the independent points X and Y fixed, and if W is on $X + Y$, then $(W)\phi$ must lie on $X + Y$, on the sphere centered at X of radius $|X, W|$, and on the sphere centered at Y of radius $|Y, W|$. Thus, $(W)\phi = W$, and ϕ leaves $X + Y$ pointwise fixed.

If ϕ leaves the independent points X, Y, and Z fixed, then by the paragraph above it leaves fixed every point on the (extended) sides of the triangle XYZ. But through every point in $X + Y + Z$, there is a line intersecting these three sides in at least two independent points. Since ϕ must leave this line pointwise fixed, it leaves $X + Y + Z$ pointwise fixed.

An analogous argument shows that $\phi = \mathbf{1}$, i.e., that ϕ leaves *every* point fixed, if ϕ leaves at least four independent points fixed.

Exercise 2.1

1. Show that if ϕ is a similarity, and K and L are lines, then:
(1) $K \perp L$ implies $(K)\phi \perp (L)\phi$.
(2) $K \mathbin{/\!\!/} L$ implies $(K)\phi \mathbin{/\!\!/} (L)\phi$.

2. Prove that a similarity which is not an isometry cannot have finite order.

3. Show that properties 1 and 2 below are equivalent for any permutation, ϕ, of euclidean space.

(1) If L is a line, then so is $(L)\phi$.

(2) If A is a plane, then so is $(A)\phi$.

Hint. First show that ϕ^{-1} shares either 1 or 2 with ϕ.

Such permutations are called collineations. Find at least one collineation that is not a similarity. Are all similarities collineations?

4. Prove theorem 2.5.

5. Show that any collineation (see problem 3 above), ϕ, of space such that $L \perp A$ implies $(L)\phi \perp (A)\phi$ is a similarity.

2.2. *Involutions in* S

An *involution* is an element of order two in a group, i.e., an element which is its own inverse but is not the identity. In this section we show that with each point, line, and plane there is a naturally associated involution in **S** and that most important geometric relationships can be characterized by algebraic relations between involutions. Once again we stress the importance of the conjugacy relation within a group, and we illustrate how conjugating an involution in **S** by a similarity has a natural geometric interpretation.

Definition. Let A be a plane, M be a line, and X be a point in euclidean space.

1. The *reflection* with *mirror* A is that transformation a such that a leaves every point in A fixed and sends any point, W, not in A to the unique point $(W)a$ such that A is the perpendicular bisector of $[W, (W)a]$.

2. The *half-turn* with *axis* M is that transformation, m, which leaves every point in M fixed and sends any point, W, not in M to the unique point $(W)m$ such that M contains the midpoint of $[W, (W)m]$ and is contained in the perpendicular bisector of $[W, (W)m]$.

3. The *inversion* with *center* X is that transformation, x, which leaves X fixed and sends any point $W \neq X$ to the unique point $(W)x$ such that X is the midpoint of $[W, (W)x]$.

Theorem 2.8. Reflections, half-turns, and inversions are all involutions in **E**. Moreover, the only involutions in **S** are the reflections, half-turns, and inversions.

Proof. Let a, m, and w be the reflection, half-turn, and inversion with mirror A, axis M, and center W, respectively. Introduce three cartesian coordinate systems, the first with A as its yz-plane, the second with M as its z-axis, and the third with W as its origin.

In the first, second, or third coordinate system, respectively, a, m, or w

sends the point with coordinates (x, y, z) to the point with coordinates $(-x, y, z)$, $(-x, -y, z)$, or $(-x, -y, -z)$, respectively. From this it is clear that a, m, w are involutions and preserve distances. Thus, reflections, half-turns, and inversions are all involutions in **E**.

Now assume that ϕ is some involution in **S**. The stretching factors of ϕ and $\phi^{-1} = \phi$ are reciprocal yet equal; hence ϕ is in **E**. If X is any point, then ϕ interchanges X and $(X)\phi$. Hence ϕ either leaves X or the midpoint of $[X, (X)\phi]$ fixed. In either case, ϕ has at least one fixed point. It follows easily from theorems 2.7 and 2.1 that ϕ is an inversion, half-turn, or reflection according as 1, 2, or 3 is the maximum number of *independent* fixed points of ϕ (exercise 2.2, problem 3).

Notational Convention. In order to exploit fully the connection among points, lines, and planes and their associated involutions, we adopt the following convention: Whenever we use a capital letter as the name of a particular point, line, or plane, we use the corresponding lowercase letter for the associated inversion, half-turn, or reflection. Lowercase Greek letters are used for similarities which may or may not be involutions.

Before considering products of similarities, we caution the reader that similarities *do not move* points, lines, or planes. Nevertheless, it is a valuable aid to the intuition to have some method of visualizing similarities in terms of motion. We suggest the following as a reasonably safe one.

To visualize a similarity, ϕ, imagine two superimposed copies of space, one fixed and the other capable of being moved, reflected through a plane, and uniformly dilated. The correspondence, $X \leftrightarrow (X)\phi$, between points (which according to our definition *is* the similarity) can be visualized by "moving" the movable copy of space such that the point in that copy originally superimposed on X is now superimposed on $(X)\phi$. There are, of course, many different ways to carry out this motion for a particular similarity, e.g., if M and N are perpendicular lines, then $mnmn = \mathbf{1}$, and hence the "motion" corresponding to the identity can be effected either by "no motion" or by rotating the movable copy 180° first about M, then about N (in the *fixed* copy!), then about M, and finally about N.

The example above illustrates two important points. *First:* Products are read from left to right! *Second:* Names for isometries or descriptions of isometries are given in terms of objects in the fixed copy! Figure 6 illustrates this in the case of a product of two inversions.

The connection between the geometry of euclidean space and the algebra of **S** depends on the fixed points, lines, and planes of the involutions above. Table 9 summarizes this information.

The usefulness of this information about fixed points is vastly increased by combining it with the geometric interpretation of conjugacy in **S**. At this

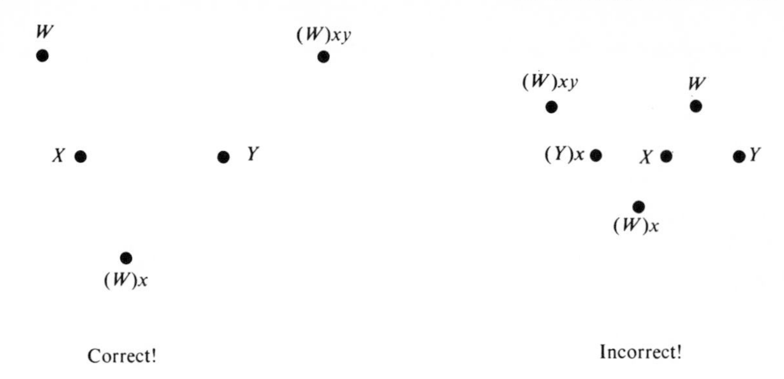

FIG. 6. Correct and incorrect ways to compute the image of a point W under the product, xy, of two inversions.

TABLE 9. Fixed Points, Lines, and Planes of Involutions

Involution	*Fixed points*	*Fixed lines*	*Fixed planes*
Reflection, a	$X \in A$	$L \subseteq A$ (pointwise) $L \perp A$ (globally)	$B = A$ (pointwise) $B \perp A$ (globally)
Half-turn, m	$X \in M$	$L = M$ (pointwise) $L \perp M$ (globally)	$B \supseteq L$ (globally) $B \perp L$ (globally)
Inversion, w	$X = W$	$W \in L$ (globally)	$W \in B$ (globally)

point, the reader should review propositions 1.9 and 1.10, examples 5 and 6 at the end of section 1.2, and the material on inner automorphisms in section 1.4.

Theorem 2.9. Let ϕ be a similarity, x the inversion with center X, m the half-turn with axis M, and b the reflection with mirror B. Then

(1) $(x)\phi^i = \phi^{-1}x\phi$ is the inversion with center $(X)\phi$.
(2) $(m)\phi^i = \phi^{-1}m\phi$ is the half-turn with axis $(M)\phi$.
(3) $(b)\phi^i = \phi^{-1}b\phi$ is the reflection with mirror $(B)\phi$.

Proof. By proposition 1.9, the fixed points of $(x)\phi^i$, $(m)\phi^i$, and $(b)\phi^i$ are $(X)\phi$, $(M)\phi$, and $(B)\phi$, respectively. Since ϕ^i is an automorphism of **S**, $(x)\phi^i$, $(m)\phi^i$, and $(b)\phi^i$ are involutions in **S**, and, therefore, by theorem 2.8, they must be (in view of their fixed points) an inversion, half-turn, and reflection, respectively.

Exploiting our notational convention concerning the capital—lowercase correspondence among points, lines, and planes and their associated involutions, we can restate theorem 2.9 as follows:

Theorem 2.9 (restatement). If ϕ is a similarity, then the following hold true:

(1) $(X)\phi = Y$ and $(x)\phi^i = y$ are equivalent statements for points and inversions.
(2) $(M)\phi = N$ and $(m)\phi^i = n$ are equivalent statements for lines and half-turns.
(3) $(A)\phi = B$ and $(a)\phi^i = b$ are equivalent statements for planes and reflections.

Corollary 2.10. No nontrivial similarity commutes with every inversion, nor with every half-turn, nor with every reflection. Thus, the centers of **E** and of **S** are both $\{\mathbf{1}\}$.

Proof. In view of theorem 2.9, a similarity, ϕ, that commutes with every inversion must satisfy $\phi^{-1}x\phi = x$, or, in other words, $(X)\phi = X$ for *every* point X. However, this means that $\phi = \mathbf{1}$. Likewise, if ϕ commutes with all half-turns (or all reflections), then it must leave every line (or every plane) fixed and hence leave every point fixed.

Since the centers of **E** and **S** are $\{\mathbf{1}\}$, each of these groups is isomorphic to its group of inner automorphisms. At the end of this chapter, we prove that **S** is isomorphic to the group of all automorphisms of **E** and that **S** is a complete group.

Using theorem 2.9, the equivalence of $\phi^{-1}\theta\phi = \theta$ and $\phi\theta = \theta\phi$, and our table of fixed objects for involutions in **S**, we can translate statements about incidence and perpendicularity into algebraic statements.

Theorem 2.11. Each of the following geometric statements about points X and Y, lines M and N, and planes A and B is equivalent to its algebraic counterpart in the second column.

(1) $X = Y$	x and y commute
(2) $X \in M$	x and m commute
(3) $X \in A$	x and a commute
(4) $M \perp N$	m and n are distinct and commute
(5) $M \perp A$ or $M \subseteq A$	m and a commute
(6) $A \perp B$	a and b are distinct and commute.

The proofs are left as problems (exercise 2.2, problem 4).

Note that whenever ϕ and θ are involutions in a group, $\phi\theta$ is also an involution if and only if ϕ and θ commute. The following theorem gives the complete story as to what type of involution in **S** is involved whenever ϕ and θ are involutions in **S** that commute. It also formalizes the intuitive observation from solid analytic geometry that two symmetries in the coordinate planes, axes, or origin invariably imply a third.

Theorem 2.12. Let A, B, and C be three mutually perpendicular planes intersecting at the point W. Let $L = B \cap C$, $M = A \cap C$, and $N = A \cap B$ be the lines of intersection. Then $\{\mathbf{1}, a, b, c, l, m, n, w\}$ is an abelian group with products determined as follows for typical lines $K \in \{L, M, N\}$ and planes $D \in \{A, B, C\}$:

(1) If $K \perp D$, then the product of any two of w, k, and d, in either order, is the third.

(2) If $D_1 \perp D_2$, then $d_1 d_2 = d_2 d_1$ is the half-turn with axis $D_1 \cap D_2$.

(3) If $K_1 \perp K_2$, then $k_1 k_2 = k_2 k_1$ is the half-turn whose axis contains $K_1 \cap K_2$ and is perpendicular to both K_1 and K_2.

(4) If $K \subseteq D$, then $kd = dk$ is the reflection whose mirror contains K and is perpendicular to D.

Proof. Introduce a cartesian coordinate system with L as x-axis, M as y-axis, and N as z-axis. Then the eight isometries send a point X with coordinates (x, y, z) to a point with coordinates $(\pm x, \pm y, \pm z)$ according to the following scheme:

$$\begin{array}{ll}
\mathbf{1} & (+, +, +) \\
a & (-, +, +) \\
b & (+, -, +) \\
c & (+, +, -) \\
l & (+, -, -) \\
m & (-, +, -) \\
n & (-, -, +) \\
w & (-, -, -).
\end{array}$$

With this information it is tedious, but trivial, to complete the proof.

Using the results of the theorem above, we can find the product of any two involutions in $\mathbf{S}$ whenever the involutions commute. In the next section we investigate products of involutions that do not commute.

Exercise 2.2

1. Find three nonisomorphic groups, $\mathbf{G}_i$, $i = 1, 2, 3$, each of which has all of the following properties: $\mathbf{G}_i \subseteq \mathbf{E}$, $\#\mathbf{G}_i = 8$, $\mathbf{G}_i$ is abelian.

2. Use Theorems 2.7, 2.11, and 2.12 to show that the three lines joining the midpoints of opposite edges of a regular tetrahedron are mutually perpendicular. Is the converse also true?

3. Complete the proof of theorem 2.8, i.e., show that if ϕ is an involution, then ϕ is an inversion, half-turn, or reflection according as 1, 2, or 3 is the maximum number of independent fixed points of ϕ.

4. Prove the six assertions in theorem 2.11.

5. Show that the set of reflections is a complete conjugacy class in **E**. Do the same for the set of half-turns and the set of inversions.

6. Let **G** be the group defined in theorem 2.12. Find all automorphisms of **G**. Are they all induced by inner automorphisms of **S**?

2.3. *The classification of isometries*

Since we know that reflections, half-turns, and inversions are all isometries, we know they generate a subgroup of **E**. In this section we show that they in fact generate **E** and show how the isometries can be classified into surprisingly few distinct types that are relatively easy to visualize. Products of two reflections lead to two of the most familiar types of isometries.

Definition. Let a and b be reflections whose mirrors are the planes A and B, respectively. If A and B are not disjoint, then ab is called a *rotation*. If, furthermore, $A \neq B$, then the line $A \cap B$ is called the *axis* of the rotation, and twice the angle from A to B is considered the *angle* of rotation. If A and B are parallel, then ab is called a *translation*. If, furthermore, $A \neq B$, then the direction perpendicular to A and B going from A to B is considered the *direction* of translation, and twice the distance from A to B is considered the *length* of translation. If $A = B$ so that A and B intersect and are parallel, then, of course, $ab = \mathbf{1}$, and thus $\mathbf{1}$ is considered to be both a translation with length zero and a rotation with angle zero.

Motivation. The above definition is rather strained, so we now try to connect it with our intuitive ideas of translation and rotation.

First assume that $A \cap B = K$ and that the angle from A to B is α. Introduce a cylindrical coordinate system with K as its z-axis, the plane A as the plane in which the polar angle θ is 0, and the plane B as the plane in which θ is α (Figure 7a). Then a and b send a point with coordinates (r, θ, z) to the points with coordinates $(r, -\theta, z)$ and $(r, 2\alpha - \theta, z)$, respectively. Thus, the product ab sends the point with coordinates (r, θ, z) first to $(r, -\theta, z)$, then on to $(r, 2\alpha + \theta, z)$. Hence ab is, in the intuitive sense, a rotation through the angle 2α. Note that ba sends (r, θ, z) to $(r, \theta - 2\alpha, z)$, so it is a rotation through the angle -2α, and thus $ab \neq ba$ unless $\alpha = 0$ or $\pi/2$.

If A and B are parallel, introduce a cartesian coordinate system in which A is the xz-plane and B is the plane $y = k$ (Figure 7b). Then ab sends the point with coordinates (x, y, z) first to $(x, -y, z)$ and then on to $(x, 2k + y, z)$, so ab is intuitively a translation through $2k$ units along the y-axis. Once again the product ba is exactly opposite ab, i.e., $(ab)^{-1} = ba$.

Caution! We are ignoring the difficulties inherent in measuring directed distances and directed angles in space. For example, how does one choose a positive direction along a given line? Similarly for angles, how does one decide

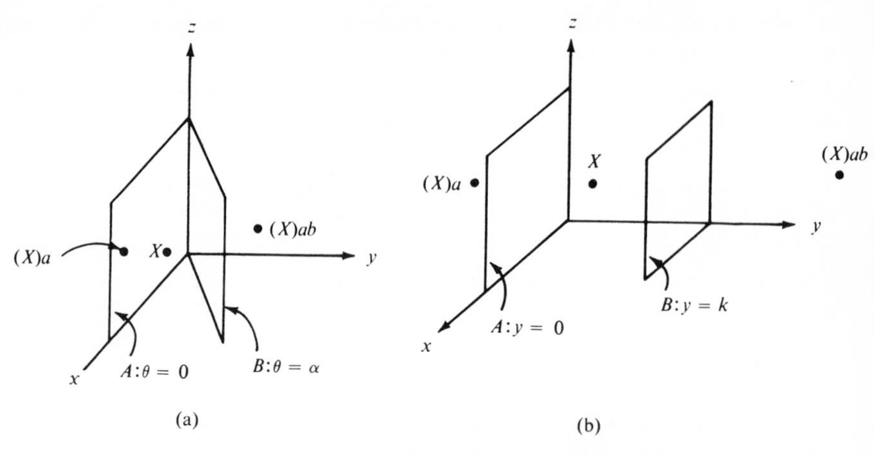

FIG. 7. (a) The rotation ab; (b) the translation ab.

which angles are positive and which are negative? See the bibliographical comments at the end of this chapter for references on this subject.

Translations, rotations, and the various involutions we have studied can be combined in simple ways to yield familiar types of isometries.

Definition. Let τ be a translation and ρ a rotation such that K, the axis of ρ, is in the direction of τ. The product $\rho\tau$ is called a *screw displacement* with *axis K.*

Definition. Let τ be a translation leaving the plane A (globally) invariant. The product τa is called a *glide reflection.*

Definition. Let ρ be a rotation with axis K and X a point in K. The product ρx is called a *rotatory inversion*. It is also called a *rotatory reflection* since [part (1) of theorem 2.12] it can be realized as the product of the rotation ρk and the reflection a if A is the plane through X perpendicular to K, i.e., $\rho x = \rho(ka) = (\rho k)a$.

Note that in each of the defining products for screw displacement, glide reflection, and rotatory inversion, the factors commute. This is especially intuitive in the case of a screw displacement where one usually visualizes the motion as a screw-type motion rather than the type of motion involved in unlocking a door (first push the key, then rotate) or opening a door (first rotate the knob, then push it). In the case of rotatory inversions, the products involved are the only *commutative* products of the type rotation followed by inversion; if ρ is a rotation and x an inversion, then $\rho x = x\rho$ if and only if $(X)\rho = X$ (theorem 2.9), i.e., if and only if X is on the axis of ρ. The analogous statement holds for glide reflections.

Certain isometries fall into more than one of the types above. We have already remarked that the identity can be viewed as either a translation or a rotation, and we add that it is also a degenerate screw displacement. Similarly, translations and rotations can be regarded as degenerate screw displacements. Inversions and reflections can be viewed as either glide reflections or rotatory inversions, e.g., a reflection is the product of the half-turn (180° rotation) about the line perpendicular to its mirror followed by the inversion in the foot of that perpendicular.

Interestingly enough, screw displacements, glide reflections, and rotatory inversions all can be represented as the product of just two involutions.

Proposition 2.13. A glide reflection, screw displacement, or a rotatory inversion can always be represented as a product of two involutions in which the first factor is a half-turn and the second is an inversion, half-turn, or reflection, respectively.

Proof. Assume τa is a glide reflection. Since τ is a translation that leaves A invariant, $\tau = bc$ where B and C are parallel planes perpendicular to A. Choose a plane D such that $D \perp (A \cap B)$, and let $K = B \cap D$. Let $X = A \cap C \cap D$ (Figure 8a), then $\tau a = bca = (bd)(dca) = kx$.

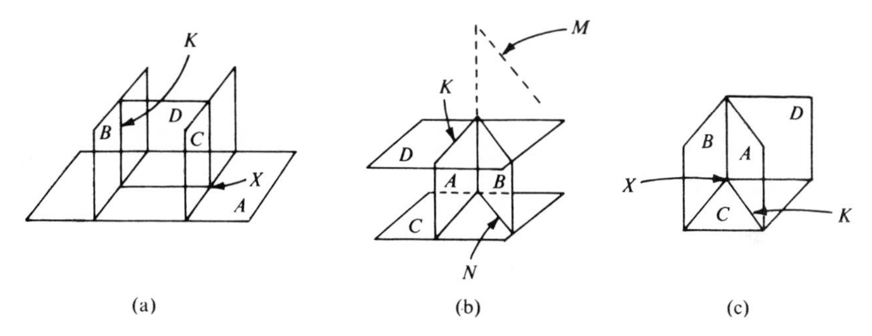

FIG. 8. (a) Glide reflection kx; (b) screw displacement km; (c) rotatory inversion kd.

Now assume $\rho\tau$ is a screw displacement with ρ a rotation, say $\rho = ab$, and τ a translation along $A \cap B$. Since τ is a translation along $A \cap B$ there are planes, C and D, both perpendicular to $A \cap B$ such that $\tau = cd$. Let $N = B \cap C$, $M = (N)d$, and $K = A \cap D$ (Figure 8b). By Theorem 2.9 $dnd = m$, i.e., $nd = dm$. Thus $\rho\tau = abcd = a(bc)d = and = adm = km$.

Finally, assume that ρx is a rotatory inversion, say, $\rho = ab$ with $X \in A \cap B$. Let C be the plane through X perpendicular to $A \cap B$, D the plane through X such that B, C, and D are mutually perpendicular, and $K = A \cap C$ (Figure 8c). Then $\rho x = ab(bcd) = acd = kd$.

In the three products above the first factor, k, is always a half-turn, and the second factor is, respectively, an inversion (x), a half-turn (m), and a reflection (d), so the proof is complete.

So far the types of isometries considered have been rather familiar ones. One would expect that by considering more complicated products of involutions, e.g., the product of the reflections in the four sides of a nonsymmetric tetrahedron, that we would discover more complicated isometries. This is not so! Our next objective is to show that *any* isometry is one of the types considered above. As a first step, we divide all similarities into two broad categories, direct or opposite, according to whether or not they preserve right- and left-handed. To do this we need some method of recognizing right- and left-handed. For the present we adopt a nonrigorous method. In Chapter 5 we give a more formal definition.

Definition. Let R_1, R_2, and R_3 be three half-lines (rays) that are mutually perpendicular and originate at a common point W. This ordered triple of rays is called *right- or left-handed,* according as the cartesian coordinate system with R_1 as its positive x-axis, R_2 as its y-axis, and R_3 as its z-axis is right- or left-handed.

For those who find the familiar mnemonic device using left or right hands difficult to remember, we present a substitute that is almost impossible to forget. To see if a coordinate system is right- or left-handed impale yourself (Ouch!) through the stomach on the positive z-axis (It usually points up, so this is the natural thing to do!), then spin your head in the direction of the positive y-axis. The "handedness" is determined by the way you must turn your head to see the positive x-axis.

Definition. A similarity ϕ is called *direct* if it transforms all right-handed triples of rays into right-handed triples. If it converts any right-handed triple into a left-handed one, then it is called *opposite.*

We *assume* for the present that a similarity is consistent in either preserving or reversing the "handedness" of all triples of rays.

The next theorem shows that the reflections generate **E**. It also provides the basis for our classification of isometries.

Theorem 2.14. If an isometry has at least k independent fixed points ($k = 0, 1, 2, 3,$ or 4), then it can be factored as the product of at most $4 - k$ reflections. Moreover, the reflections can be chosen such that their mirrors all contain the fixed points.

Proof. Assume that ϕ is an isometry and that $X_1, X_2, \ldots, X_k$ are k independent points left fixed by ϕ ($k = 0, 1, 2, 3,$ or 4). If $k = 4$, then $\phi = \mathbf{1}$ by theorem 2.7. We view the identity as a product of zero reflections in order to simplify the statement of the theorem. If $\phi \neq \mathbf{1}$, then $k \leq 3$, and we can choose a point Y such that $(Y)\phi \neq Y$. Let A be the perpendicular bisector of $[Y, (Y)\phi]$. Since $X_i = (X_i)\phi$ and ϕ preserves distances, $|X_i, Y| = |X_i, (Y)\phi|$, so the X_i are in A. It follows easily that the isometry ϕa leaves all the X_i and the point Y fixed. By theorem 2.7, these $k + 1$ points are independent. Carrying out this process at most $4 - k$ times, we find that ϕ followed by at most $4 - k$ reflections is the identity, i.e., that ϕ is the product (in the reverse order!) of these reflections.

Corollary 2.15. There is a unique homomorphism of **E** *onto* a group of order two. The kernel of this homomorphism consists of all the direct isometries and does not include any reflections. Thus, an isometry is direct if and only if it can be factored as the product of 0, 2, or 4 reflections and is opposite if and only if it can be factored as the product of 1 or 3 reflections.

Proof. Let $\mathbf{G} = \{e, p\}$ be a two-element group with identity e. There is only one possible multiplication table for **G**, namely, a product of two elements in **G** is p if and only if exactly one of the two factors is p. The product of two isometries is opposite if and only if exactly one factor is opposite; hence the mapping, f, sending direct isometries to e and opposite isometries to p is a homomorphism of **E** onto **G**.

Now assume that g is any homomorphism of **E** onto **G** and that **H** is the kernel of g. Since g maps **E** onto **G**, we must have $\mathbf{H} \neq \mathbf{E}$. Reflections are all conjugate in **E** (exercise 2.2, problem 5), and **H** is a normal subgroup of **E**. Hence if one reflection is in **H**, then all reflections are in **H**. But the group **E** is generated by reflections and $\mathbf{H} \neq \mathbf{E}$, so we cannot have all reflections in **H**. Thus, there are no reflections in **H**, and the homomorphisms, f and g, both send reflections to p. The reflections generate **E**; hence $f = g$.

Clearly the natural homomorphism mapping direct isometries to e and opposite isometries to p can be extended to a homomorphism of **S** onto a two-element group $\{e, p\}$. We do not yet claim that this is the only homomorphism of **S** onto a two-element group.

Definition. The group of direct similarities is denoted by $\mathbf{S}^+$. The group, $\mathbf{S}^+ \cap \mathbf{E}$, consisting of all direct isometries is called $\mathbf{E}^+$.

Recall that in Chapter 1 we agreed to let $[\mathbf{G}, x]$ be the subgroup consisting of those elements in **G** leaving x fixed.

Corollary 2.16. A direct isometry with at least one invariant point, X, is a rotation (possibly **1**) whose axis contains X. Consequently, the rotations with axis through X form a group, namely, the group $[\mathbf{E}^+, X]$. This group is not abelian.

Proof. By theorem 2.14, any ϕ in $[\mathbf{E}, X]$ is a product of at most three reflections all of whose mirrors contain X. If, in addition, ϕ is direct, then by corollary 2.15 ϕ is either $\mathbf{1}$ or ab with $X \in A \cap B$. In other words, ϕ is either the identity or a rotation with axis through X. If M and N are lines through X which are distinct and not perpendicular, then both m and n are in $[\mathbf{E}^+, X]$, but $mn \neq nm$ [part (4) of theorem 2.11]. Thus, $[\mathbf{E}^+, X]$ is not abelian and contains only rotations with axis through X. Clearly, any rotation with axis through X is in $[\mathbf{E}^+, X]$, so the group $[\mathbf{E}^+, X]$ is just the set of all rotations with axis through X.

The group $[\mathbf{E}^+, X]$ is isomorphic to that group in linear algebra known as O_3^+, the group of proper orthogonal transformations. Some other terms used in the literature for direct isometries are motions, proper motions, rigid motions, or displacements. Similarly, improper motions, reversals, or improper congruences are often used for opposite isometries.

We now have enough information to prove that any isometry is one of the types mentioned above.

Theorem 2.17 (classification of isometries). A direct isometry is either a translation, rotation, or screw displacement. An opposite isometry is either a glide reflection or a rotatory reflection.

Proof. We use proposition 2.13 as our guide and first show (lemmas 3 and 4 below!) that any isometry can be factored as $m\theta$ with m a half-turn and θ a half-turn or reflection, according as the isometry is direct or opposite. In order to avoid repeated consideration of two related cases we introduce the concept of a pencil of planes.

Definition. Let A and B be distinct planes. The *pencil of planes determined by A and B* is the set of all planes parallel to both A and B if $A \mathbin{/\!/} B$ or the set of all planes containing the line $A \cap B$ if $A \cap B \neq \varnothing$.

Thus, pencils of planes come in two varieties: all planes sharing a common line or all planes perpendicular to a common line. Note that the planes in the pencil containing a line M and those in the pencil of planes perpendicular to M are the mirrors of all reflections that commute with m. There is a similar (but more useful since only one pencil is involved) algebraic characterization of the pencil determined by the planes A and B. This is our first lemma in the proof.

Lemma 1. The plane C is in the pencil of planes determined by A and B ($A \neq B$) if and only if abc is a reflection.

Proof. Assume that $A \cap B \neq \varnothing$. If $abc = d$, then $ab = dc$, i.e., ab and dc are the same rotation. Since $A \neq B$, $ab \neq \mathbf{1}$. Hence the set of fixed points of ab is the line $A \cap B$. The same must be true for dc; hence $A \cap B = D \cap C$, and C and D are therefore both in the pencil determined by A and B. Conversely, if C is in the pencil, then we can choose a plane D in that pencil such that the directed angle from A to B equals the directed angle from D to C. Then $ab = dc$, and hence $abc = d$.

The proof in case the planes A and B are parallel is similar and is left as a problem (exercise 2.3, problem 2).

Lemma 2. If X is a point and ϕ is an opposite isometry, then there are planes A, B, and C, with $X \in B \cap C$ such that $\phi = abc$.

Proof. By corollary 2.15, either $\phi = a$ or $\phi = abc$ (but perhaps $X \notin B \cap C$). If $\phi = a$, then choose a plane B with $X \in B$. Then $\phi = a = abb$ is the desired factorization of ϕ. In the other case, $\phi = abc$, choose a plane C_1 such that $X \in C_1$ and C_1 is in the pencil determined by B and C. Then, by lemma 1, $\phi = abc = a(bcc_1)c_1 = ab_1c_1$ for some reflection b_1 with mirror B_1. Now choose a plane B_2 containing X and contained in the pencil determined by A and B_1. Again, by lemma 1, $\phi = ab_1c_1 = (ab_1b_2)b_2c_1 = a_1b_2c_1$. X is in $B_2 \cap C_1$ so the proof of lemma 2 is complete.

Lemma 3. If X is a point and ϕ is an opposite isometry, then $\phi = md$ for some half-turn m and some reflection d whose mirror contains X.

Proof. By lemma 2 there are planes A, B, and C such that $X \in B \cap C$ and $\phi = abc$. Let K be the line $B \cap C$ (or any line through X and in $B \cap C$ if $B = C$), and let D be the plane containing K and perpendicular to A. Then, by lemma 1, $\phi = abc = ad(dbc) = add_1$ for some reflection d_1 whose mirror, D_1, contains K and hence contains X. Since $A \perp D$, $ad = m$, the half-turn about $M = A \cap D$. Thus, $\phi = md_1$ and $X \in D_1$ as claimed.

Lemma 4. If X is a point and ϕ is a direct isometry, then $\phi = mn$ for some half-turns m and n such that $X \in N$.

Proof. Since ϕ is direct, ϕx is opposite. Hence, by lemma 3, $\phi x = md$ with $X \in D$. However, then $\phi = mdx = mn$ where N is the line through X perpendicular to D.

Lemmas 3 and 4 show that any isometry can be factored in exactly the same way that glide reflections, screw displacements, and rotatory inversions were factored in proposition 2.13. We leave to the reader (exercise 2.3, problem 3) the relatively simple problem of reversing the arguments in proposition 2.13 and showing that:

(1) A product of the form md is a reflection, glide reflection, or rotatory inversion, according as $M \subseteq A$, $M \parallel A$, or M meets A in one point.
(2) A product of the form mn is a translation, rotation, or screw displacement, according as $M \parallel N$, M intersects N, or M and N are skew lines.

The proofs of (1) and (2) above complete the proof of theorem 2.17.

The lemmas used in proving theorem 2.17 are interesting in their own right. Lemmas 2, 3, and 4 tell how much freedom is allowed in the factorization of isometries as products of reflections and half-turns. Note also that lemma 4 states that $\mathbf{E}^+$ is generated by the half-turns. The analog of lemma 1 for plane geometry, called the *Satz von den drei Spiegelungen*, has been used extensively (see axiom 3, p. 24 of [3] or axiom 3, p. 236 of [10]) as an axiom in the formal development of plane geometry based on assumptions about the group of symmetries of that geometry. It is not difficult to show that lemma 1, for pencils of planes containing a common line, is equivalent to the fact that rotations with a fixed axis commute (exercise 2.3, problem 4).

Exercise 2.3

1. Show that a rotation is of order n if and only if the angle between the two mirrors is $\pi m/n$ with m and n relatively prime.

2. Assume that $A \mathbin{/\!/} B$. Show that abc is a reflection if and only if the plane C is parallel to both A and B.

3. Complete the proof of theorem 2.17, i.e.,

(1) Show that md is a reflection, glide reflection, or rotatory inversion, according as $M \subseteq A$, $M \mathbin{/\!/} A$, or M meets A in one point.

(2) Show that mn is a translation, rotation, or screw displacement, according as $M \mathbin{/\!/} N$, M intersects N, or M and N are skew lines.

Hint. Reverse the steps in the proof of proposition 2.13.

4. Let a, b, c, and d be involutions in a group $\mathbf{G}$. Show that $(ab)(cd) = (cd)(ab)$ and $(da)(bd) = (bd)(da)$ if and only if $abc = cba$ and $bad = dab$. Apply this to relate lemma 1 and the statement that rotations with the same axis commute.

5. Let X, Y, and Z be any three points, not necessarily distinct.

(1) Show that xy is a translation, i.e., find parallel planes, A and B, such that $xy = ab$.

(2) Show that xyz is an inversion.

(3) Use parts (1) and (2) to show that translations form a commutative group.

Hint. Consider problem 4.

6. Let K, M, and N be three lines with a common perpendicular plane or line. Show that kmn is also a half-turn.

7. Let $\mathbf{G}$ be the group of rotations whose axes pass through X, and let g be an automorphism of $\mathbf{S}$. Show that g maps $\mathbf{G}$ onto the group of rotations whose axes pass through the center of $(x)g$.

8. Let ϕ and θ be rotations with parallel axes but opposite angles. Show that $\phi\theta$ is a translation and is nontrivial if the axes are distinct.

9. Review the proof of theorem 2.14 and then prove that if X_1, X_2, X_3, X_4 are four points such that $|X_1X_2| = |X_3X_4|$ there is a direct isometry, ϕ, such that $(X_1)\phi = X_3$, $(X_2)\phi = X_4$.

2.4. Geometric implications of conjugacy in S

In section 2.2 we showed that conjugating an inversion, half-turn, or reflection by a similarity yields an inversion, half-turn, or reflection, respectively. We now generalize this result and show that if ϕ is a similarity and θ an isometry of a certain type, then conjugating θ by ϕ yields an isometry, $(\theta)\phi^i = \phi^{-1}\theta\phi$, of the same type as θ.

Theorem 2.18. Let θ be an isometry and ϕ a similarity with stretching factor α.

(1) If θ is a screw displacement with axis K, then $\phi^{-1}\theta\phi = (\theta)\phi^i$ is also a screw displacement. Moreover, the axis of $(\theta)\phi^i$ is $(K)\phi$, the angle of $(\theta)\phi^i$ is the same as that of θ, and the length of $(\theta)\phi^i$ is α times the length of θ.

(2) If θ is a rotatory inversion with center X and axis L, then $(\theta)\phi^i$ is a rotatory inversion through the same angle about the axis $(L)\phi$ and with center $(X)\phi$.

(3) If θ is a glide reflection along the line L and with mirror A ($L \mathbin{/\!/} A$), then $\phi^{-1}\theta\phi$ is a glide reflection along $(L)\phi$ with mirror $(A)\phi$.

Proof. The proofs of the three parts are similar so we prove part (1) only, leaving the rest to the reader (exercise 2.4, problem 1). If θ is a screw displacement with axis K, then there are planes A, B, C, and D such that $A \cap B = K$, $C \perp K$, $D \perp K$, and $\theta = (ab)(cd)$, i.e., θ can be factored as a rotation about K followed by a translation along K. The mapping ϕ^i is an inner automorphism of **S**; hence

$$(\theta)\phi^i = (a)\phi^i(b)\phi^i(c)\phi^i(d)\phi^i .$$

Since

$$(A)\phi \cap (B)\phi = (K)\phi,\ (C)\phi \perp (K)\phi \quad \text{and} \quad (D)\phi \perp (K)\phi,$$

it follows that $(\theta)\phi^i$ is a screw displacement along $(K)\phi$. Moreover, the angle from A to B is congruent to the angle from $(A)\phi$ to $(B)\phi$ (corollary 2.3), and the distance from $(C)\phi$ to $(D)\phi$ is α times the distance from C to D. Thus, the angle of $(\theta)\phi^i$ is the same as that of θ, and the length is α times that of θ.

By specializing the above result to degenerate screw displacements, we find two interesting results about rotations and translations.

Corollary 2.19. If θ is a rotation through the angle β, then the conjugacy class of θ in **S** or in **E** consists of all rotations through the angle β. If τ is a translation of length $\alpha \neq 0$, then the conjugacy class of τ in **E** consists of all translations of length α, and the conjugacy class in **S** consists of *all* nontrivial translations.

Note that any rotation or translation is conjugate in **E** to its own inverse. In the case of a rotation, ρ, whose angle is not 0 nor 180°, one must be cautious about orientation when conjugating, for conjugating by an opposite similarity reverses the direction of the rotation. For example, let ρ be a 90° rotation about the line K, X a point on K, and M a line perpendicular to K. Then, although both x and m flip K end for end, $(\rho)x^i = \rho$, whereas $(\rho)m^i = \rho^{-1}$. From another viewpoint, one can predict that $(\rho)x^i = \rho$, but $(\rho)m^i \neq \rho$ by noting that $(x)\rho^i = x$, whereas $(m)\rho^i \neq m$.

Theorem 2.18 is the tool that we promised (at the end of section 1.2) to develop as an efficient method to find conjugacy classes in symmetry groups (exercise 2.4, problem 2). For example, in the symmetry group of a cube, the conjugacy class of a 90° rotational symmetry must be contained in its conjugacy class in **S** and hence contains *only* 90° rotations. On the other hand, we can carry the axis of any 90° rotational symmetry to any of the three facial axes of a cube or flip the axis end for end. Hence the 90° rotational symmetries must all be in the same conjugacy class.

Theorem 2.18 also leads to algebraic criteria for geometric relations. We introduced a few of these in theorem 2.11 which for convenience we repeat here.

(1) $xm = mx$	The point X is on the line M.
(2) $xa = ax$	The point X is on the plane A.
(3) $ma = am$	The line M is contained in or is perpendicular to the plane A.
(4) $ab = ba \neq \mathbf{1}$	$A \perp B$.
(5) $mn = nm \neq \mathbf{1}$	$M \perp N$.
(6) $kmn = \mathbf{1}$	The lines K, M, and N are mutually perpendicular.
(7) $wxyz = \mathbf{1}$	W, X, Y, and Z are vertices of a parallelogram, or, if collinear, the distance from W to X equals the distance from Z to Y.
(8) $(kmn)^2 = \mathbf{1}$, but $(kmn) \neq \mathbf{1}$	The lines K, M, and N are parallel or have a common perpendicular line.
(9) $(abc)^2 = \mathbf{1}$	The planes A, B, and C are mutually perpendicular ($abc = x$) or else lie in the same pencil of planes ($abc = d$).
(10) $xzyz = \mathbf{1}$	Z is the midpoint of $[X, Y]$.

(11) $xaya = \mathbf{1}$	A is the perpendicular bisector of $[X, Y]$.
(12) $acbc = \mathbf{1}$	C is an angle bisector of the planes A and B, or, if $A \parallel B$, C is the plane midway between A and B.
(13) There is a $\phi \in \mathbf{E}$ such that $(wx)\phi^i = (yz)$	$\|WX\| = \|YZ\|$.

Exercise 2.4

1. Prove parts (2) and (3) of theorem 2.18.

2. Let **G** be the group of all 120 isometries that are symmetries of a dodecahedron. Find the conjugacy classes in **G**, and determine the number of isometries in each class. *Hint.* Use theorems 2.18 and 1.17. Review example 3 in section 1.5.

3. Prove some of statements (1)–(13) concerning the equivalences between algebraic statements and geometric relations.

4. Let ϕ be a rotatory inversion, say, $\phi = \rho x$ where ρ is a rotation with axis K and $X \in K$. Find an isometry θ such that $(\phi)\theta^i = \phi^{-1}$.

5. Determine the conjugacy class in **E** and in **S** of a glide reflection.

6. Find at least two additional algebraic statements characterizing geometric relations.

2.5. An exercise: Isometries in the plane

The study of isometries in the euclidean plane is analogous to the study carried out in the preceding sections. The definitions of similarity and isometry are exactly the same, except that now one considers permutations of the points in the plane. There are only two types of involutions: *half-turns* about a *point* and *reflections* in a *line*. Once again *rotations* and *translations* are products of two reflections whose axes intersect or are parallel. The unique fixed point of a nontrivial rotation is called its center. Note that, *in the plane*, reflections reverse orientation and half-turns preserve it, even though both can be viewed as restrictions to the plane of direct isometries *of space*. Theorems 2.1 through 2.18 all have obvious analogs that the reader should state for himself. In particular, he should state and prove the analogs of the central theorems: theorem 2.14 (reflections generate the group of isometries), theorem 2.17 (classification of isometries), and theorem 2.18 (conjugating isometries). Do not use results about isometries in space to derive the results for the plane. Instead, review the proofs and construct their analogs in the plane.

2.6. Dilatations and spiral similarities

In this section we first consider those symmetries of space that preserve "direction" and not necessarily relative distances. We show that these symmetries are all familiar and in fact are all similarities. Finally we show that any similarity is either an isometry or can be visualized as a spiral-type motion.

Definition. A permutation, ϕ, of the points of space is called a *dilatation* if and only if $X + Y \mathbin{/\!/} (X)\phi + (Y)\phi$ for *all* independent points X and Y. *Caution!* Remember that we consider a line to be parallel to itself. Clearly the dilatations form a subgroup of the group of all permutations of space. This group is denoted by **D**.

The only isometries that are also dilatations are the translations and inversions. There is one other obvious type of dilatation in **S**.

Definition. Let X be a point and α a nonzero real number. We introduce a cartesian coordinate system with origin at X and define the mapping x_α as follows. If Y is the point with coordinates (x, y, z), then $(Y)x_\alpha$ is that point with coordinates $(\alpha x, \alpha y, \alpha z)$. Clearly x_α is a similarity with stretching factor $|\alpha|$ and is an isometry if and only if $\alpha = \pm 1$ ($x_\alpha = \mathbf{1}$ if $\alpha = 1$ and $x_\alpha = x$ if $\alpha = -1$). The mapping x_α is called a *central dilatation* (many authors say *homothetic transformation*), and X is called the *center*.

The definition of x_α is independent of the particular cartesian coordinate system chosen, and in fact x_α can be described in a way that does not appeal to a coordinate system (exercise 2.6, problem 1). Note that x_α is direct if $\alpha > 0$ and opposite if $\alpha < 0$. The analogue in plane geometry is not exact since *in the plane* x_α is direct, regardless of the sign of α. One way to visualize x_α is to imagine a balloon with the source of air at point X. If $\alpha < 0$, then one first turns the balloon inside out. One then blows air into the balloon if $|\alpha| > 1$, or lets air out if $|\alpha| < 1$. For any two points Y and Z, the vector from $(Y)x_\alpha$ to $(Z)x_\alpha$ is α times the vector from Y to Z. Hence x_α is a dilatation.

Proposition 2.20. A dilatation is uniquely determined by its action on two points.

Proof. Assume that we know ϕ is a dilatation, that X and Y are independent points, and that $(X)\phi$ and $(Y)\phi$ are known. We must show that for any point Z, $(Z)\phi$ is determined.

Case 1. Z is not on $X + Y$. Let K and M be the lines through $(X)\phi$ and $(Y)\phi$ parallel to $X + Z$ and $Y + Z$, respectively (Figure 9). Then $(Z)\phi$ is the unique point at the intersection of K and M.

Case 2. Z is on $X + Y$. First determine $(W)\phi$ for some point W not on

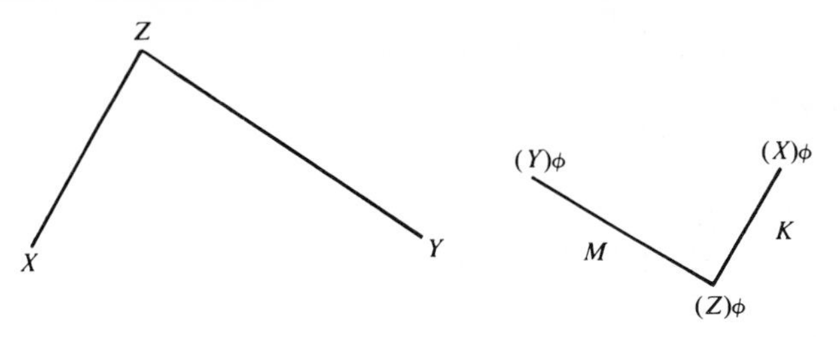

FIG. 9. Dilatation determined by action on two points.

$X + Y$. Then repeat the above construction of $(Z)\phi$ using the points W and X instead of X and Y.

This proposition can be rephrased as follows. If X, Y and X_1, Y_1 are two pairs of independent points such that $X + Y \mathbin{/\!\!/} X_1 + Y_1$, then there is at most one dilatation carrying X to X_1 and Y to Y_1. In the next theorem we reverse the inequality and show that there is at least one such dilatation.

Theorem 2.21. If X, Y and X_1, Y_1 are two pairs of independent points (perhaps $X = X_1$ or $Y = Y_1$) such that $X + Y \mathbin{/\!\!/} X_1 + Y_1$, then there is either a central dilatation or a translation carrying X to X_1 and Y to Y_1.

Proof. Since the fixed point of a central dilatation must always be on the line joining any other point to its image, we first consider the case in which $X \neq X_1$, $Y \neq Y_1$, and $X + X_1$ meets $Y + Y_1$ in one point.

Case 1. $X + X_1 \not\mathbin{/\!\!/} Y + Y_1$. Since $X + Y \mathbin{/\!\!/} X_1 + Y_1$, we know that X, Y, X_1, and Y_1 are coplanar. Hence there is a unique point W on both $X + X_1$ and $Y + Y_1$. We choose a nonzero number, α, as follows:

$$\alpha = \begin{cases} \dfrac{|WX_1|}{|WX|} & \text{if } W \text{ is not between } X \text{ and } X_1\,, \\ -\dfrac{|WX_1|}{|WX|} & \text{if } W \text{ is between } X \text{ and } X_1\,. \end{cases}$$

Now consider the central dilatation w_α which leaves W fixed and carries X to X_1. Since w_α is a dilatation, it carries Y to a point on the line through X_1 parallel to $X + Y$, i.e., the line $X_1 + Y_1$. But w_α must also carry Y to a point on $Y + W = Y + Y_1$ since it is a *central* dilatation with center W. Thus, $(Y)w_\alpha \doteq Y_1$, and w_α is the desired central dilatation carrying X to X_1 and Y to Y_1.

Case 2. $X = X_1$ or $Y = Y_1$. If both $X = X_1$ and $Y = Y_1$, then **1** is the desired translation. If only one equality holds, then we can assume it is $Y = Y_1$. Since $X + Y \mathbin{/\!\!/} X_1 + Y_1$ and $Y = Y_1$, it follows that Y, X, and X_1 are collinear. Choose α as in case 1 with Y replacing W, then y_α is the desired central dilatation.

Case 3. $X + X_1 \mathbin{/\!/} Y + Y_1$, but $X + X_1 \neq Y + Y_1$. The four points X, X_1, Y, and Y_1 are vertices of a parallelogram; hence the translation carrying X to X_1 also carries Y to Y_1.

Case 4. $X + X_1 = Y + Y_1$. Choose any point Z not on $X + X_1$. Through X_1 and Y_1 construct the two lines parallel to the lines $X + Z$ and $Y + Z$, respectively. Let Z_1 be the intersection of these two lines, and let ϕ be the translation or central dilatation (found as in cases 1, 2, or 3) carrying X to X_1 and Z to Z_1 (Figure 10). This ϕ then carries Y to Y_1.

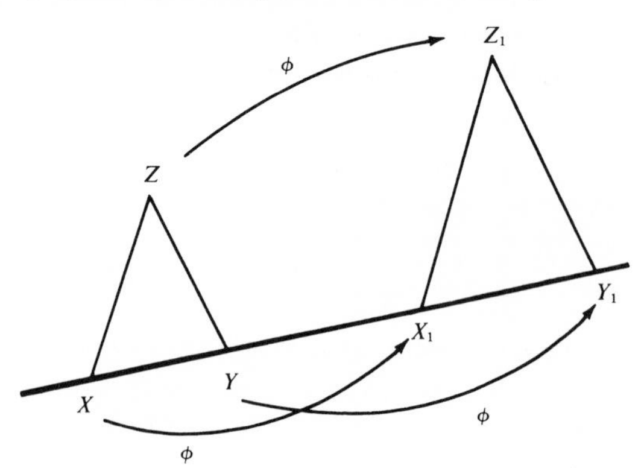

FIG. 10. The dilatation mapping one line segment onto another on the same line.

This exhausts the possibilities for points X, Y, X_1 and Y_1 such that $X \neq Y, X_1 \neq Y_1$, and $X + Y \mathbin{/\!/} X_1 + Y_1$. In each case we have found a translation or central dilatation carrying X to X_1 and Y to Y_1, so the proof is complete.

For any dilatation ϕ, we can consider the action of ϕ on two independent points. By theorem 2.21, there is a translation or central dilatation with the same action on these two points. However, by proposition 2.20, this means that ϕ equals the translation or central dilatation, and thus we arrive at the following result.

Corollary 2.22. Every dilatation is a translation or a central dilatation. Thus, a product of central dilatations and/or translations is always a central dilatation or translation. In other words, **D**, the group of dilatations, is the subgroup of **S** consisting of all translations and all central dilatations.

The reader should experiment with various products of dilatations. The cases $x_\alpha y_\beta$ with $\beta = \pm 1/\alpha$ are especially interesting (exercise 2.6, problem 2).

We remarked earlier that the only isometries that are dilatations are the translations and inversions. Since any translation can be factored as the product of two inversions, we see that $\mathbf{D} \cap \mathbf{E}$ is the subgroup of **S** generated by the inversions.

With central dilatations available, we can classify all similarities very easily and show that those that are not isometries are "spirals." This classification is *not* a classification into conjugacy classes in **S.**

Definition. A similarity, ϕ, is called a *spiral* if $\phi = \rho x_\alpha$ where x_α is a central dilatation with center X, and ρ is a rotation whose axis contains X.

Note that if $\phi = \rho x_\alpha$ is a spiral, then $\rho x_\alpha = x_\alpha \rho$, i.e., we can first rotate and then expand or first expand and then rotate. Moreover, ρx_α is direct or opposite, according as α is positive or negative.

Theorem 2.23 (classification of similarities). A similarity is either a spiral or else an isometry without fixed points (translation, screw displacement, or glide reflection).

This is an obvious consequence of theorem 2.6 and corollary 2.16, so the proof is left to the reader (exercise 2.6, problem 3).

Exercise 2.6

1. Describe the central dilatation x_α in terms that do not involve coordinate systems. *Hint.* Consider the cases $\alpha > 0$ and $\alpha < 0$ separately, and review case 1 in the proof of theorem 2.21.

2. Prove that the product of two central dilatations, $x_\alpha y_\beta$, is a translation if and only if $\alpha\beta = 1$. Derive as a corollary that the product of a central dilatation and a translation, $x_\alpha \tau$, is a central dilatation, y_β, with $\beta = \alpha$. Note that if $\alpha\beta \neq 1$, then $x_\alpha y_\beta$ must be a central dilatation. Show how to find its center.

3. Prove theorem 2.23.

4. Let X be the orthocenter, Y the centroid, and Z the circumcenter of a triangle. Using the central dilatation $y_{(-1/2)}$, show that $Y \in [X, Z]$ and that $2|XY| = |YZ|$. Moreover, show that the circle with center at the midpoint of $[X, Z]$ and radius one half of the radius of the circumscribed circle hits all nine of the following points: the midpoints of the sides, the bases of the altitudes, and the midpoints of the line segments joining X and the vertices.

5. Let X have coordinates $(0, -1, 0)$ and Y have coordinates $(0, 1, 0)$. Describe the products:

(1) $x_{-1}y_{-1}$
(2) $x_2 y_{1/2}$
(3) $x_2 y_{-1/2}$
(4) $x_2 y_{-1}$
(5) $x_3 y_4$.

6. Show that the various isometries with fixed points are all spirals, ρx_α, with $\alpha = \pm 1$.

7. Let ρx_α be a spiral (i.e., X is on the axis of the rotation ρ), and let ϕ be a similarity. Describe $\phi^{-1}\rho x_\alpha \phi$ and its relation to ρx_α. Use this result to describe as a spiral the product σx_α when σ is a rotation whose axis does *not* contain X, i.e., $\sigma x_\alpha \neq x_\alpha \sigma$.

8. Classify similarities of the euclidean plane. *Caution!* x_α is always a direct similarity in the plane.

2.7. *Automorphisms of* **E** *and* **S**

One of the guiding principles in mathematics today is that in order to fully understand a mathematical structure one should investigate its group of symmetries or automorphisms (see [11], p. 144). In this chapter we are primarily concerned with the two groups of symmetries of euclidean space: **E**, the group of symmetries leaving distances invariant, and **S**, the group of symmetries leaving equal distances invariant. In this section we go one step farther and consider the symmetries of **E** and of **S**. We prove that **S** is a complete group and, furthermore, that **S** is, in a natural way, isomorphic to the group of symmetries of **E**. Once again the involutions in **S** play an important role.

First let us review several important normal subgroups of **S**. It is easy to show directly from the definition of **E** and **S** that **E** is a normal subgroup of **S**. Alternatively, we could appeal to the fact that **E** is generated by the reflections to prove the same result (exercise 2.7, problem 1). In section 2.3 we defined direct and opposite symmetries and showed that $\mathbf{S}^+$, the group of direct symmetries, is the kernel of a homomorphism of **S**. As such, it must be normal in **S**. Since **E** is also normal, it follows that $\mathbf{E}^+ = \mathbf{E} \cap \mathbf{S}^+$ is a normal subgroup of **S**. Finally, in section 2.6, we considered **D**, the group of dilatations, and showed that it is a subgroup of **S**. Once again it is easy to show from the definition that **D** is a normal subgroup of **S**. An alternative proof is given in theorem 2.25 below. Since **D**, $\mathbf{S}^+$, **E**, and $\mathbf{E}^+$ are all normal in **S**, it follows that $\mathbf{D}^+ = \mathbf{D} \cap \mathbf{S}^+$, $\mathbf{D} \cap \mathbf{E} =$ the group of all inversions and translations, and $\mathbf{D} \cap \mathbf{E}^+ = \mathbf{T} =$ the group of all translations, are all normal subgroups of **S**. The subgroup relationships for these groups are diagramed in Figure 11. Note that **E** is generated by the reflections, $\mathbf{E}^+$ by the half-turns, and $\mathbf{D} \cap \mathbf{E}$ by the inversions. Our first objective is to show that these three groups, and the group **D**, are not only normal but also characteristic subgroups of **S**.

Definition. Let $\mathcal{I}$, $\mathcal{H}$, and $\mathcal{R}$, respectively, denote the set of inversions, half-turns, and reflections.

Notational Convention. Automorphisms of **E** or **S** are denoted by lowercase Latin letters, especially f, g, and h.

Theorem 2.24. The three sets of involutions, $\mathcal{I}$, $\mathcal{H}$, and $\mathcal{R}$, are all characteristic subsets of **S**, and hence $\mathbf{D} \cap \mathbf{E}$, $\mathbf{E}^+$, and **E** are all characteristic subgroups of **S**.

Proof. By theorem 2.8, $\mathcal{I} \cup \mathcal{H} \cup \mathcal{R}$ is the set of all involutions in **S** and hence is characteristic in **S**. By theorem 2.9 each of the three sets $\mathcal{I}$, $\mathcal{H}$, and $\mathcal{R}$ is a complete conjugacy class in **S**. Since an automorphism of **S** must preserve conjugacy classes in **S**, an automorphism must permute $\mathcal{I}$, $\mathcal{H}$, and $\mathcal{R}$

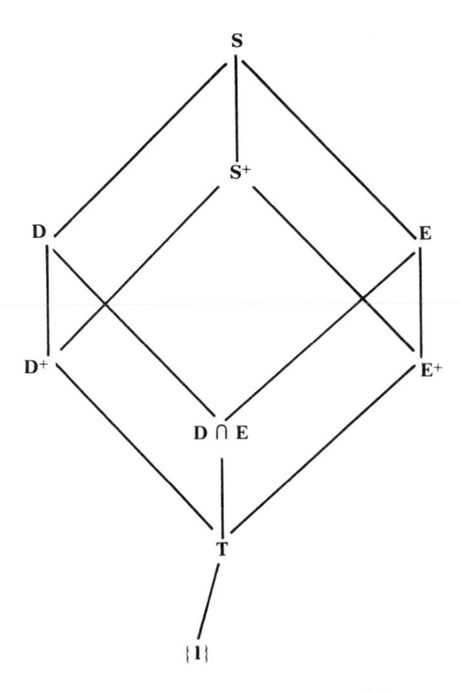

FIG. 11. Normal subgroups of **S**.

as sets. However, $\mathcal{I}$, $\mathcal{H}$, and $\mathcal{R}$ can be distinguished algebraically as follows:

(1) Inversions are unique in that the product of any two has infinite order, whereas we can find two half-turns or two reflections whose product is of order two.

(2) Half-turns are unique in that they are perfect squares (of 90° rotations) in **S**, whereas inversions and reflections are opposite and therefore cannot be perfect squares in **S**.

(3) Reflections are distinguishable as the one remaining conjugacy class of involutions.

An automorphism of **S** preserves all of these algebraic properties; hence $\mathcal{I}$, $\mathcal{H}$, and $\mathcal{R}$ are all characteristic subsets of **S**.

Theorem 2.25. The group of dilatations, **D**, is a characteristic subgroup of **S**.

Proof. Let θ be a dilatation, g an automorphism of **S**, and $\phi = (\theta)g$. We need to show that if X and Y are independent points, then $X + Y \mathbin{/\!\!/} (X)\phi + (Y)\phi$. Since g is an automorphism of **S** and $\mathcal{I}$ is a characteristic subset of **S**, g must permute $\mathcal{I}$, and thus we can choose (independent) points X_1 and Y_1 such that $(x_1)g = x$ and $(y_1)g = y$. The basic idea of the proof is to express the hypothesis ($X_1 + Y_1 /\!/ (X_1)\theta + (Y_1)\theta$) in an algebraic form and then show that g transfers this algebraic relationship (and hence also transfers the geometric analog) over to ϕ, X, and Y.

Since θ is a dilatation, $X_1 + Y_1 /\!/ (X_1)\theta + (Y_1)\theta$, i.e., the lines $M_1 = X_1 + Y_1$ and $N_1 = (X_1)\theta + (Y_1)\theta$ are parallel. The geometric statement, $M_1 = X_1 + Y_1$, has as its algebraic analog: m_1 is the unique half-turn commuting with the inversions x_1 and y_1. The geometric relationship $M_1 /\!/ N_1$ can be translated: there is an inversion, w_1, such that $w_1 m_1 w_1 = n_1$.

Clearly, if $w = (w_1)g$, $m = (m_1)g$, and $n = (n_1)g$, then m commutes with x and y and $wmw = n$. Moreover, by the previous theorem, m and n are half-turns, and w is an inversion. Thus, if we can show that $N = (X)\phi + (Y)\phi$, it follows that $X + Y /\!/ (X)\phi + (Y)\phi$, and the proof is complete. To do this we consider the inversions with centers $(X)\phi$ and $(Y)\phi$. By theorem 2.9, these inversions are $\phi^{-1}x\phi$ and $\phi^{-1}y\phi$, respectively. However,

$$\phi^{-1}x\phi = ((\theta)g)^{-1}(x_1)g(\theta)g = (\theta^{-1}x_1\theta)g\,,$$

and similarly, $\phi^{-1}y\phi = (\theta^{-1}y_1\theta)g$. Since n_1 is the half-turn commuting with the inversions $\theta^{-1}x_1\theta$ and $\theta^{-1}y_1\theta$, it follows that $n = (n_1)g$ is the half-turn commuting with the inversions $\phi^{-1}x\phi = (\theta^{-1}x_1\theta)g$ and $\phi^{-1}y\phi = (\theta^{-1}y_1\theta)g$, i.e., $N = (X)\phi + (Y)\phi$, and the proof is complete.

The observation, already used in the proof above, that any automorphism of **S** induces a permutation of $\mathscr{I}$ is the basis of our proof that **S** and **A(S)** are naturally isomorphic.

Theorem 2.26. The mapping sending each automorphism of **S**, g, to the permutation of points, γ, defined by $(X)\gamma = Y$ if and only if $(x)g = y$, is an isomorphism of **A(S)** onto **S**.

Proof. The mapping sending g to γ is clearly a permutation representation of **A(S)** in the group of permutations of the points in space. Hence to complete the proof we need only show that this representation is faithful and maps **A(S)** onto **S**.

Assume that $\gamma = \mathbf{1}$, i.e., that the automorphism g leaves the inversions pointwise fixed. Given a similarity ϕ, let us compare the action of the similarities ϕ and $(\phi)g$ on an arbitrary point X. Assume that $(X)\phi = Y$. Since $(x)g = x$,

$$((\phi)g)^{-1}x(\phi)g = ((\phi)g)^{-1}(x)g(\phi)g = (\phi^{-1}x\phi)g = (y)g = y\,.$$

But this implies $(X)(\phi)g = Y$, so ϕ and $(\phi)g$ have the same action on any point X, i.e. $\phi = (\phi)g$. This is true for an arbitrary similarity, ϕ, so $g = \mathbf{1}$ is the only automorphism in the kernel of the permutation representation $g \rightarrow \gamma$. Thus, this representation of **A(S)** is faithful.

To show that $\gamma \in \mathbf{S}$ we must show that $|W,X| = |Y,Z|$ if and only if $|(W)\gamma(X)\gamma| = |(Y)\gamma(Z)\gamma|$. If $|W,X| = |Y,Z|$, then there is an isometry, σ, carrying W to Y and X to Z. The similarity $(\sigma)g$ is also in **E** since **E** is characteristic in **S**. Moreover, $(\sigma)g$ carries $(W)\gamma$ to $(Y)\gamma$ since

$$((\sigma)g)^{-1}(w)g((\sigma)g) = (\sigma^{-1}w\sigma)g = (y)g\,.$$

Similarly, $(\sigma)g$ carries $(X)\gamma$ to $(Z)\gamma$. Hence $|(W)\gamma(X)\gamma| = |(Y)\gamma(Z)\gamma|$. Conversely, if $|(W)\gamma(X)\gamma| = |(Y)\gamma(Z)\gamma|$, the same argument, using g^{-1} in place of g, shows $|W,X| = |Y,Z|$. Thus, γ is a similarity, and the mapping $g \rightarrow \gamma$ is a faithful representation of **A(S)** in **S**. In this representation a similarity ϕ is clearly the similarity induced by the automorphism ϕ^i. Hence the mapping $g \rightarrow \gamma$ maps **A(S)** *onto* **S**, and the proof is complete.

Corollary 2.27. If two automorphisms of **S** agree for inversions, then they are equal.

Proof. If two automorphisms agree for inversions, then they induce the same similarity. Hence they are equal since the correspondence $g \rightarrow \gamma$ is an isomorphism.

Corollary 2.28. **S** is a complete group.

Proof. By corollary 2.10, the center of **S** is $\{\mathbf{1}\}$; hence distinct similarities induce distinct inner automorphisms. If g is any automorphism, and if g induced the similarity γ according to the method of theorem 2.26, then clearly g and γ^i agree for inversions, and thus $g = \gamma^i$ by corollary 2.27.

Corollary 2.29. The group of automorphisms of **E**, **A(E)**, is isomorphic to **S** in a natural way. Moreover, restricting automorphisms of **S** to **E** maps **A(S)** onto **A(E)**.

Proof. Since inversions are isometries, it follows from corollary 2.27 that restricting automorphisms of **S** to **E** is one to one. Since **E** is a characteristic subgroup of **S**, the restriction process maps **A(S)** into **A(E)**. To complete the proof we need only show that no additional automorphisms of **E** are possible. Probably the easiest way to show this is to observe that the proof of theorem 2.26 can be repeated almost verbatim to show that any automorphism of **E** is induced by a similarity and thus by an automorphism of **S**.

Exercise 2.7

1. Give two proofs that **E** is a normal subgroup of **S**, the first directly from the definitions of **E** and **S** and the second (quoting appropriate theorems) using the result that **E** is generated by reflections.

2. Show that reflections can be characterized algebraically as follows: If $\phi \in \mathbf{S}$, then ϕ is a reflection if and only if all three of the following hold:

(1) $\phi^2 = \mathbf{1}$.
(2) ϕ is not a perfect square in **S**.
(3) There is a $\sigma \neq \phi$, $\sigma \in \mathbf{S}$, such that $\sigma^2 = (\phi\sigma)^3 = \mathbf{1}$.

3. Let $\mathbf{E}_2$ and $\mathbf{S}_2$ be the groups of isometries and similarities, respectively, of the euclidean plane. Diagram the subgroup relations between the two-dimensional analogs of the nine normal subgroups of **S** shown in Figure 11. Discuss the differences between your diagram and the diagram in Figure 11.

4. In Chapter 1 we gave an example of a subgroup of the cube group which was of index two but was not a characteristic subgroup. Show that $\mathbf{S}^+$ is not in this unhappy predicament by showing that $\mathbf{S}^+$ consists of all the perfect squares in **S**. Similarly, show that $\mathbf{D}^+ = \{\phi^2 | \phi \in \mathbf{D}\}$.

2.8. *Homomorphisms of* E *and* S

In this final section on the symmetries of euclidean space, we investigate the homomorphisms of **E** and of **S**. Since a homomorphism is "essentially" determined when we know its kernel, this problem is equivalent to finding the normal subgroups of **E** or **S**. We shall show that there are very few normal subgroups of **E** and that the homomorphisms of **E** are all very natural. The analogous theorems do not hold for **S**, and we shall give some examples to show how complicated the situation is.

We have already used to great advantage one homomorphism of **S**, namely, the homomorphism of **S** onto a group of order two with kernel $\mathbf{S}^+$, the subgroup of direct similarities. Restricting this to any subgroup, **H**, of **S** yields a homomorphism of **H** which is trivial if $\mathbf{H} \subseteq \mathbf{S}^+$ and otherwise has a kernel of index two in **H**, i.e., a kernel that is "very large" in **H**. This homomorphism can be described intuitively as "ignore everything about the similarities except direct or opposite."

At the other extreme, we have a homomorphism of **S** (or subgroups of **S**) which has a very small kernel but is not an isomorphism. This homomorphism can be described intuitively as ignoring or "factoring out" the translational aspect of each similarity. It is described carefully in the following theorem for **S** and in a corollary for subgroups of **S**.

Theorem 2.30. Given a point, X, any similarity, ϕ, can be factored uniquely in the form $\phi_X\tau$ with ϕ_X a similarity leaving X fixed and τ a translation. The mapping $\phi \to \phi_X$ is a homomorphism of **S** onto $[\mathbf{S}, X]$, the group of all similarities leaving X fixed. The kernel of this homomorphism is **T**, the group of translations.

Proof. Given X and ϕ, let τ be the translation carrying X to $(X)\phi$. Then $\phi\tau^{-1} = \phi_X$ is a similarity leaving X fixed, and $\phi = \phi_X\tau$ is a factorization of the desired type. If $\phi = \phi_X^1\tau_1$ is also a factorization of this type, then τ and τ_1 are both translations sending X to $(X)\phi$; hence $\tau = \tau_1$. This in turn implies that $\phi_X^1 = \phi\tau^{-1} = \phi_X$, and thus the factorization is unique.

Let $\phi = \phi_X\tau$ and $\theta = \theta_X\tau_1$ be the factorizations of the similarities ϕ and θ. Then

$$\phi\theta = (\phi_X\tau)\theta = \phi_X\theta\theta^{-1}\tau\theta = \phi_X(\theta_X\tau_1)(\theta^{-1}\tau\theta) = (\phi_X\theta_X)(\tau_1\theta^{-1}\tau\theta),$$

but $\phi_X\theta_X$ is a similarity leaving X fixed, and $\tau_1(\theta^{-1}\tau\theta)$ is a product of two translations (**T** is normal in **S**!). Thus, $\phi\theta = (\phi_X\theta_X)(\tau_1\theta^{-1}\tau\theta)$ is a factorization of $\phi\theta$ in the desired form. Since this factorization is unique, $(\phi\theta)_X = \phi_X\theta_X$, and the mapping is a homomorphism. Clearly, the homomorphism has kernel **T** and maps **S** onto $[\mathbf{S}, X]$.

When we restrict this homomorphism to a subgroup, **G**, of **S**, then the image may not be $[\mathbf{G}, X]$ for any point X. For example, if **G** is the group generated by a screw displacement with angle 90° and length 1 then the image of **G** is a cyclic group of order 4, whereas the only nontrivial subgroups of **G** are cyclic of infinite order. The following corollary clarifies this situation. It is one of the principal tools in mathematical crystallography.

Corollary 2.31. If **G** is a subgroup of **S** and X is a point, then the homomorphism $\phi \rightarrow \phi_X$ restricted to **G** has kernel $\mathbf{G} \cap \mathbf{T}$ and maps **G** onto a group, $\mathbf{G}_X$, called the *point group* of **G**. The various point groups of **G** are all conjugate in **S** and hence isomorphic. The group $[\mathbf{G}, X]$ is a subgroup of $\mathbf{G}_X$, and $[\mathbf{G}, X] = \mathbf{G}_X$ if and only if the **G**-orbit of X equals the $\mathbf{G} \cap \mathbf{T}$-orbit of X.

Proof. It follows directly from the definition of ϕ_X and theorem 2.30 that the homomorphism $\phi \rightarrow \phi_X$ restricted to **G** has kernel $\mathbf{G} \cap \mathbf{T}$ and that $[\mathbf{G}, X] \subseteq \mathbf{G}_X \subseteq [\mathbf{S}, X]$. We leave to the reader (exercise 2.8, problem 1) the details of the proof that $\tau^{-1}\mathbf{G}_X\tau = \mathbf{G}_Y$ if τ is the translation carrying X to Y. This shows that the various point groups for **G** are all conjugate in **S** (or even in **E** if $\mathbf{G} \subseteq \mathbf{E}$).

If the **G**-orbit of X coincides with the $\mathbf{G} \cap \mathbf{T}$-orbit of X, then for each $\phi \in \mathbf{G}$ the translation, τ, carrying X to $(X)\phi$ is in **G**, and hence $\phi_X = \phi\tau^{-1}$ is in $[\mathbf{G}, X]$, i.e., $\mathbf{G}_X \subseteq [\mathbf{G}, X]$. The reverse inclusion is obvious, so $\mathbf{G}_X = [\mathbf{G}, X]$.

Conversely, if $\mathbf{G}_X = [\mathbf{G}, X]$ and if $\phi \in \mathbf{G}$, then $\phi = \phi_X\tau$ implies $\phi_X \in \mathbf{G}$, and hence $\tau = \phi_X^{-1}\phi \in \mathbf{G} \cap \mathbf{T}$. Thus, $(X)\phi$ is in the $\mathbf{G} \cap \mathbf{T}$-orbit of X for any $\phi \in \mathbf{G}$, i.e., the **G**-orbit of X is contained in the $\mathbf{G} \cap \mathbf{T}$-orbit of X. The reverse inclusion is obvious so the orbits must coincide.

Note that even if $\mathbf{G}_X = [\mathbf{G}, X]$ for one point X, the same need not be true for some other point Y. In fact, if **G** does not have too many elements, then we can often find a point Y such that $[\mathbf{G}, Y] = \{\mathbf{1}\}$. Such a point is said to be in *general position* with respect to **G**.

The inversion x commutes with every similarity in $[\mathbf{S}, X]$; hence $\{\mathbf{1}, x\}$ is a normal subgroup of $[\mathbf{S}, X]$ and, in fact, is the center of $[\mathbf{S}, X]$. This leads us to expect a homomorphism of $[\mathbf{S}, X]$ onto some group with kernel $\{\mathbf{1}, x\}$. Since x is an opposite isometry, this homomorphism would "ignore direct and opposite." There is a natural homomorphism of $[\mathbf{S}, X]$ onto $[\mathbf{S}^+, X]$ with these properties, namely, for each $\phi_X \in [\mathbf{S}, X]$ let $\phi_X^+ = \phi_X$ if ϕ_X is direct or $x\phi_X$ if ϕ_X is opposite. In either case, ϕ_X^+ is direct. The composite map $\phi \to \phi_X \to \phi_X^+$ is a homomorphism of $\mathbf{S}$ onto $[\mathbf{S}^+, X]$ with kernel $\mathbf{D} \cap \mathbf{E}$. Of course, there is an analog to corollary 2.31 when this homomorphism is restricted to subgroups of $\mathbf{S}$.

Finally, the mapping sending each similarity to its stretching factor is a natural homomorphism of $\mathbf{S}$ onto the multiplicative group of nonzero real numbers and has kernel $\mathbf{E}^+$. When followed by the absolute value mapping, we have a homomorphism with kernel $\mathbf{E}$.

This completes the catalog of natural homomorphisms of $\mathbf{S}$ corresponding to the various normal subgroups of $\mathbf{S}$ presented at the beginning of section 2.7. Our next objective is to show that the restrictions of these natural homomorphisms to $\mathbf{E}$ are essentially the *only* homomorphisms of $\mathbf{E}$.

Theorem 2.32. The only normal subgroups of $[\mathbf{E}^+, X]$ are $\{\mathbf{1}\}$ and $[\mathbf{E}^+, X]$, i.e., $[\mathbf{E}^+, X]$ is a *simple* group.

Proof. The only elements of $[\mathbf{E}^+, X]$ are the rotations with axis through X. Thus, it is enough to show that if ρ is a nontrivial rotation with axis through X, then ρ and its conjugates in $[\mathbf{E}^+, X]$ generate $[\mathbf{E}^+, X]$. The following two lemmas help us to do this.

Lemma 1. If ϕ is a rotation with axis through X, and if for some other point, Y, $(Y)\phi = (Y)x$, then ϕ is a half-turn whose axis contains X.

Proof. ϕx is an isometry with two independent, fixed points, X and Y. Hence there are one or two reflections, a and b, such that $\phi x = a$ or $\phi x = ab$ (theorem 2.14). Since ϕx is opposite, we must have $\phi x = a$. Thus, $\phi = ax$, and since $X \in A$, this implies that ϕ is a half-turn.

Lemma 2. Let ρ be a nontrivial rotation with axis through X, let S be a sphere centered at X with north and south poles on the axis of ρ, and let $\alpha = |Y, (Y)\rho|$ for some point Y on the equator of S. If Z_1 and Z_2 are any two points on S such that $|Z_1, Z_2| \leq \alpha$, then some conjugate of ρ in $[\mathbf{E}^+, X]$ carries Z_1 to Z_2.

Proof. As the point W varies along the shortest path on S joining Y to the north pole, the distance $|W, (W)\rho|$ decreases monotonically from α down to 0. Thus, we can choose a point W such that $|W, (W)\rho| = |Z_1, Z_2|$. Now let ϕ be a rotation in $[\mathbf{E}^+, X]$ sending W to Z_1 and $(W)\rho$ to Z_2 (exercise 2.8, problem 3). Then $\phi^{-1}\rho\phi$ is the desired conjugate of ρ in $[\mathbf{E}^+, X]$ sending Z_1 to Z_2.

Now let ρ be a nontrivial rotation in $[\mathbf{E}^+, X]$ and let S, Y, and α be as in lemma 2. The ρ-orbit of Y consists of points aroung the equator of S with $(Y)\rho^k$ and $(Y)\rho^{k+1}$ exactly α units apart. Of these points at least one, say $Z_1 = (Y)\rho^n$, is at most α units from $Z_2 = (Y)x$. By lemma 2 some conjugate, $\phi^{-1}\rho\phi$, of ρ in $[\mathbf{E}^+, X]$ sends $(Y)\rho^n$ to $(Y)x$. Thus, by lemma 1, $\rho^n\phi^{-1}\rho\phi$ is a half-turn. Since all half-turns in $[\mathbf{E}^+, X]$ are conjugate in $[\mathbf{E}^+, X]$, this shows that ρ and its conjugates in $[\mathbf{E}^+, X]$ generate $[\mathbf{E}^+, X]$. Thus, $[\mathbf{E}^+, X]$ is a simple group.

The proof presented above is essentially the proof appearing in [2], p. 178. If the emphasis is on linear transformations or matrices, and X is the origin, then the group $[\mathbf{E}^+, X]$ is called the (proper) orthogonal group in 3-space and is denoted O_3^+. One can investigate generalizations in two directions; consider O_n^+ and/or consider $O_n^+(\mathbf{F})$ for fields, $\mathbf{F}$, other than the field of real numbers. See Chapter 5 in [2] for an introduction to these problems.

Since the group $\mathbf{T}$, the group of all translations, is abelian, it is not a simple group. We can show, however, that the only subgroups of $\mathbf{T}$ that are normal in $\mathbf{E}^+$ are $\{\mathbf{1}\}$ and $\mathbf{T}$.

Theorem 2.33. Translations are conjugate in $\mathbf{E}^+$ if and only if they have the same length. The conjugates in $\mathbf{E}^+$ of a nontrivial translation generate $\mathbf{T}$.

Proof. Let $\tau = xy$ be a translation of length $2|X, Y|$. For any isometry σ, $|(X)\sigma, (Y)\sigma| = |X, Y|$. Thus, $\sigma^{-1}\tau\sigma = (\sigma^{-1}x\sigma)(\sigma^{-1}y\sigma)$ also has length $2|X, Y|$ since the centers of the inversions $\sigma^{-1}x\sigma$ and $\sigma^{-1}y\sigma$ are $(X)\sigma$ and $(Y)\sigma$. Conversely, given two translations, $\tau_1 = x_1y_1$ and $\tau_2 = x_2y_2$, of the same length $(2|X_1, Y_1| = 2|X_2, Y_2|)$, choose (exercise 2.3, problem 9) a direct isometry, ϕ, carrying X_1 to X_2 and Y_1 to Y_2. Then $\phi^{-1}\tau_1\phi = \tau_2$, so τ_1 and τ_2 are conjugate in $\mathbf{E}^+$.

Now let τ be a nontrivial translation, and let α be a positive number. Choose an integer n such that the length of τ^n is greater than $\alpha/2$. If τ^n sends X to Y, then, since $|XY| > \alpha/2$, we can choose a rotation, ρ, with axis through X such that $|Y, (Y)\rho| = \alpha$. The translation $\tau^{-n}\rho^{-1}\tau^n\rho$ sends Y to $(Y)\rho$ and hence has length α. Thus, there is among the products of τ and its conjugates in $\mathbf{E}^+$ a translation of length α. However (by the first half of this theorem), this implies that any nontrivial translation is conjugate in $\mathbf{E}^+$ to some product of the conjugates of τ, i.e., *is* a product of conjugates of τ.

Now we have enough information about conjugates in $\mathbf{E}^+$ to find *all* normal subgroups of $\mathbf{E}^+$ or $\mathbf{E}$.

Theorem 2.34. The only normal subgroups of $\mathbf{E}^+$ are $\{\mathbf{1}\}$, $\mathbf{T}$, and $\mathbf{E}^+$. The only additional normal subgroups of $\mathbf{E}$ are $\mathbf{D} \cap \mathbf{E}$ and $\mathbf{E}$.

Proof. We have exhibited **T**, $\mathbf{D} \cap \mathbf{E}$, $\mathbf{E}^+$, and **E** as kernels of homomorphisms of **S**. Hence they are all normal in **S**, and, therefore, each is normal in any subgroup of **S** in which it is contained. Note that these groups are, in fact, characteristic subgroups of **E** or of **S**.

Now let **G** be a normal subgroup of $\mathbf{E}^+$. If $\mathbf{G} \subseteq \mathbf{T}$, then $\mathbf{G} = \{\mathbf{1}\}$ or **T** by theorem 2.33. Assume that **G** contains a direct isometry, ϕ, that is not a translation. We consider three cases.

Case 1. ϕ is a nontrivial rotation. Let X be on the axis of ϕ, then $\mathbf{G} \cap [\mathbf{E}^+, X]$ is a nontrivial normal subgroup of $[\mathbf{E}^+, X]$, and hence, by theorem 2.32, $\mathbf{G} \supseteq [\mathbf{E}^+, X]$. In particular, **G** contains half-turns about lines through X and thus contains all of the half-turns since it is normal in $\mathbf{E}^+$. The half-turns generate $\mathbf{E}^+$, so $\mathbf{G} = \mathbf{E}^+$.

Case 2. ϕ is a screw displacement whose rotational part is a half-turn, i.e., $\phi = xym = mxy$ with $X + Y = M$. Choose a line K lying in the perpendicular bisector of $[X, Y]$ and such that K and M are skew lines. Let $N = (M)k$, then N and M are parallel lines, and $k\phi k = n(xy)^{-1}$. (Check this!) Hence $\phi k \phi k = mn$ is a nontrivial translation in **G**. By theorem 2.33, all translations, in particular, $(xy)^{-1}$, are in **G**; hence $\phi(xy)^{-1} = m$ is in **G** and (as in case 1) $\mathbf{G} = \mathbf{E}^+$.

Case 3. ϕ is a screw displacement whose rotational part is neither **1** nor a half-turn, i.e., $\phi = xy\rho = \rho xy$ with $X + Y =$ axis of ρ and $\rho^2 \neq \mathbf{1}$. Choose K as in case 2. As in case 2, $\phi k \phi k = \rho(k\rho k)$. Since ρ and $k\rho k$ are rotations with parallel axes but opposite angles, the product $\rho(k\rho k)$ is a nontrivial translation. Since this product is in **G**, all translations are in **G**. Hence $\rho = \phi(xy)^{-1}$ is in **G** and thus, as in case 1, $\mathbf{G} = \mathbf{E}^+$.

Thus, $\mathbf{E}^+$ is the only normal subgroup of $\mathbf{E}^+$ containing direct isometries that are not translations.

If **G** is a normal subgroup of **E**, then $\mathbf{G} \cap \mathbf{E}^+$ is a normal subgroup of $\mathbf{E}^+$; hence $\mathbf{G} \cap \mathbf{E}^+$ is either $\{\mathbf{1}\}$, **T**, or $\mathbf{E}^+$.

Case 1. $\mathbf{G} \cap \mathbf{E}^+ = \{\mathbf{1}\}$. The only possible opposite isometries in **G** are inversions or reflections since $\sigma \in \mathbf{G}$ implies $\sigma^2 \in \mathbf{G} \cap \mathbf{E}^+$. The group generated by an inversion (respectively a reflection) and its conjugates in **E** is $\mathbf{D} \cap \mathbf{E}$ (respectively **E**) which is too large to satisfy $\mathbf{G} \cap \mathbf{E}^+ = \{\mathbf{1}\}$. Thus, in this case, $\mathbf{G} = \{\mathbf{1}\}$.

Case 2. $\mathbf{G} \cap \mathbf{E}^+ = \mathbf{T}$. The only possible opposite isometries in **G** are inversions or glide reflections (exercise 2.8, problem 4). No glide reflection could be in **G** since we could cancel the translational part of the glide ($\mathbf{T} \subseteq \mathbf{G}$!!) and find a reflection in **G** which would imply that $\mathbf{G} = \mathbf{E}$ and, therefore, that $\mathbf{G} \cap \mathbf{E}^+ = \mathbf{E}^+$, instead of **T**. Thus, $\mathbf{G} \cap \mathbf{E}^+ = \mathbf{T}$ implies $\mathbf{G} = \mathbf{D} \cap \mathbf{E}$ or **T**, according as **G** does or does not contain an inversion.

Case 3. $\mathbf{G} \cap \mathbf{E}^+ = \mathbf{E}^+$. The only subgroups (normal or not) of **E** containing $\mathbf{E}^+$ are $\mathbf{E}^+$ and **E**; hence $\mathbf{G} = \mathbf{E}^+$ or **E**.

This theorem enables us to classify the homomorphisms of **E**. Recall that in our discussion of direct and opposite similarities we considered (corollary 2.15) homomorphisms of **S** onto *any* group of order 2. Since any two groups of order 2 are isomorphic—and in a unique way—there was no need to pin down a specific group of order 2. Similarly, in corollary 2.31, we called $\mathbf{G}_X$ *the* point group of **G**, even though (except in special cases) $\mathbf{G}_X \neq \mathbf{G}_Y$ if $X \neq Y$. The reason for ignoring the difference between $\mathbf{G}_X$ and $\mathbf{G}_Y$ is that there is a natural isomorphism (simply restrict to $\mathbf{G}_X$ the inner automorphism of **S** induced by the translation carrying X to Y) of $\mathbf{G}_X$ onto $\mathbf{G}_Y$. In this case, as opposed to the case for groups of order 2, there are often (exercise 2.8, problem 2) other isomorphisms of $\mathbf{G}_X$ onto $\mathbf{G}_Y$. Note that the kernel of the homomorphism of **G** onto its point group $\mathbf{G}_X$ is always $\mathbf{G} \cap \mathbf{T}$, regardless of what point X is chosen.

The two examples in the paragraph above are special cases of a general theorem which asserts that if f_1 and f_2 are homomorphisms of a group **G** onto groups $\mathbf{H}_1$ and $\mathbf{H}_2$, respectively, and if f_1 and f_2 have the same kernel, then $\mathbf{H}_1$ and $\mathbf{H}_2$ are isomorphic in a natural way, namely, $h_1 \leftrightarrow h_2$ if and only if the pre-images of h_1 under f_1 and the pre-images of h_2 under f_2 form the same coset of the kernel of f_1 and f_2. Thus, a homomorphism is essentially determined when we know its kernel. Since the kernel of a homomorphism is a normal subgroup, we know all of the possible kernels of homomorphisms of **E**, and we arrive at the following result.

Corollary 2.35. Essentially the only homomorphisms of **E** are:
(1) Isomorphisms (kernel $= \{\mathbf{1}\}$).
(2) The homomorphism of **E** onto its point group $\mathbf{E}_X = [\mathbf{E}, X]$ (kernel $=$ **T**) obtained by factoring out the translational aspect of each isometry.
(3) The homomorphism of **E** onto $\mathbf{E}_X^+ = [\mathbf{E}^+, X]$ (kernel $= \mathbf{E} \cap \mathbf{D}$) obtained by first factoring out the translational aspect of each isometry and then ignoring direct or opposite.
(4) The homomorphism of **E** onto a group of order 2 (kernel $= \mathbf{E}^+$) in which we ignore everything except direct or opposite.
(5) The trivial homomorphism of **E** onto $\{\mathbf{1}\}$.

Keep in mind the word "essentially" which appears in this corollary. We have *not* proved that these five homomorphisms are all that are possible, but only that any homomorphism of **E** corresponds in a natural way to one of them. More precisely, if ϕ is any homomorphism of **E**, although ϕ need not be any of the five homomorphisms above, there is an isomorphism, θ, defined on the range of ϕ such that $\phi\theta$ is one of the homomorphisms above.

Determining the homomorphisms of **S** is much more difficult since there are many normal subgroups of **S** which are rather complicated to describe. The simplest are of the following type.

Definition. Let **F** be a subgroup of the multiplicative group, **P**, of positive real numbers. Then **S(F)** denotes the group of all similarities with stretching factor in **F**. Elements of **S(F)** are called **F**-*similarities*. Thus, $\{\mathbf{1}\}$-similarities are isometries, and **P**-*similarities* are simply similarities.

The group, **P**, is an abelian group; hence all of its subgroups are normal. Thus, if we combine theorem 2.5 and problem 5 in exercise 1.3, we find that **S(F)** is always a normal subgroup of **S**. Since $\mathbf{S}^+$, **D**, and $\mathbf{D}^+$ are also normal in **S**, we have proved the following proposition.

Proposition 2.36. If **F** is a subgroup of **P**, the multiplicative group of positive reals, then the following four groups are all normal in **S**:

(1) **S(F)**, the group of all **F**-similarities
(2) $\mathbf{S(F)} \cap \mathbf{S}^+$, the group of all direct **F**-similarities
(3) $\mathbf{S(F)} \cap \mathbf{D}$, the group of all **F**-dilatations
(4) $\mathbf{S(F)} \cap \mathbf{D}^+$, the group of all direct **F**-dilatations.

In the special case in which $\mathbf{F} = \{\mathbf{1}\}$, these are just the nontrivial normal subgroups of **E**. If **F** is a larger group, then we can define still more complicated normal subgroups of **S**. For example, if **F** is the subgroup of **P** generated by 2, i.e., $\mathbf{F} = \{\ldots, \frac{1}{4}, \frac{1}{2}, 1, 2, 4, \ldots\}$, then the set $\{x_2 a \mid X$ any point, and A any plane, such that $X \in A\}$ is a normal sub*set* of **S** and hence generates a normal sub*group* of **S**. A similarity in this group is direct or opposite, according as its stretching factor is an even or odd power of 2. Thus, this group is not one of the four types above.

Even the groups **F** themselves may be very complicated. For example, in [4] the additive groups of rational numbers are examined. These, of course, are only a few of the possible additive groups of real numbers, and (since any logarithm function is an isomorphism from **P** onto the additive group of reals) each such group corresponds to a possible **F**.

Exercise 2.8

1. Let τ be the translation carrying X to Y, and let **G** be a subgroup of **S**. Show that the inner automorphism, τ^i, of **S** maps $\mathbf{G}_X$ onto $\mathbf{G}_Y$, and thus all of the point groups of **G** are conjugate in **S**. Note (see problem 2 below) that this isomorphism between $\mathbf{G}_X$ and $\mathbf{G}_Y$ need not be unique.

2. Let **G** be the group of isometries leaving a given cube globally invariant. Find a point X such that $\mathbf{G} = \mathbf{G}_X = [\mathbf{G}, X]$ and a point Y such that $[\mathbf{G}, Y] \neq \mathbf{G}_Y \neq \mathbf{G}$. Find at least two distinct isomorphisms of $\mathbf{G}_X$ onto $\mathbf{G}_Y$.

3. Prove that if W, X, Y, and Z are points on a sphere such that $|WX| = |YZ|$, then there is a rotation sending W to Y and X to Z which leaves the sphere globally fixed.

4. Let **G** be a subgroup (not necessarily normal!) of **E** for which $\mathbf{G} \cap \mathbf{E}^{+} = \mathbf{T}$. Show that the only possible opposite isometries in **G** are inversions or glide reflections (including reflections). Find a group **G** such that $\mathbf{G} \neq \mathbf{E}$, and yet **G** does contain reflections and $\mathbf{G} \cap \mathbf{E}^{+} = \mathbf{T}$.

5. In Chapter 1 we considered the two-dimensional symmetry group of a square and found that it was a group of order 8. Show that the group of all isometries of space leaving a given square fixed is a group of order 16.

6. (continuation of 5) Let X be, respectively, the centroid, a vertex, or a point $\frac{1}{3}$ of the way along a side of the square. Find $[\mathbf{G}, X]$ in each case, where **G** is the group of all 16 isometries leaving the square globally invariant.

7. (continuation of 5) Find (in the spirit of corollary 2.35) all homomorphisms of the group of 16 isometries leaving a given square globally invariant.

8. Show that the 5 subgroups of $\mathbf{E}_2$, the group of isometries of the plane considered in exercise 2.7, problem 3, are the only normal subgroups of $\mathbf{E}_2$.

9. Describe intuitively a homomorphism of **S** with kernel **S(F)** if **F** is a subgroup of **P**. You may first want to do the same problem with the kernel $\mathbf{S(F)} \cap \mathbf{S}^{+}$, and then speak in terms of a composite mapping.

10. Prove that if **G** is a nontrivial, normal subgroup of **S**, then $\mathbf{T} \subseteq \mathbf{G}$. *Hint.* If $\mathbf{G} \subseteq \mathbf{E}$, then use results in this section; otherwise use theorem 2.23, and show that a spiral followed by suitable conjugates is a nontrivial translation.

Bibliography and suggestions for further reading

[1] Allendoerfer, Carl B., "Angles, Arcs, and Archimedes," *The Math. Teacher* **58,** 82–88 (1965).

[2] Artin, E., *Geometric Algebra*, Wiley, New York, 1957.

[3] Bachmann, Friedrich, *Aufbau der Geometrie aus dem Spiegelungsbegriffe*, in *Die Grundlehren der Mathematischen Wissenschaften in Einzeldarstellungen*, Vol. 96, Springer, Berlin, 1959.

[4] Beaumont, R. A., and H. S. Zuckerman, "Subgroups of the additive group of rationals," *Pacific J. Math.* **1,** 169–177 (1951).

[5] Coxeter, H. S. M., *Introduction to Geometry*, Wiley, New York, 1961.

[6] Guggenheimer, Heinrich W., *Plane Geometry and its Groups*, Holden-Day, San Francisco, 1966.

[7] Kolmogorov, A. N., and S. V. Fomin, *Functional Analysis*, Vol. 1, Graylock, Rochester, N.Y., 1957.

[8] MacGillavry, Caroline H., *Symmetry Aspects of M. C. Escher's Periodic Drawings* (published for the International Union of Crystallography), A. Oosthoek, Utrecht, 1965.

[9] Modenov, P. S., and A. S. Parkhomenko, *Geometric Transformations*, Vol. 1, Academic Press, New York, 1965.

[10] Thomsen, Gerhard, "The treatment of elementary geometry by a group-calculus," *The Math. Gazette* **17,** 230–242 (1933). This is a translation of the article appearing in *Mathematische Zeitschrift* **34,** 668–720 (1932).

[11] Weyl, Hermann, *Symmetry*, Princeton Univ. Press, Princeton, N.J., 1952.

Of the many introductions to geometric transformations, we consider [5] and [9] to be two of the best. Chapters 2, 3, 5, and 7 of Coxeter's text cover the material in the first six sections of this chapter from a slightly different point of view. Some of the remarkable artwork of M. Escher is reproduced there to show how various groups of isometries in the plane can be represented as symmetry groups. The symmetry aspects of Escher's work are analyzed in [8]. The paperback translation, [9], of a book by two Russian authors is an excellent, elementary, and inexpensive introduction to geometric transformations. Be careful, for products of mappings read from right to left in [9] and from left to right in [5].

For those who wish to go further into the formal study of euclidean geometry, we strongly recommend [6], in which euclidean plane geometry is developed in terms of axioms about points, lines, and reflections. In particular, angles are carefully defined and analyzed, a problem we have avoided in our consideration of rotations. See [1] for a discussion of the difficulties that must be overcome in a careful treatment of angles.

Review problems for Chapters 1 and 2

These problems are presented in random order. Some are easy, others are very difficult, and a few are more in the way of suggestions for term-paper topics rather than problems.

1. Determine the centralizer and normalizer in **E** of each of the following sets:
(1) $\{x, y\}$, two cases, according as $X = Y$ or $X \neq Y$.
(2) $\{k, m\}$, assuming $K \neq M$.
(3) $\{a, b\}$, assuming A and B are not parallel.

2. Show that $|W, X| = |Y, Z|$ if and only if there is an isometry ϕ such that $(wx)\phi^i = (yz)$.

3. Show that the only abelian subgroup of **E** that is transitive on points is **T**.

4. In exercise 2.3, problem 3, we examined products of the type *md*. Describe products of the type *dm*.

5. Let ϕ be an isometry, L a line, and A a plane. Prove that there is a line M and a plane B such that:
(1) $\{Y|Y = \text{midpoint of } [X, (X)\phi], X \in L\} \subseteq M$.
(2) $\{Z|Z = \text{midpoint of } [X, (X)\phi], X \in A\} \subseteq B$.

6. Find all similarities of finite order.

7. Look up the topic of eulerian angles, and show how they can be used to describe the product of two rotations whose axes intersect. We have proved that such a product is a rotation, but we have not shown how to find the axis and angle of rotation.

8. Let **F**, **G**, and **H** be groups such that **G** is a normal subgroup of **F** and **H** is a characteristic subgroup of **G**. Show that **H** is a normal subgroup of **F**. Find an example, with $\mathbf{F} = \mathbf{S}_4$, showing that if **H** is merely normal in **G**, then it need not be normal in **F**.

9. Let M be the plane with equation $ax + by + cz = d$. Find equations describing m.

10. The search for homomorphisms of **S** onto two element groups is equivalent to the search for normal subgroups of index 2 in **S**. Show that if **F** were a subgroup of index 2 in **P**, the multiplicative group of positive reals, then the **F**-similarities would be such a group. Then show that no such **F** exists.

11. Is there a homomorphism of **S** onto a two-element group with kernel $\neq \mathbf{S}^+$?

12. Prove that if g and h are, respectively, 90° and 120° rotational symmetries of a cube such that the axis of h hits one of the vertices of a face left invariant by g, then g and h generate the group of rotational symmetries of that cube.

13. Let **G** be the symmetry group of a regular triangular prism ($\#G = 12$).
(1) Show that **G** does not have a faithful representation in $\mathbf{S}_4$.
(2) Find a faithful representation of **G** in $\mathbf{S}_5$.
(3) Find the normalizer in **G** of each of its 7 involutions. *Hint*. First show that if h is the conjugate of g by ϕ in **G**, then $(N(g))\phi^i = N(h)$.
(4) Find an outer automorphism of **G**, and show that **G** is isomorphic to its group of automorphisms. *Hint*. Find faithful representations of **G** and **A(G)** in $\mathbf{S}_6$.

14. Find an involution in $\mathbf{S}_8$ that commutes with (1, 2, 3, 4)(5, 6, 7, 8).

15. How many different 4 by 4 Latin squares are there?

Chapter 3

An Introduction to Crystallography

In this chapter we present an elementary introduction to an important field of applied group theory and geometry. The subject of crystallography is one which affects many of the sciences, and, in fact, one can argue that it is one of the most important subjects in theoretical chemistry or physics, *cf.* [10]. We exploit the tools that have been developed in Chapter 2 to find all finite subgroups of **E** and to classify the crystallographic groups.

Throughout this chapter we restrict ourselves to isometries of space. There are several good treatments of crystallographic groups in the plane which will be of help to the reader as he reads this book. Two of these are [7], chapter 3, and [5], chapter 4. In section 3.6 we present many helpful hints for reading the *International Tables of X-Ray Crystallography*, [8]. These tables provide many examples of the general theory discussed here.

3.1. Discrete groups of isometries

The adjective "discrete" suggests "spread apart" or "discontinuous." It is often used in the phrase "discrete group" to imply that the group is "almost finite" in some sense. In this section we define carefully what *we* mean when we use the phrase "discrete group." We also prove that this definition is equivalent to two others that are often used, and that "discrete" is *not* the same as "countable."

Before stating the formal definition, we recall a few basic facts about bounded sets in euclidean space. First of all, a set of points in euclidean space is *bounded* if it is contained in some ball or, equivalently, if there is an upper bound to the distance between pairs of points in the set. We often use the basic property that any infinite bounded set of points in euclidean space has a limit point, i.e., if S is an infinite bounded set, then there is a point X (which need not be in S) such that, for any $\epsilon > 0$, there is a point $Y \in S$ for which $0 < |XY| < \epsilon$. An easy consequence of the definition of limit point is that if X is a limit point of the set S, then there is a sequence of *distinct* points in S that converges to X. In particular, a finite set cannot have any limit points.

We often need to refer to the **G**-orbit of a point X in this chapter, so let us agree to denote it by $(X)\mathbf{G}$, i.e., $(X)\mathbf{G} = \{(X)g | g \in G\}$.

Definition. A group, **G**, is a *discrete group* if and only if **G** is a group of isometries such that for any point X and any bounded set S the set S contains only finitely many points in the **G**-orbit of X, i.e, $S \cap (X)\mathbf{G}$ is finite. *Caution!* This definition for discrete groups is fairly standard in the context of geometry but may not be equivalent to other definitions given in other contexts, e.g., abstract group theory.

Clearly, any finite subgroup of **E** is a discrete group. We classify the finite subgroups of **E** in section 3.2. In section 3.3, we consider the discrete groups generated by translations. The simplest of these is the group generated by a single nontrivial translation. This is also the simplest example of a discrete group that is not finite. It is, however, *countable*, i.e., there is a one-to-one correspondence between this group and some subset of the integers. The relationship between "discrete" and "countable," although not crucial to the rest of this chapter, is interesting, so we pause to present it here.

Proposition 3.1. A discrete group is countable, but not all countable subgroups of **E** are discrete.

Proof. Let **G** be a discrete group, and let X_1, X_2, X_3, and X_4 be four independent points. It follows easily from theorem 2.7 that there is a one-to-one correspondence between **G** and the set of **G**-images of the ordered quadruple of points (X_1, X_2, X_3, X_4). Let $\mathcal{B}_n$ be the ball of radius n centered at X_1. Since **G** is discrete, each X_i has only a finite number of images in $\mathcal{B}_n$, and hence there are only a finite number of **G**-images of the ordered quadruple (X_1, X_2, X_3, X_4) in which the image of each X_i is in $\mathcal{B}_n$. The union of the $\mathcal{B}_n$, $n = 1, 2, 3, \ldots$ is the entire space; hence the set of **G**-images of (X_1, X_2, X_3, X_4) is a countable union of finite sets. Such a set is itself countable; hence **G** is countable.

To see that the converse is not true consider the group of all translations of rational length, leaving a given line globally fixed or the group of all rotations of finite order with a given axis. Neither of these groups is discrete, yet both are countable—the first since the rationals are countable, and the second as a countable union of finite sets.

Essentially the same technique as that used in the first part of the proof above can be used to show that in a discrete group only a finite number of elements can leave a given point fixed.

Theorem 3.2. If $\mathbf{G}$ is a discrete group and W is a point, then $[\mathbf{G}, W]$ is a finite subgroup of $\mathbf{G}$.

Proof. Choose points X, Y, and Z such that W, X, Y, and Z are independent, and then choose a ball, $\mathcal{B}$, centered at W and containing X, Y, and Z. Since each element of $[\mathbf{G}, W]$ is an isometry and leaves W fixed, the $[\mathbf{G}, W]$-images of X, Y, or Z are all in $(X)\mathbf{G} \cap \mathcal{B}$, $(Y)\mathbf{G} \cap \mathcal{B}$, or $(Z)\mathbf{G} \cap \mathcal{B}$, respectively. These are finite sets so there are only a finite number of possible $[\mathbf{G}, W]$-images of the ordered quadruple (W, X, Y, Z), and thus $[\mathbf{G}, W]$ is finite.

The analog of this result in the euclidean plane can be used to prove that a discrete but infinite group of plane isometries must contain a translation (exercise 3.1, problem 7). This result is not true in space—consider the discrete group generated by a proper screw displacement whose angle α (in radians) is incommensurable with π (i.e., π/α is irrational). This discrete group is infinite yet contains no translations.

The following theorem explains the connection between "discrete" and "spread apart."

Theorem 3.3. Let $\mathbf{G}$ be a group of isometries. $\mathbf{G}$ is a discrete group if and only if for each $\mathbf{G}$-orbit there is an $\epsilon > 0$ such that any two points in that $\mathbf{G}$-orbit are at least ϵ units apart.

Proof. Assume $\mathbf{G}$ is a discrete group, and consider a $\mathbf{G}$-orbit, $(X)\mathbf{G}$. If $(X)\mathbf{G} = \{X\}$, then any $\epsilon > 0$ will do. If not, let $\mathcal{B}$ be any ball centered at X containing at least one other point in $(X)\mathbf{G}$. Among the finite points in $\mathcal{B} \cap (X)\mathbf{G}$, choose one that is closest to X, and let ϵ be the distance between X and this point. If Y and Z are two distinct points in $(X)\mathbf{G}$, then we can choose $g, h \in \mathbf{G}$ such that $(X)g = Y$ and $(X)h = Z$. Let $W = (Y)h^{-1} = (X)gh^{-1}$. Since $W \neq X$ and $W \in (X)\mathbf{G}$, it follows that $|W, X| \geq \epsilon$. However, $|W, X| = |Y, Z|$ since h is an isometry sending W to Y and X to Z. Thus, the points in this $\mathbf{G}$-orbit are spread apart by at least ϵ.

Now assume that $\mathbf{G}$ is not discrete. Then there is at least one $\mathbf{G}$-orbit and one bounded set containing infinitely many points in that orbit. By the remarks at the beginning of this section, there is a sequence of distinct points in this $\mathbf{G}$-orbit that converges. Since this sequence converges, it must be a Cauchy sequence, i.e., given any $\epsilon > 0$, if we go far enough out in the sequence, the points are less than ϵ units apart. Thus, for this $\mathbf{G}$-orbit, it is impossible to find a satisfactory ϵ. Taking the contrapositive, we see that if it is possible to find a satisfactory ϵ for each $\mathbf{G}$-orbit, then $\mathbf{G}$ must be discrete.

Note carefully the "order of the quantifiers" in the theorem above. We claim that for each $\mathbf{G}$-orbit there is a satisfactory $\epsilon > 0$, but *not* that there is an $\epsilon > 0$ that is satisfactory for all $\mathbf{G}$-orbits. For example, consider the discrete group consisting of the 48 symmetries of a cube centered at W. If X

is a point on one of the rays joining W to a vertex, then $(X)\mathbf{G}$ consists of the eight vertices of a cube. The amount by which these eight points are spread apart depends on $|WX|$. Clearly, no one ϵ will be satisfactory for all of these orbits.

Corollary 3.4. If **G** is a group of isometries, then **G** is discrete if and only if no **G**-orbit has a limit point.

Proof. If **G** is discrete, then each **G**-orbit is "spread apart" and hence cannot have a limit point. On the other hand, if **G** is not discrete, then we have shown in the proof above that at least one **G**-orbit has a limit point.

Exercise 3.1

1. Prove that any subgroup of a discrete group is discrete.

2. Prove, or disprove by giving a counterexample, that if **G** is discrete and **H** is a subgroup of **G** containing a nontrivial translation, then $[\mathbf{G}:\mathbf{H}]$ is finite.

3. Devise a pattern in the plane whose group of symmetries is a discrete group but is not finite.

4. Prove or disprove that the symmetry group of a countable pattern of points is discrete.

5. Assume that $\mathcal{S}$ is a set of points in space such that $\mathcal{S} \cap \mathcal{B}$ is finite for every bounded set $\mathcal{B}$. Show that the symmetry group of $\mathcal{S}$ is discrete if and only if $\mathcal{S}$ contains at least three independent points.

6. Give an example of a discrete group containing at least one proper screw displacement *and* at least one glide reflection.

7. Prove that a discrete, but not finite, group of isometries of the plane must contain a nontrivial translation. *Hint.* Show that the direct isometries in the group cannot all be rotations with the same center.

8. Let **G** be a subgroup of **S** such that $(X)\mathbf{G} \cap \mathcal{B}$ is finite for any point X and any bounded set $\mathcal{B}$. Prove that $\mathbf{G} \subseteq \mathbf{E}$, and hence **G** is discrete.

3.2. *Finite groups of isometries*

Our objective in this section is to classify all finite subgroups of **E**. There are several natural ways to classify these groups. We could, for example, classify them into classes of isomorphic groups, a crude classification which ignores their connections with geometry and merely identifies their algebraic structure. Another classification is into sets of conjugate groups within **S**. This more restrictive classification takes into account the geometric nature of the group, e.g., it distinguishes the cyclic group of order 2 generated by a half-turn from the isomorphic group generated by a reflection. We shall focus on these two classifications in this section. We already have the term "isomorphic" to assert that two groups are of the same type in the first classification. In the following definition we establish an analogous term for the second classification.

Definition. Subgroups **G** and **H** of **E** are called *similar groups* if and only if they are conjugate subgroups in **S**.

By corollary 2.29 similar subgroups of **E** are groups that are isomorphic via an isomorphism that can be extended to an automorphism of **E**.

We have considered several finite subgroups of **E** in the previous two chapters, e.g., the symmetry groups of various solids. Let us now change our outlook and look for restrictions on finite subgroups of **E**. After doing this, we shall complete the list of examples, catalog them, and exhibit each as the symmetry group of a simple solid figure.

Theorem 3.5. If **G** is a finite subgroup of **E**, then there is at least one point left fixed by all of the isometries in **G**.

Proof. Let $\mathcal{S}$ be a finite set of points. If g is an isometry, X is the centroid of $\mathcal{S}$, then $(X)g$ is the centroid of $(\mathcal{S})g$. However, if $\mathcal{S}$ is a **G**-orbit, then $\mathcal{S}$ is finite since **G** is finite, and $(\mathcal{S})g = \mathcal{S}$ for each $g \in \mathbf{G}$. Thus, the centroid of any **G**-orbit is a fixed point for each isometry in **G**.

We saw in Chapter 2 that the only isometries leaving a point X fixed are the rotations and rotatory inversions with axes through X. Thus, for any finite group, **G**, of isometries, there is a point X such that **G** contains only rotations and rotatory inversions with axes through X. The inversion x is an opposite isometry and commutes with all isometries in **G**. Hence there are only three possible relationships between the group, **G**, and the subgroup, $\mathbf{G}^+$, of all direct isometries in **G**. These are stated in the following theorem.

Theorem 3.6. If **G** is a finite group of isometries all of which leave the point X fixed, and if $\mathbf{G}^+$ is the group of direct isometries in **G**, then exactly one of the following three possibilities holds:

(1) $\mathbf{G} = \mathbf{G}^+$
(2) $\mathbf{G} = \mathbf{G}^+ \cup x\mathbf{G}^+$
(3) $\mathbf{G} \neq \mathbf{G}^+$ and $x \notin \mathbf{G}$.

Moreover, in the third case,

$$\mathbf{G}^+[\mathbf{G} = \mathbf{G}^+ \cup \{xg | g \in \mathbf{G}, g \notin \mathbf{G}^+\}$$

is a group of direct isometries isomorphic to **G** and containing $\mathbf{G}^+$ as a subgroup of index 2.

Proof. **G** fits into case (1) if and only if it contains no opposite isometries. If it does contain opposite isometries, then it fits into cases (2) or (3) according as x does or does not belong to **G**. Thus, these three cases are exclusive and exhaust all possibilities. Clearly if $\mathbf{G} \neq \mathbf{G}^+$ and $x \notin \mathbf{G}$, then $\mathbf{G}^+[\mathbf{G}$ is a *set* of direct isometries. It is an easy exercise in group theory (exercise 3.2, problem 1) to show that $\mathbf{G}^+[\mathbf{G}$ is a group isomorphic to **G** and containing $\mathbf{G}^+$ as a subgroup of index 2.

From the result above, it is clear that in order to find all finite groups of isometries we need only find all finite groups of *direct* isometries and look for all subgroups of index 2 in these groups. The technique used below to search for finite groups of direct isometries is based on the ideas of P. Curie of the famous husband and wife team of pioneers in radioactivity (*cf.* [6] or Appendix A in [13]).

Definition. Let **G** be a group of direct isometries all leaving the point W fixed. Let $\mathcal{S}$ be a sphere centered at W. A point X on $\mathcal{S}$ is a *pole* if some nontrivial rotation in **G** leaves X fixed.

The isometries in **G** permute the poles on a given sphere, and thus the set of poles on that sphere is partitioned into one or more **G**-orbits. For example, the symmetry group of a cube leads to three orbits of poles, those on the axes through the centers of opposite faces, opposite vertices, or the midpoints of opposite sides. The symmetry group of a pyramid (with a regular base but excluding a regular tetrahedron), on the other hand, leads to two orbits of poles, each containing just one of the two points at opposite ends of the axis of rotational symmetry. This division into just two or three orbits is typical, as we now show.

Theorem 3.7. If **G** is a nontrivial, finite group of direct isometries all of which leave the point W fixed, and if $\mathcal{S}$ is a sphere centered at W, then the number of **G**-orbits of poles on $\mathcal{S}$ is either 2 or 3.

Proof. Let X_i, $i = 1, 2, \ldots, k$ be a set of poles on $\mathcal{S}$, one from each **G**-orbit of poles on $\mathcal{S}$. We are to show that $k = 2$ or 3. Let $\alpha_i = \#[\mathbf{G}, X_i]$ be the number of isometries in **G** leaving the ith pole fixed. By definition of "pole," $\alpha_i \geq 2$. Let $\beta_i = \#(X_i)\mathbf{G}$ be the number of poles in the ith **G**-orbit, and let $n = \#(\mathbf{G})$. It follows directly from corollary 1.18 (see [13], pp. 150–151 if you want to see a proof that starts from basic principles) that $\alpha_i\beta_i = n$ for $i = 1, 2, \ldots, k$.

Since $\alpha_i = [\mathbf{G}, X]$ for any pole X in $(X_i)\mathbf{G}$, and there are β_i poles in $(X_i)\mathbf{G}$, it follows that $\beta_i(\alpha_i - 1)$ is the number (counting repetitions) of *nontrivial* isometries in **G** leaving at least one pole in the ith orbit fixed. Summing over the k orbits, we find that

$$\sum_{i=1}^{k} \beta_i (\alpha_i - 1) = \sum_{i=1}^{k} (n - \beta_i)$$

is the total number of nontrivial elements in **G** leaving at least one pole fixed, again counting each isometry once for each pole it leaves fixed. Each nontrivial element in **G**, however, is a rotation leaving exactly two poles on $\mathcal{S}$ fixed; hence

$$2(n - 1) = \sum_{i=1}^{k} (n - \beta_i) . \qquad (1)$$

Dividing both sides of this equation by n, and using the fact that $n = \alpha_i \beta_i$, we arrive at

$$2 - (2/n) = \sum_{i=1}^{k} (1 - (1/\alpha_i)) . \qquad (2)$$

Since $n \geq 2$ (**G** is nontrivial), $1 \leq 2 - (2/n) < 2$. Since each $\alpha_i \geq 2$, we also have $\frac{1}{2} \leq 1 - (1/\alpha_i) < 1$, $i = 1, 2, \ldots, k$. Summing over the k orbits and using eq. (2) this leads to $k/2 \leq 2 - (2/n) < k$. These two inequalities on $2 - (2/n)$ are compatible only if $k = 2$ or 3.

Corollary 3.8. If **G** is a group of order n, $n > 1$, consisting entirely of direct isometries leaving the sphere $\mathcal{S}$ globally fixed, and if there are only two **G**-orbits of poles on $\mathcal{S}$, then **G** is cyclic and is generated by a rotation through $(360/n)°$.

Proof. Let X_i, α_i, and β_i, $i = 1, 2$, be as in the proof above. Then eq. (1) simplifies to $2 = \beta_1 + \beta_2$. Since β_1 and β_2 are integers, each of the two **G**-orbits contains just one pole. Obviously these two inequivalent poles must lie on a diameter of $\mathcal{S}$, and, therefore, **G** consists entirely of rotations about this diameter. Let Y be any point on $\mathcal{S}$ except one of the two poles. The **G**-orbit of Y is spread apart and contained in the circle on $\mathcal{S}$ passing through Y and in a plane perpendicular to the common axis of the rotations in **G**. Let $Z \neq Y$ be one of the two points in the **G**-orbit of Y closest to Y, and let ρ be the rotation in **G** carrying Y to Z. Let γ be the order of ρ. Then Y, $(Y)\rho$, $(Y)\rho^2, \ldots, (Y)\rho^{\gamma-1}$ determine γ congruent arcs of the circle containing the **G**-orbit of Y. If $\phi \in$ **G**, and if ϕ is not a power of ρ, then $(Y)\phi$ must be in the interior of one of these arcs, say $(Y)\phi$ is between $(Y)\rho^j$ and $(Y)\rho^{j+1}$. But then $\phi\rho^{1-j}$ is in **G** and sends Y to a point inside the arc from Y to $Z = (Y)\rho$. This contradicts our choice of Z, so $\mathbf{G} \subseteq \{\mathbf{1}, \rho, \ldots, \rho^{\gamma-1}\}$. The reverse inclusion is obvious, so **G** is generated by ρ and $\gamma = n$. In view of the way ρ was chosen, its angle is $(360/n)°$.

In the following corollary we meet the direct symmetry groups of the regular polyhedra. Since the cube and the octahedron are *dual figures*, i.e., either one can be obtained from the other by using the centers of the faces as vertices, their direct symmetry groups are clearly similar. Likewise, the direct symmetry groups of the dodecahedron and the icosahedron are similar. It is customary to refer to these groups as **O**, the *octahedral group*, and **I**, the *icosahedral group*, respectively. Similarly **T** is the customary name for the *tetrahedral group*. Unfortunately we have an alphabetic collision at this point for we have used **T** as the name for the group of all translations. The context *always* determines which of these two groups is under consideration so we shall use the same name for both groups.

Corollary 3.9. If **G** is a group of order n, $n > 1$, consisting entirely of direct isometries leaving the sphere $\mathcal{S}$ invariant, and if there are three **G**-orbits of poles on $\mathcal{S}$, then **G** is similar to exactly one of the following groups:

(1) the dihedral group, $\mathbf{D}_q$, generated by two half-turns whose axes intersect at an angle of $(180/q)°$, $q = 2, 3, 4, \ldots$
(2) the group, **T**, of direct symmetries of a tetrahedron
(3) the group, **O**, of direct symmetries of an octahedron
(4) the group, **I**, of direct symmetries of an icosahedron.

Proof. Again let X_i, α_i, and β_i, $i = 1, 2, 3$, be as in the proof of theorem 3.7, and assume that $\alpha_1 \leq \alpha_2 \leq \alpha_3$. Corollary 3.8 applied to $[\mathbf{G}, X_i]$ asserts that $[\mathbf{G}, X_i]$ is cyclic and is generated by a rotation of order α_i, $i = 1, 2, 3$.

Equations (1) and (2) in the proof of theorem 3.7 reduce to $\beta_1 + \beta_2 + \beta_3 = n + 2$ and $(1/\alpha_1) + (1/\alpha_2) + (1/\alpha_3) = 1 + (2/n)$. An easy analysis of these two equations, using the information that $2 \leq \alpha_1 \leq \alpha_2 \leq \alpha_3$ and that $\alpha_i\beta_i = n$, shows that the only integral solutions are as shown in Table 10.

TABLE 10. Possible Orders of Poles, α_i, and Numbers of Poles, β_i, in Three **G**-Orbits

	α_1	α_2	α_3	β_1	β_2	β_3	n
Case 1	2	2	q	q	q	2	$2q$
Case 2	2	3	3	6	4	4	12
Case 3	2	3	4	12	8	6	24
Case 4	2	3	5	30	20	12	60

Case 1. $[\mathbf{G}, X_3]$ is cyclic of order q. Let ρ be a generator. Since $\alpha_1 = \alpha_2 = 2$, all rotations in **G**, but not in $[\mathbf{G}, X_3]$, are half-turns. Let m be one of these half-turns, then $m\rho$ is not in $[\mathbf{G}, X_3]$. Hence $m\rho = k$ is a half-turn. Clearly m and ρ generate **G**, and since $\rho = km$ is a rotation through $(360/q)°$, it follows that k and m are half-turns whose axes intersect at an angle of $(180/q)°$ and that they generate **G**.

Case 2. Let $Y_1 = X_2$, Y_2, Y_3, Y_4 be the four poles in the second **G**-orbit. The rotations in **G** permute these four points among themselves transitively. Since there are rotations in **G** of orders 2 and 3, it is easy to see that **G** is transitive on the line segments $[Y_1, Y_j]$. Thus, the Y_i fall at the vertices of a regular tetrahedron, and **G** is a subgroup of the group of direct symmetries of this tetrahedron. Both groups have 24 elements, so they are equal. All tetrahedra are similar in euclidean space, so all of their groups of direct symmetries are similar.

Case 3. Consider the six poles in the third **G**-orbit. A rotation of order 4 in **G** must leave fixed two of these poles and permute the remaining four poles in this orbit cyclicly. This holds for all three possible axes so the twelve line segments joining poles on distinct axes all have the same length, and the six poles lie at the vertices of a regular octahedron. **G** is the group of direct symmetries of this octahedron since it is a subgroup with just as many elements. Again all regular octahedra are similar so all these groups of direct symmetries are similar.

Case 4. Consider the 12 poles in the third orbit lying on six axes of rotations of order 5. Choose a rotation, ρ, in **G** and of order 5. View the two poles of ρ as the north and south poles of the sphere $\mathbb{S}$. The remaining 10 poles in the third **G**-orbit split into two ρ-orbits of 5 poles each. These 10 poles do not all lie on the equator for if they did some rotations in **G** of order 5 would have both poles on the equator and would send the other 8 poles on the equator to poles of order 5 which are neither on the equator nor at the north or south poles. Thus each pole of order 5 must have five near neighbors, five remote neighbors, and its antipode. Since all 12 poles are in one **G**-orbit the distance from a pole to its 5 near neighbors is invariant and therefore the 12 poles lie at the vertices of a regular icosahedron. The group **G** leaves this set of 12 poles globally invariant and hence is a subgroup of the group of direct symmetries of an icosahedron. The orders of the two groups are the same, hence they are equal.

Summarizing these last three results, we find that any finite group of direct isometries is similar to one of the groups in the following list:

- $\mathbf{C}_q$, a cyclic group of order q generated by a rotation through $(360/q)°$, $q = 1, 2, 3, \ldots$
- $\mathbf{D}_q$, a dihedral group of order $2q$ generated by two half-turns whose axes intersect at an angle of $(180/q)°$, $q = 2, 3, 4, \ldots$ (*Note!* $q \neq 1$.)
- **T**, the group of direct symmetries of a tetrahedron, $\#\mathbf{T} = 12$
- **O**, the group of direct symmetries of a cube or of an octahedron, $\#\mathbf{O} = 24$
- **I**, the group of direct symmetries of an icosahedron or a dodecahedron, $\#\mathbf{I} = 60$.

No two groups in this list are isomorphic (exercise 3.2, problem 2) so this is the desired classification into isomorphism classes and into similarity classes for finite groups of direct isometries. As an immediate consequence of this result and theorem 3.6, we see that any finite group of isometries that contains an inversion, x, is similar to exactly one of the following groups:

$$\mathbf{C}_q \cup x\mathbf{C}_q, q = 1, 2, \ldots; \quad \mathbf{D}_q \cup x\mathbf{D}_q, q = 2, 3, 4, \ldots;$$
$$\mathbf{T} \cup x\mathbf{T}, \mathbf{O} \cup x\mathbf{O}, \mathbf{I} \cup x\mathbf{I}.$$

The last two groups are, of course, the complete symmetry groups of the octahedron and icosahedron, respectively. No group containing an inversion can be similar to a group without an inversion so no two groups in the combined lists above can be similar. There are some isomorphic pairs of groups, however. These are: $\mathbf{C}_q \cup x\mathbf{C}_q \cong \mathbf{C}_{2q}$ if q is odd, $\mathbf{C}_2 \cup x\mathbf{C}_2 \cong \mathbf{D}_2$, and $\mathbf{D}_q \cup x\mathbf{D}_q \cong \mathbf{D}_{2q}$ if q is odd ($q \neq 1$) (exercise 3.2, problem 7).

The only other possible finite groups of isometries must contain opposite isometries but no inversion. According to theorem 3.6, the only candidates for such groups arise from pairs, $\mathbf{G}^+$ and $\mathbf{H}$, of groups of direct isometries in which $\mathbf{G}^+$ is a subgroup of index 2 in $\mathbf{H}$. Examining our list of finite groups of direct isometries, we find the following are the only such pairs: $\mathbf{C}_q$ in $\mathbf{C}_{2q}$, $\mathbf{C}_q$ in $\mathbf{D}_q$, $\mathbf{D}_q$ in $\mathbf{D}_{2q}$, and $\mathbf{T}$ in $\mathbf{O}$. If $\mathbf{G}^+$ and $\mathbf{H}$ are any such pair, then $\mathbf{G}^+[\mathbf{H} = \mathbf{G}^+ \cup \{xh | h \in \mathbf{H}, h \notin \mathbf{G}^+\}$ is a group containing opposite isometries but no inversion, provided X, the center of x, is a point left fixed by all isometries in $\mathbf{H}$ (exercise 3.2, problem 1). The groups $\mathbf{G}^+[\mathbf{H}$ and $\mathbf{H}$ are isomorphic. Hence we do not gain any new isomorphism classes in this sequence of groups. We do, however, gain new similarity classes, for these groups, containing opposite isometries but no inversion, clearly are not similar to any of our previous groups. As for isomorphisms between these groups, we observe that since $\mathbf{G}^+[\mathbf{H}$ and $\mathbf{H}$ are isomorphic, the only isomorphic pairs in this list are $\mathbf{C}_{2q}[\mathbf{D}_{2q}$ and $\mathbf{D}_q[\mathbf{D}_{2q}$, $q = 2, 3, \ldots$. These pairs of groups are not similar since their subgroups of direct isometries are not similar. Hence each new group corresponds to a distinct similarity class.

To conclude this section on finite groups of isometries we present a solid for each finite group, $\mathbf{G}$ such that $\mathbf{G}$ is either the direct symmetry group or the complete symmetry group of that figure. In order to include the case $q = 2$ we define a *regular 2-gon* to be a plane figure shaped like (), i.e., like the cross section of a convex lens. With this convention we define a *q-prism* as a right cylinder whose top and bottom are regular q-gons and whose height is not equal to the length of one side of the top, $q = 2, 3, 4, \ldots$. Similarly a *q-pyramid* is a right pyramid based on a regular q-gon, $q = 2, 3, 4, \ldots$. In the case $q = 3$ avoid a regular tetrahedron. The direct symmetry groups and complete symmetry groups of these and the regular polyhedra include most of the finite subgroups of $\mathbf{E}$. See Table 11.

TABLE 11. Symmetry Groups for Irregular Solids, Pyramids, Prisms, and the Regular Polyhedra

Values of q	*Solid*	*Direct symmetry group*	*Complete symmetry group*
1	completely irregular	$\mathbf{C}_1$	$\mathbf{C}_1$
1	irregular except for a center of symmetry	$\mathbf{C}_1$	$\mathbf{C}_1 \cup x\mathbf{C}_1$
1	irregular except for a plane of symmetry	$\mathbf{C}_1$	$\mathbf{C}_1[\mathbf{C}_2$
2, 3, 4, . . .	q-pyramid	$\mathbf{C}_q$	$\mathbf{C}_q[\mathbf{D}_q$
2, 4, 6, . . .	q-prism	$\mathbf{D}_q$	$\mathbf{D}_q \cup x\mathbf{D}_q$
3, 5, 7, . . .	q-prism	$\mathbf{D}_q$	$\mathbf{D}_q[\mathbf{D}_{2q}$
	tetrahedron	$\mathbf{T}$	$\mathbf{T}[\mathbf{O}$
	cube	$\mathbf{O}$	$\mathbf{O} \cup x\mathbf{O}$
	octahedron	$\mathbf{O}$	$\mathbf{O} \cup x\mathbf{O}$
	dodecahedron	$\mathbf{I}$	$\mathbf{I} \cup x\mathbf{I}$
	icosahedron	$\mathbf{I}$	$\mathbf{I} \cup x\mathbf{I}$

The only remaining groups all contain opposite isometries and (with the one exception, $\mathbf{T} \cup x\mathbf{T}$) can all be realized as complete symmetry groups of modified prisms. These modified prisms can all be easily constructed with scissors, cardboard, and paste. A piece of cardboard cut in the shape of a regular q-gon makes a very short but satisfactory q-prism. Pasting two of them together yields a slightly taller q-prism. If (before we paste them together) we rotate the top piece through an angle of $(180/q)°$, we have what we call a *twisted q-prism*. Top views of twisted q-prisms are shown in Figure 12. A

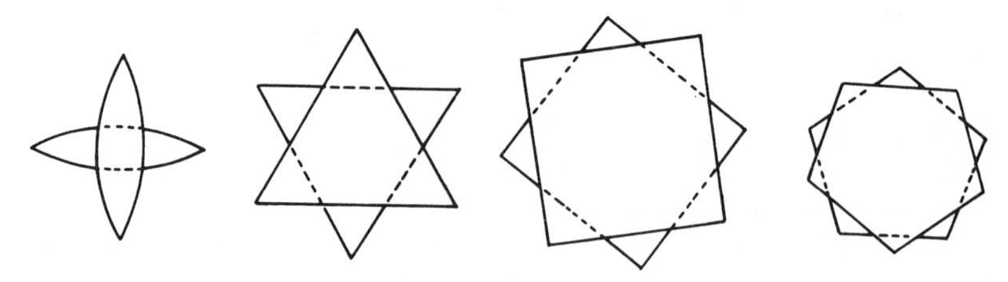

FIG. 12. Top views of twisted q-prisms, $q = 2, 3, 4, 5$.

prism can be given a clockwise or counterclockwise orientation by shaving off the corners, as shown in Figure 13. We call a prism modified in this way a *shaved prism*. If we start with two $2q$-prisms and cut off alternate corners from the top and bottom prism, as shown in Figure 14, then we have an *alter-*

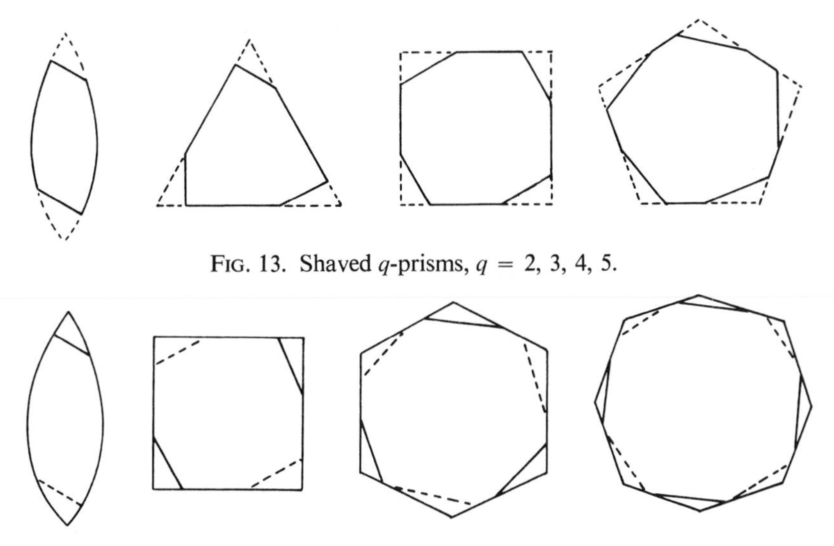

FIG. 13. Shaved q-prisms, $q = 2, 3, 4, 5$.

FIG. 14. Alternating $2q$-prisms, $q = 1, 2, 3, 4$.

nating 2q-prism. The complete symmetry groups of these three types of modified prisms are in Table 12.

TABLE 12. Symmetry Groups of Modified Prisms

Values of q	*Solid*	*Complete symmetry group*
3, 5, 7, . . .	alternating $2q$-prism	$\mathbf{C}_q \cup x\mathbf{C}_q$
2, 4, 6, . . .	alternating $2q$-prism	$\mathbf{C}_q[\mathbf{C}_{2q}$
2, 4, 6, . . .	shaved q-prism	$\mathbf{C}_q \cup x\mathbf{C}_q$
3, 5, 7, . . .	shaved q-prism	$\mathbf{C}_q[\mathbf{C}_{2q}$
3, 5, 7, . . .	twisted q-prism	$\mathbf{D}_q \cup x\mathbf{D}_q$
2, 4, 6, . . .	twisted q-prism	$\mathbf{D}_q[\mathbf{D}_{2q}$

The only remaining group, $\mathbf{T} \cup x\mathbf{T}$, is a subgroup of $\mathbf{O}$ so we ought to be able to modify a cube to obtain a solid with $\mathbf{T} \cup x\mathbf{T}$ as its symmetry group. One way to do this is as follows: inscribe congruent rectangles (not square) on the six faces of a cube as in Figure 15, and then delete those portions of the cube outside of the planes joining parallel sides of adjacent rectangles. In order to simplify drawing this solid we have also deleted those portions of the cube outside of the planes joining the three closest vertices of the three rectangles around each vertex of the cube. The resulting solid, shown in Figure 15, clearly has three half-turns (all of which are conjugate) in its symmetry

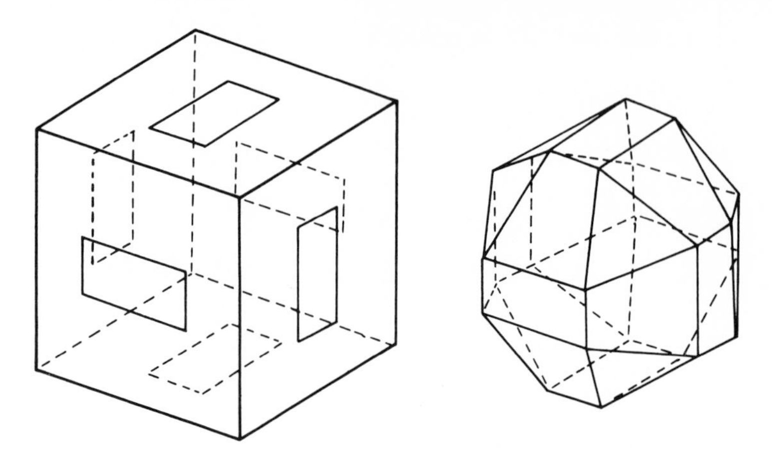

FIG. 15. Modified cube. Complete symmetry group $= \mathbf{T} \cup x\mathbf{T}$.

group. It also has rotations of order 3, but none of orders 4 or 5; thus, (corollary 3.9) the group of direct symmetries must be **T.** Since the inversion in the centroid is a symmetry, the complete symmetry group must be $\mathbf{T} \cup x\mathbf{T}$.

Exercise 3.2

1. Let **H** and **G,** $\mathbf{H} \subseteq \mathbf{G}$ be subgroups of a group **F** such that $[\mathbf{G}:\mathbf{H}] = 2$. Let x be an involution in **F** but not in **H,** and assume that x commutes with all elements of **G.** Show that:

(1) $\mathbf{H}[\mathbf{G} = \mathbf{H} \cup \{xg | g \in \mathbf{G}, g \notin \mathbf{H}\}$ is a subgroup of **F.**
(2) $\mathbf{H} = \mathbf{H}[\mathbf{G}$ or is of index 2 in $\mathbf{H}[\mathbf{G}$ according as x is or is not in **G.**
(3) If x is not in **G,** then **G** and $\mathbf{H}[\mathbf{G}$ are isomorphic in a natural way.
(4) If x is not in **G,** then $\mathbf{G} = \mathbf{H}[(\mathbf{H}[\mathbf{G})$.

2. Prove that no two groups of distinct types in our list of finite groups of direct isometries are isomorphic.

3. Find all subgroups of **O,** the octahedral group, and show how each one fits into our list of similarity classes of finite subgroups of **E.**

4. Show that the complete symmetry group of a twisted q-prism ($q > 1$) is $\mathbf{D}_q \cup x\mathbf{D}_q$ if q is odd and $\mathbf{D}_q[\mathbf{D}_{2q}$ if q is even.

5. Describe the isometries in $\mathbf{C}_3[\mathbf{C}_6$, $\mathbf{C}_4[\mathbf{C}_8$, $\mathbf{C}_3 \cup x\mathbf{C}_3$, and $\mathbf{C}_4 \cup x\mathbf{C}_4$.

6. Show that $\mathbf{D}_q \cup x\mathbf{D}_q$ has $q + 3$ conjugacy classes if q is odd and $q + 6$ if q is even.

7. Prove that $\mathbf{C}_q \cup x\mathbf{C}_q \cong \mathbf{C}_{2q}$ and $\mathbf{D}_q \cup x\mathbf{D}_q \cong \mathbf{D}_{2q}$ if q is odd.

8. Show that neither **T** nor **I** contains a subgroup of index 2 and that **T** is the only subgroup of index 2 in **O.**

9. Prove that **I** is a simple group.

3.3. Lattices and lattice groups

The term "lattice" is used in at least two distinct subjects in mathematics. The term "lattice theory" usually refers to the study of a special type of partially ordered set (*cf.* [1]). We shall not study this type of lattice. The other use of the word "lattice" occurs in geometry and refers to a set of points obtained by stacking copies of a parallelepiped (Figure 16). We shall define this type of

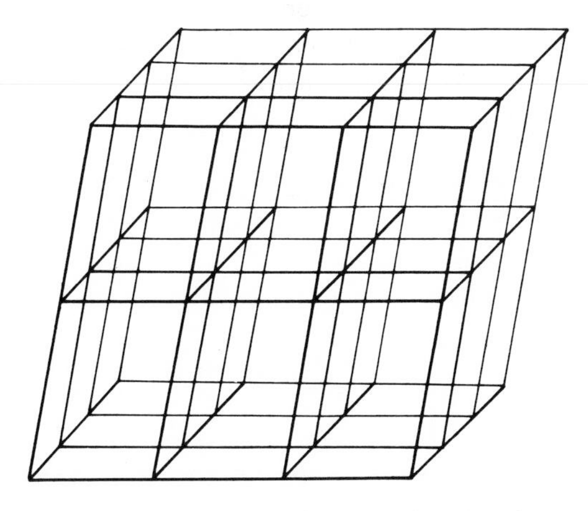

FIG. 16. Part of a three-dimensional lattice.

lattice more accurately in a moment, but first let us remark that the theory of (geometric) lattices is well developed and is useful in many branches of mathematics, especially number theory (*cf.* [3]).

Some of the results in this section are stated without proof. We return to the subject of lattices in Chapter 5, show how the concept may be generalized to n-dimensions, and present the proofs that are omitted in this section.

Definition. A *lattice group* is a nontrivial, discrete subgroup of $\mathbf{T}$, the group of translations, i.e., $\mathbf{G}$ is a lattice group, if $\mathbf{G} \neq \{\mathbf{1}\}$, $\mathbf{G}$ is discrete, and every element of $\mathbf{G}$ is a translation. If $\mathbf{G}$ is a lattice group, then any $\mathbf{G}$-orbit is called a *lattice*, and $\mathbf{G}$ (or one of its lattices) is said to be *k-dimensional* if $k + 1$ is the maximal number of independent points in any one $\mathbf{G}$-orbit.

The importance of lattice groups in crystallography arises from the fact that if $\mathbf{G}$ is a discrete group containing a nontrivial translation, then the translations in $\mathbf{G}$ form a subgroup, $\mathbf{G} \cap \mathbf{T}$, that is clearly a lattice group. This subgroup is the kernel of the homomorphism of $\mathbf{G}$ onto its point group (corollary 2.31). Thus, the definition we have chosen for lattices is very convenient for our purposes. Unfortunately it does not demonstrate, a priori, any connection with the intuitive idea of "stacking parallelepipeds." The connection is established in the following theorem in which we show, in effect, that a lattice is one-, two-, or three-dimensional if and only if it consists of all points with integral coordinates in some coordinate system (which need not be a cartesian system!) on a line, on a plane, or in space, respectively.

Theorem 3.10. A group of translations is a k-dimensional lattice group if and only if it is generated by k-translations with the property that any point and its k-images under these translations are independent.

Proof. We defer the proof of the "if" half of the theorem until Chapter 5 where we generalize lattices to n-dimensions and exploit the techniques of affine geometry to simplify the proof. Theorem 5.35 is the generalization if you are curious and want to see the proof now.

To prove the "only if" half assume that **G** is a three-dimensional lattice group. Obvious simplifications of the proof for this case cover the one- or two-dimensional cases. We must find three translations g_1, g_2, g_3 that generate **G.** The technique used here to pick the generators is not the same as that used by crystallographers to find a primitive cell, since that technique, while more useful in applications, poses difficult theoretical problems which we defer until Chapter 5 (*cf.* theorem 5.36).

In order to simplify notation and emphasize the fact that translations and vectors are equivalent in some sense we temporarily shift to the additive notation for products of translations. Since **T** is abelian this does not violate our convention on using the additive notation. Our task is to find g_1, g_2, g_3 such that $g \in \mathbf{G}$ if and only if there are integers n_1, n_2, n_3 such that $g = n_1g_1 + n_2g_2 + n_3g_3$.

Let X be a point. Choose a translation, g_1, of minimal length among the nontrivial translations in **G,** and let L be the line determined by X and$(X)g_1$. Next choose a translation, g_2, of minimal length among those translations in **G** whose directions are not along L, and let A be the plane determined by X, $(X)g_1$, and $(X)g_2$. Finally choose a translation, g_3, of minimal length among the translations in **G** whose directions are not along A. The point X and its seven images under g_1, g_2, g_3, $g_1 + g_2$, $g_1 + g_3$, $g_2 + g_3$, and $g_1 + g_2 + g_3$ are the vertices of a *closed* parallelepiped, $\mathcal{P}$, containing no other point in the **G**-orbit of X. The reader should verify that any other point in both $\mathcal{P}$ and the **G**-orbit of X would lead to a contradiction in the choice of g_1, g_2, and g_3.

Let g be any translation in **G.** Since g_1, g_2, and g_3 are independent, there are real numbers, a_1, a_2, and a_3, such that $g = a_1g_1 + a_2g_2 + a_3g_3$. (Here we use a result from linear algebra, viewing the translations as vectors.) Let $[a_i]$ be the largest integer less than or equal to a_i, and consider

$$h = g - ([a_1]g_1 + [a_2]g_2 + [a_3]g_3) .$$

The translation h is in **G,** and $(X)h$ is in the closed parallelepiped, $\mathcal{P}$, since $0 \leq a_i - [a_i] < 1$. This implies that h is trivial, and $g = \Sigma[a_i]g_i$ is an integral combination of the g_i. Thus, the g_i generate **G.** Clearly a point and its three images under g_1, g_2, and g_3 are independent.

If **G** is a three-dimensional lattice group then the parallelepiped formed by a point X and its images under the simplest seven combinations of the generators is called a *primitive cell* or *basic parallelepiped* of **G** or of the **G**-orbit (lattice) containing X. By deleting three suitable closed faces of a primitive cell one obtains a *fundamental domain* for **G,** i.e., a set containing exactly one point from each **G**-orbit. Although the generators (and hence primitive cells) of a lattice group are not unique, the volume of the various primitive cells is invariant and is called the *volume* or *determinant* of the lattice group. The method of choosing g_1, g_2, and g_3 used in the proof above may be refined to show that a primitive cell for a lattice may be chosen such that one edge is the line segment joining any two points in the lattice such that one is "visible" from the other. Moreover, we can insist that a face containing this edge be contained in any plane joining these two points of the lattice to a third independent point in the lattice. In Chapter 5 we prove this and show that in spite of the diversity of primitive cells they all have the same volume, i.e., the volume of the lattice is not affected by the way the primitive cell is chosen, *cf.* theorems 5.36 and 5.37.

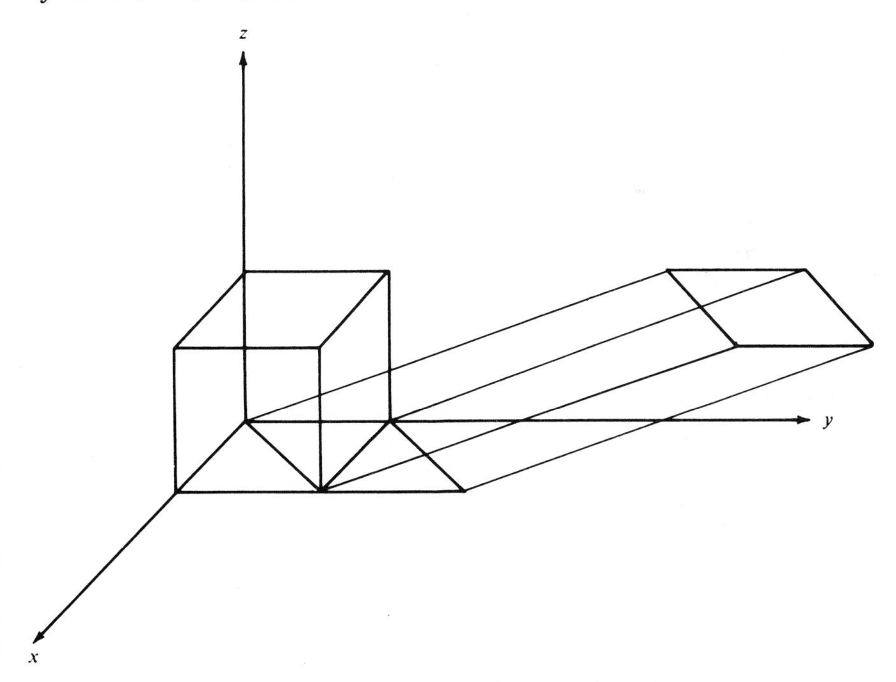

FIG. 17. Two primitive cells of a primitive cubic lattice.

Example 1. Let g_1, g_2, and g_3 be the translations one unit long along the x, y, and z axes, respectively, in a standard cartesian coordinate system. The lattice group, **P**, generated by g_1, g_2, and g_3 is called *primitive cubic* since one of its primitive cells is a cube. Figure 17 illustrates part of the **P**-orbit of the origin, the obvious cubic primitive cell, and another (less obvious) primitive cell corresponding to the generators g_2, $g_1 + g_2$, and $3g_2 + g_3$. Let **F** be the lattice group generated by $(1/2)(g_2 + g_3)$, $(1/2)(g_1 + g_3)$, $(1/2)(g_1 + g_2)$. **F**, or any of its orbits, is called *face-centered cubic* since none of its primitive cells is a cube. However, there is a cube such that its vertices and the centers of all six faces are lattice points. Part of the **F**-orbit of the origin is shown in Figure 18. For analogous reasons, **I**, the lattice group generated by g_1, g_2, and $(1/2)(g_1 + g_2 + g_3)$, is called *body-centered cubic* (Figure 19). Note that all

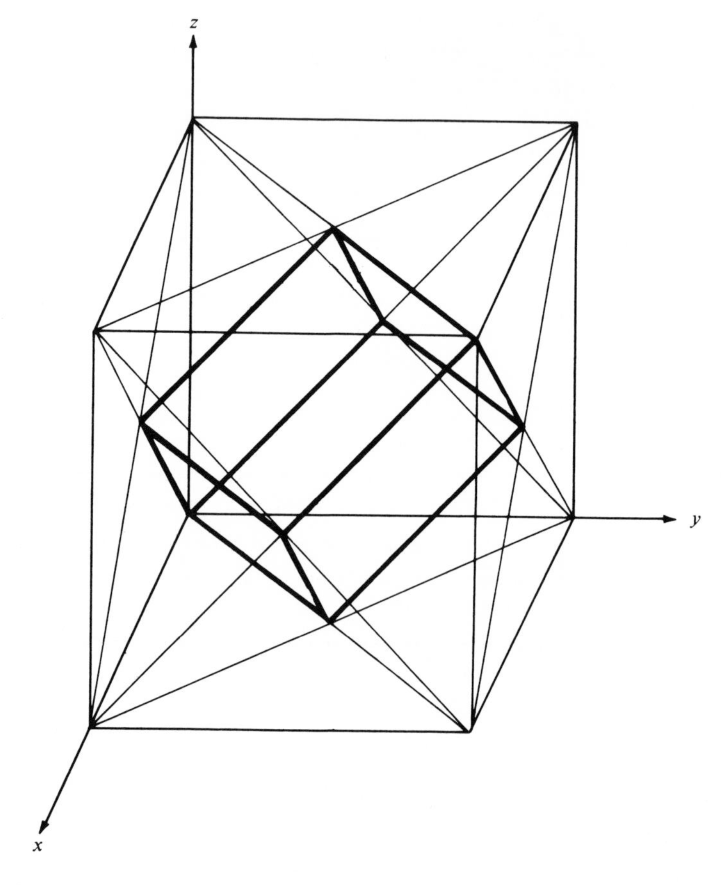

FIG. 18. Primitive cell and unit cell of a face-centered cubic lattice.

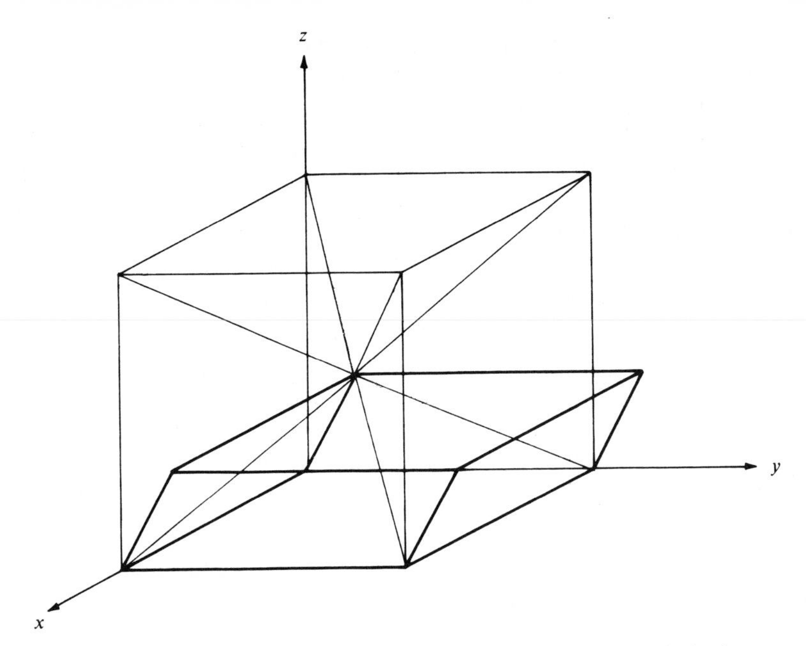

FIG. 19. Primitive cell and unit cell for a body-centered cubic lattice.

three lattices (the **P**, **F**, or **I**-orbits of the origin) are left invariant by all 48 symmetries of the unit cube.

Exercise 3.3

1. In choosing generators for a lattice group (proof of theorem 3.10), we chose translations in **G** satisfying conditions of the type "minimal length." Show how the assumption that the lattice group is discrete justifies this procedure.

2. Let g_1 and g_2 be generators of a two-dimensional lattice group, **G.** Show that the point, $(X)g$, $g \in \mathbf{G}$, is visible from X in the lattice $(X)\mathbf{G}$ if and only if $g = n_1 g_1 + n_2 g_2$ with n_1 and n_2 relatively prime integers.

3. Show that any two lattices for a particular lattice group are not only congruent but are translational images of each other.

3.4. Crystallographic point groups

Definition. A subgroup of **E** is called a *crystallographic point group* if it leaves a point X and some three-dimensional lattice containing X fixed.

In this section we prove that there are *at most* 32 similarity classes of crystallographic point groups. In the next section we classify these groups into the seven crystal systems and show that each of these 32 classes does in fact consist of crystallographic point groups.

Clearly a subgroup, **G**, of **E** is a crystallographic point group if and only if at least one **G**-orbit contains only one point, and there is a three-dimensional lattice group, **P**, such that **G** is contained in the normalizer of **P**. *Caution!* Review exercise 1.5, problem 7 to be sure you understand what the normalizer of a subgroup is. Note that we did not include as part of the definition that a crystallographic point group must be discrete. However, using the same technique as in the proof of theorem 3.2, the reader can easily prove (exercise 3.4, problems 1 and 4) the following theorem.

Theorem 3.11. A crystallographic point group is finite.

This theorem implies that any crystallographic point group is similar to exactly one of the groups listed in section 3.2. However, not every one of the finite groups is a crystallographic point group, for the order of a rotation in such a group must meet the following restriction.

Theorem 3.12 (the crystallographic restriction). Let g be a nontrivial rotation in a crystallographic point group. The order of g is 2, 3, 4, or 6.

Proof. Some power of a rotation of order n is a rotation through $(360/n)°$. Hence we need only show that no rotations through $(360/n)°$ are possible in crystallographic point groups if $n \notin \{2, 3, 4, 6\}$. The following lemma is the key to the proof.

Lemma. Assume that **P** is a two- or three-dimensional lattice group and that ρ is a rotation through $(360/n)°$, $n > 2$, with axis K such that ρ^i leaves **P** globally fixed, i.e., that ρ leaves globally fixed any **P**-orbit (lattice) of any point on K. Then **P** contains at least one nontrivial translation in a direction perpendicular to K.

Proof. Since **P** is not one-dimensional it contains at least one translation, τ, whose direction is not along K. Since ρ is a rotation leaving **P** invariant, $g = \rho^{-1}\tau\rho\tau^{-1}$ is in **P**. Since $n > 2$, g is nontrivial, and since ρ has axis K, the direction of g is perpendicular to K. (The reader may verify this by showing that $k^{-1}gk = g^{-1}$.)

Returning to the proof of the theorem let ρ be a rotation in a crystallographic point group. If the three-dimensional lattice left invariant is the **P**-orbit of the point X and if ρ is a rotation through $(360/n)°$, $n > 2$, then by the lemma, the group, $\mathbf{P}_K$, of all translations in **P** whose directions are perpendicular to K, the axis of ρ, is a nontrivial subgroup of **P** and is therefore itself a lattice group. Choose a translation, τ, in $\mathbf{P}_K$, of minimal length among the nontrivial translations in $\mathbf{P}_K$. The translation $\tau^{-1}\rho^{-1}\tau\rho$ is in $\mathbf{P}_K$, but is shorter (Figure 20a) than τ if $n > 6$. However, if $n = 5$ then the translation $(\rho^{-1}\tau\rho)^{-1}$

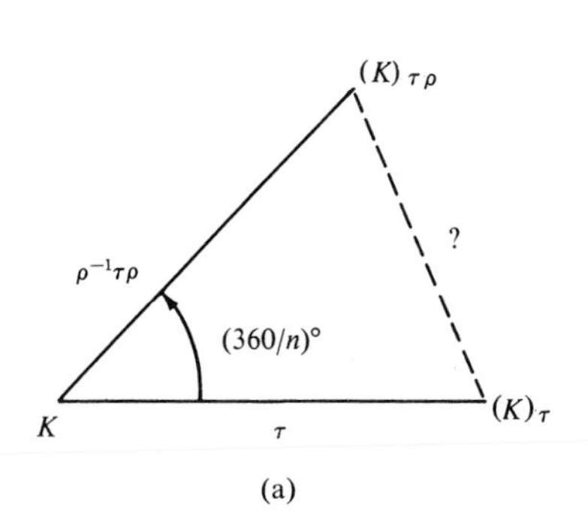

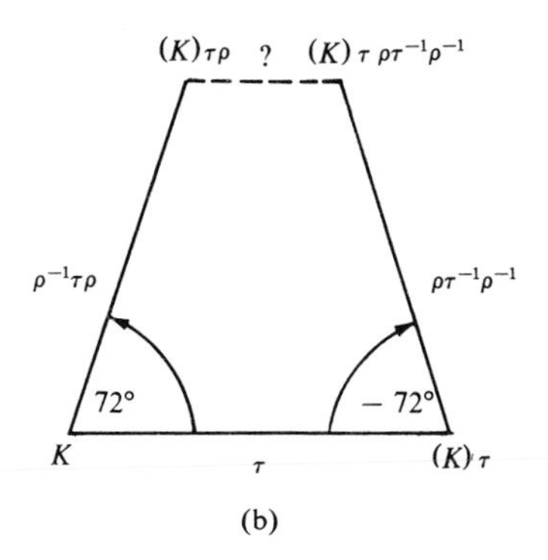

FIG. 20. The crystallographic restriction. (a) $n \leq 6$; (b) $n \neq 5$.

$\tau(\rho\tau^{-1}\rho^{-1})$ is too short (Figure 20b). Thus $n = 3$, 4, or 6 are the only possibilities for $n > 2$.

In Chapter 4 we give another proof (corollary 4.22) of this result, exploiting the trace of a matrix.

Corollary 3.13. If g is a rotatory inversion in a crystallographic point group, then the rotational part of g is of order 1, 2, 3, 4, or 6.

Proof. If the rotational part of g has order greater than 2, then g has only one fixed point X. This point X must therefore be in the lattice left invariant by the crystallographic point group. The inversion x leaves any lattice containing X invariant. Hence xg = the rotational part of g also leaves the lattice invariant and must therefore satisfy the crystallographic restriction. The order of xg is the same as that of g unless the order of xg is three and the order of g is six.

With these results we need only examine our list of finite groups of isometries and select those whose rotations and rotatory inversions meet the requirements above. This examination yields the 32 candidates for point groups listed in Table 13 below. We demonstrate in the next section that each of these is actually a point group, i.e., we give a three-dimensional lattice for each of these 32 groups.

In Table 13 we give both the name used in section 3.2 and the symbol used in the *International Tables for X-Ray Crystallography*, [8], for each of the 32 groups. The International symbols were chosen to give some idea about the generators of the group. For example, **3** indicates that a rotation of order 3 is one of the generators, $\bar{\mathbf{3}}$ that a rotatory inversion involving a rotation of order 3 is a generator, and m that a reflection (m for mirror) is one of the generators. The column labeled "crystal system" is explained in the next section. The groups are listed in the order that they appeared in Section 3.2.

Thus, the first 11 contain only direct isometries, the next 11 contain an inversion, and the last ten contain opposite isometries but no inversion. In Figure 21 we illustrate the subgroup relations for the 32 point groups.

The 32 classes of similar point groups are usually referred to as the 32 *crystal classes*. The crystal class to which a crystal belongs determines its "macroscopic" properties such as relative position of cleavage planes and optical properties. This is true since the point group is determined by ignoring the translational aspect of any symmetry of the atomic lattice (the distances

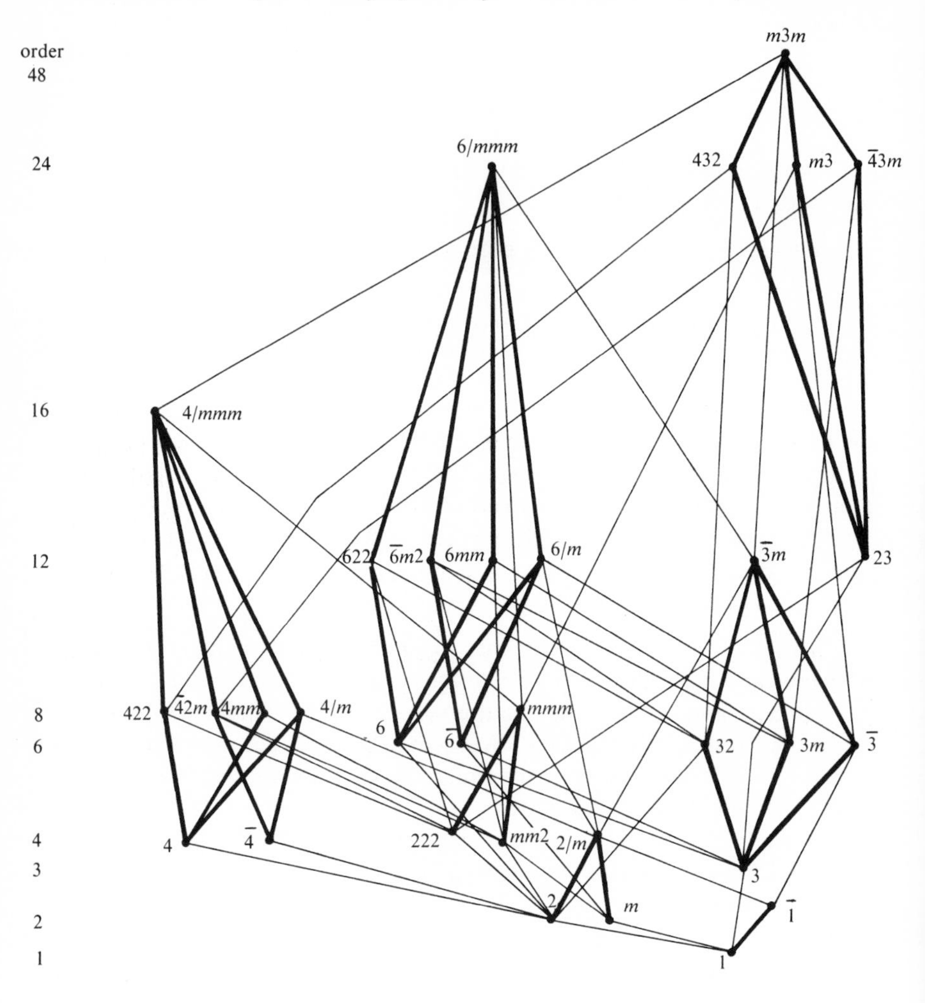

FIG. 21. Subgroup relations for the 32 point groups. Heavy lines connect groups in the same crystal system.

TABLE 13. The Crystallographic Point Groups

Temporary name from section 3.2	*International*	*Generators*	*Crystal system*
1. $\mathbf{C}_1$	**1**		triclinic
2. $\mathbf{C}_2$	**2**	k = a half-turn with axis K	monoclinic
3. $\mathbf{C}_3$	**3**	a rotation of order 3	trigonal
4. $\mathbf{C}_4$	**4**	a rotation of order 4	tetragonal
5. $\mathbf{C}_6$	**6**	a rotation of order 6	hexagonal
6. $\mathbf{D}_2$	**222**	two half-turns about perpendicular axes	orthorhombic
7. $\mathbf{D}_3$	**32**	ρ and k; order $\rho = 3$, $K \perp$ (axis ρ)	trigonal
8. $\mathbf{D}_4$	**422**	ρ and k; order $\rho = 4$, $K \perp$ (axis ρ)	tetragonal
9. $\mathbf{D}_6$	**622**	ρ and k; order $\rho = 6$, $K \perp$ (axis ρ)	hexagonal
10. $\mathbf{T}$	**23**	k and ρ; $\rho^3 = \mathbf{1}$, $1/\sqrt{3} = \cos(K, \text{axis } \rho)$	cubic
11. $\mathbf{O}$	**432**	ρ and σ; $\rho^4 = \mathbf{1}$, $\mathbf{1} = \sigma^3$, cos between axes $= 1/\sqrt{3}$	cubic
12. $\mathbf{C}_1 \cup x\mathbf{C}_1$	$\bar{\mathbf{1}}$	x = an inversion with center X	triclinic (holohedry)
13. $\mathbf{C}_2 \cup x\mathbf{C}_2$	**2/m**	k and a; $K \perp A$	monoclinic (holohedry)
14. $\mathbf{C}_3 \cup x\mathbf{C}_3$	$\bar{\mathbf{3}}$	$x\rho$; $\rho^3 = \mathbf{1}$	trigonal
15. $\mathbf{C}_4 \cup x\mathbf{C}_4$	**4/m**	ρ and a; (axis ρ) $\perp A$	tetragonal
16. $\mathbf{C}_6 \cup x\mathbf{C}_6$	**6/m**	ρ and a; (axis ρ) $\perp A$	hexagonal
17. $\mathbf{D}_2 \cup x\mathbf{D}_2$	**mmm**	a, b, and c; A, B, C mutually $\perp$ planes	orthorhombic (holohedry)
18. $\mathbf{D}_3 \cup x\mathbf{D}_3$	$\bar{\mathbf{3}}$**m**	$x\rho$ and a; $\rho^3 = \mathbf{1}$, axis $\rho \subseteq A$	trigonal (holohedry)
19. $\mathbf{D}_4 \cup x\mathbf{D}_4$	**4/mmm**	ρ, a, and b; $\rho^4 = \mathbf{1}$, axis $\rho \subseteq A$, axis $\rho \perp B$	tetragonal (holohedry)
20. $\mathbf{D}_6 \cup x\mathbf{D}_6$	**6/mmm**	ρ, a, and b; $\rho^6 = \mathbf{1}$, axis $\rho \subseteq A$, axis $\rho \perp B$	hexagonal (holohedry)
21. $\mathbf{T} \cup x\mathbf{T}$	**m3**	ρ and a; $\rho^3 = \mathbf{1}$, sine between A and axis $\rho = 1/\sqrt{3}$	cubic
22. $\mathbf{O} \cup x\mathbf{O}$	**m3m**	ρ, a, and b; ρ, a, and b as in **m3** and $\bar{\mathbf{4}}$**3m**	cubic (holohedry)
23. $\mathbf{C}_1[\mathbf{C}_2$	**m**	a reflection	monoclinic
24. $\mathbf{C}_2[\mathbf{C}_4$	$\bar{\mathbf{4}}$	$x\rho$, order $\rho = 4$	tetragonal

(*cont.*)

TABLE 13 (*cont.*)

Temporary name from section 3.2	*International*	*Generators*	*Crystal system*
25. $\mathbf{C_3[C_6}$	$\mathbf{\bar{6}}$	$x\rho$; $\rho^6 = \mathbf{1}$	hexagonal
26. $\mathbf{C_2[D_2}$	**mm2**	a and b; $A \perp B$	orthorhombic
27. $\mathbf{C_3[D_3}$	**3m**	ρ and a; $\rho^3 = \mathbf{1}$, axis $\rho \subseteq A$	trigonal
28. $\mathbf{C_4[D_4}$	**4mm**	ρ and a; $\rho^4 = \mathbf{1}$, axis $\rho \subseteq A$	tetragonal
29. $\mathbf{C_6[D_6}$	**6mm**	ρ and a; $\rho^6 = \mathbf{1}$, axis $\rho \subseteq A$	hexagonal
30. $\mathbf{D_2[D_4}$	$\mathbf{\bar{4}2m}$	$x\rho$ and a; $\rho^4 = \mathbf{1}$, axis $\rho \subseteq A$	tetragonal
31. $\mathbf{D_3[D_6}$	$\mathbf{\bar{6}m2}$	$x\rho$ and a; $\rho^4 = \mathbf{1}$, axis $\rho \subseteq A$	hexagonal
32. $\mathbf{T[O}$	$\mathbf{\bar{4}3m}$	ρ and b; $\rho^3 = \mathbf{1}$, sine between B and axis $\rho = 2/\sqrt{6}$	cubic

involved are of the order of 10^{-8} centimeters) and considering only the rotational aspect. When one takes into account the translational aspect also, then the space group of the crystal is the appropriate measure of its (atomic) symmetry. See the bibliography and suggestions for further reading at the end of this chapter for sources of more information on the crystallographic point groups.

Exercise 3.4

1. Prove theorem 3.11.

2. Find all conjugacy classes in $\bar{\mathbf{6}}m2 = \mathbf{D_3[D_6}$.

3. Prove that if ρ, k, and a are as specified in items 10 and 21, Table 13, then ρ and k generate $\mathbf{T}$, and ρ and a generate $\mathbf{T} \cup x\mathbf{T}$.

4. Let $\mathbf{G}$ be a subgroup of $\mathbf{E}$ leaving a two-dimensional lattice invariant. Show that $\mathbf{G}$ is finite. Give an example showing that $\mathbf{G}$ need not be finite or even discrete if the lattice is only one-dimensional.

3.5. *The seven crystal systems*

Three-dimensional lattices are classified into crystal systems determined by the largest crystallographic point group leaving the lattice invariant. In this section we prove that there are exactly seven such systems, show how the crystallographic point groups are related to these seven systems, and demonstrate that each of the 32 groups in the last section is actually a crystallographic point group.

Definition. Let **G** be a three-dimensional lattice group. The *holohedry* of **G** at the point X is the group of all isometries, h, such that $(X)h = X$ and $(\mathbf{G})h^i = \mathbf{G}$. In other words it is the largest crystallographic point group leaving X and the **G**-orbit of X invariant (exercise 3.5, problem 1). Two lattice groups (or lattices) are in the same *crystal system* if their holohedries are similar groups.

The definition would be very unsatisfactory if the holohedries of **G** at various points were not essentially the same, i.e., were not similar point groups. This unfortunate phenomenon does not occur (exercise 3.5, problem 2), so there is no harm in speaking of *the* holohedry of a lattice.

Theorem 3.14. Let **H** be the holohedry at X of the three-dimensional lattice group **G**. Then the inversion, x, with center X is in **H**, and thus **H** can only be one of the eleven crystallographic point groups containing an inversion.

Proof. For any translation g, $xgx = (g)x^i$ is the inverse of g. Since **G** is a *group* of translations, this implies that $(\mathbf{G})x^i = \mathbf{G}$. Clearly $(X)x = X$, so $x \in \mathbf{H}$.

Theorem 3.15. If **H** is the holohedry of the three-dimensional lattice group **G**, and if h is a rotation in **H** of order 3, 4, or 6 with axis L, then there is a reflection in **H** whose mirror contains L.

Proof. As in the proof of theorem 3.10, we shift to the additive notation for products of translations. In this notation $(if + jg)\phi^i = i(\phi^{-1}f\phi) + j(\phi^{-1}g\phi)$ for any integers i and j, translations f and g, and similarity ϕ.

Using the same lemma that was used in the proof of the crystallographic restriction (theorem 3.12) we can assert that **F**, the group of translations in **G** with directions perpendicular to L, is nontrivial and hence we can choose a translation, f, in **F** of minimal length among the nontrivial translations in **F**.

Lemma 1. The translations f and $f_1 = h^{-1}fh$ generate **F**.

Proof. Let X be a point on L and consider the parallelogram whose vertices are X and its images under f, f_1, and $f + f_1$ (Figure 22). In each of the three cases (order h = 3, 4, or 6), it is easy to verify that $\mathbf{1}, f, f_1$, and $f + f_1$ are the only translations in **F** sending X to some point in this closed parallelogram. This implies (review the proof of theorem 3.10) that f and f_1 generate **F**.

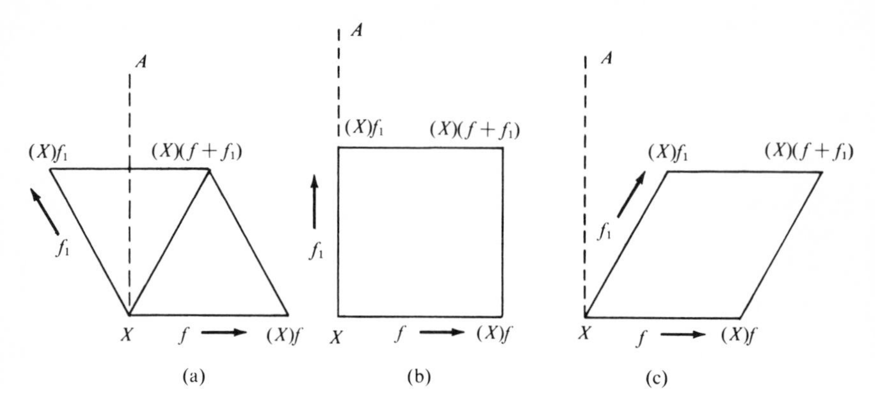

FIG. 22. (a) order $h = 3$; (b) order $h = 4$; (c) order $h = 6$.

Returning now to the main proof, let A be the plane containing L and perpendicular to the direction of f. In order to prove that the reflection a with mirror A is in **H** we need to show that $aga \in \mathbf{G}$ whenever $g \in \mathbf{G}$. To do this it is enough to show that $aga - g \in \mathbf{G}$ whenever $g \in \mathbf{G}$.

Any translation g can be "factored" in the form $g = g_1 + g_2$, with g_1 a translation whose direction is perpendicular to L and g_2 a translation along L. Since $L \subseteq A$, and therefore $ag_2a = g_2$, we find that $aga - g = ag_1a - g_1$. Thus, our task is reduced to the following: If $g \in \mathbf{G}$, and $g = g_1 + g_2$ as above, then $ag_1a - g_1 \in \mathbf{G}$. This is clear if $g_1 \in \mathbf{F}$ since in all three cases afa and af_1a are in **F** and $\mathbf{F} \subseteq \mathbf{G}$. Unfortunately this is not always the case, e.g., the face-centered cubic lattice (Figure 18) contains translations whose projections in the xy-plane are not among the lattice translations perpendicular to the z-axis. We do know, however, that $h^{-1}gh \in \mathbf{G}$; moreover, $h^{-1}g_2h = g_2$, so

$$h^{-1}gh - g = h^{-1}g_1h - g_1$$

is in **F** and therefore is an integral combination of f and f_1, say,

$$hg_1h^{-1} - g_1 = if + jf_1 .$$

Case 1. The order of h is 3. Since g_1, $h^{-1}g_1h$, and hg_1h^{-1} are 120° to each other,

$$hg_1h^{-1} - g_1 = if + jf_1, \; hfh^{-1} = -(f + f_1) ,$$

and $hf_1h^{-1} = f$, we have:

$$g_1 + h^{-1}g_1h + hg_1h^{-1} = 0$$
$$g_1 - h^{-1}g_1h = -if - jf_1$$
$$g_1 - hg_1h^{-1} = h(h^{-1}g_1h - g_1)h^{-1} = h(if + jf_1)h^{-1} = (j - i)f - if_1 .$$

Adding these three equations, we find that

$$3g_1 = (j - 2i)f - (i + j)f_1 .$$

Applying the reflection a to both sides,

$$3ag_1a = (i - 2j)f - (i + j)f_1 .$$

Finally, combining these last two equations,

$$3(ag_1a - g_1) = 3(i - j)f .$$

Thus, $(ag_1a - g_1) = (i - j)f$ is in **G** whenever $g \in$ **G** and this case is closed.

Case 2. The order of h is 4. Now we have $h^{-1}fh = f_1$ and $h^{-1}f_1h = -f$ to combine with $h^{-1}g_1h - g_1 = if + jf_1$. This yields:

$$g_1 - h^{-1}g_1h = -if - jf_1$$
$$g_1 - hgh^{-1} = jf - if_1$$
$$hg_1h^{-1} + h^{-1}g_1h = 0 .$$

Adding and then applying a leads to the equations

$$2g_1 = (j - i)f - (i + j)f_1$$
$$2ag_1a = (i - j)f - (i + j)f_1 .$$

Hence $ag_1a - g_1 = (i - j)f$ and this case also is closed.

Case 3. The order of h is 6. Then h^2 is a rotation of order 3 in **H** so, by case 1, $a \in$ **H**, thus closing this case and completing the proof of the theorem.

Among the eleven classes of crystallographic point groups containing an inversion, theorem 3.15 eliminates $\bar{\mathbf{3}}$, **4/m**, **6/m**, and **m3** as possibilities

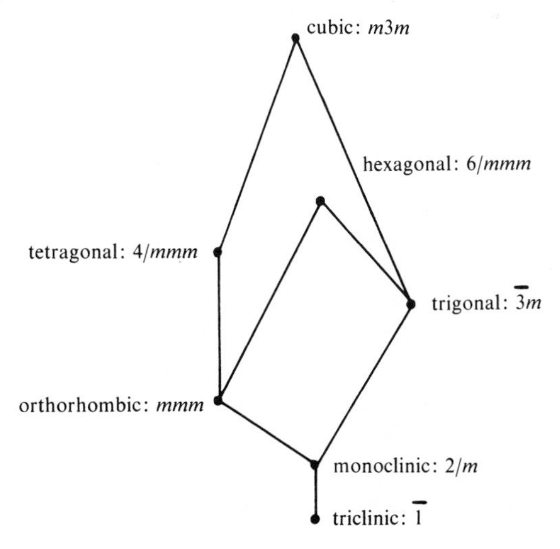

FIG. 23. Subgroup relations for the seven holohedry.

for holohedry. For each of the remaining seven classes, there is at least one lattice whose holohedry is in that class. The simplest of these lattices are given in the list below. These are referred to as the *primitive lattices* in each class. Following the system used in the International Tables [8], we describe each lattice by specifying relations for the lengths and for the angles between the three generating translations. Except in the case of the trigonal system, their positions are chosen such that the direction of the third generator is along the axis of the rotation of highest order. a, b, and c denote the lengths of the first, second, and third generators, respectively, and α, β, and γ denote the angles between generators 2 and 3, 1 and 3, or 1 and 2, respectively, as shown in Table 14.

TABLE 14. Primitive Lattices and the Holohedry for the Seven Crystal Systems

Name of crystal system	*Holohedry*	*Relations for the lengths of generators*	*Relations for the angles between generators*
triclinic	$\bar{\mathbf{1}}$	a, b, c distinct	α, β, γ distinct
monoclinic	**2/m**	a, b, c distinct	$\alpha = \beta = 90° \neq \gamma$
orthorhombic	**mmm**	a, b, c distinct	$\alpha = \beta = \gamma = 90°$
tetragonal	**4/mmm**	$a = b \neq c$	$\alpha = \beta = \gamma = 90°$
hexagonal	**6/mmm**	$a = b \neq c$	$\alpha = \beta = 90°, \gamma = 120°$
trigonal	$\bar{\mathbf{3}}$**m**	$a = b = c$	$90° \neq \alpha = \beta = \gamma < 120°$
cubic	**m3m**	$a = b = c$	$\alpha = \beta = \gamma = 90°$

Theorem 3.16. Any group in any of the 32 classes of groups given in section 3.4 is a crystallographic point group.

Proof. If **G** is a group in one of the 32 classes, then it can be realized as a subgroup of a group in one of the seven holohedry classes (Figure 21). If **P** is the primitive lattice group for this holohedry, then **G** clearly leaves **P** invariant.

Crystallographic point groups are classified into the seven crystal systems according to the *smallest* holohedry containing the point groups. The crystal system for each of the 32 classes was stated in the list of crystal classes in section 3.4. It is also demonstrated in Figure 21 by grouping classes according to their system. Note that a crystallographic point group is always of index 1, 2, or 4 in its holohedry (*holohedry*, *hemihedry*, or *tetartohedry*).

With the exception of the trigonal system where R is used, the primitive lattices described above are denoted by P in the International Tables. There are seven more types of lattices in addition to the seven primitive lattices. They fall into three types of relationships with the primitive lattices. To describe these relationships let g_1, g_2, and g_3 generate a primitive lattice and call a primitive cell of this primitive lattice a *unit cell.* The descriptions above of the seven primitive lattice groups can then be reinterpreted as descriptions of the seven types of unit cells. The nonprimitive lattices are as follows.

1. *Body-centered.* In the orthorhombic, tetragonal, and cubic systems, g_1, g_2, and $\frac{1}{2}(g_1 + g_2 + g_3)$ generate a new type of lattice (exercise 3.5, problem 3). These three types of lattices are denoted by **I** in the International Tables and are called body-centered since the center of each unit cell is a new lattice point. Any primitive cell for the body-centered lattice has $\frac{1}{2}$ the volume of the corresponding unit cell.

2. *Face-centered.* In the cubic and orthorhombic systems, $\frac{1}{2}(g_2 + g_3)$, $\frac{1}{2}(g_3 + g_1)$, and $\frac{1}{2}(g_1 + g_2)$ generate a lattice whose primitive cells are all $\frac{1}{4}$ the size of the corresponding unit cells. In each unit cell there are six new lattice points, one at the center of each of the six faces. These lattices are denoted by **F** in the International Tables. The face-centered cubic lattice was described in example 1 at the end of section 3.3.

3. *End-centered.* In the monoclinic and orthorhombic systems, g_1, g_2, and $\frac{1}{2}(g_2 + g_3)$ generate a lattice with two new lattice points at the centers of two opposite faces (ends) of the unit cell. Any primitive cell is $\frac{1}{2}$ the size of the unit cell. These two types of lattices are denoted *A*, *B*, or *C* in the International Tables, depending on which pair of opposite faces contain the two new lattice points.

This completes the description of the 14 types of lattices (seven primitive and seven nonprimitive) called the 14 Bravais lattices. For a proof that these are the only types possible, see [2], pp. 82–90.

Exercise 3.5

1. Let **G** be a lattice group and S the **G**-orbit of the point X. Assume that h is an isometry such that $(X)h = X$. Show that $(\mathbf{G})h^i = \mathbf{G}$ if and only if $(S)h = S$. *Hint.* Let $T = \{Y | Y$ is a midpoint of two points in $S\}$. First show that $(S)h = S$ if and only if $(T)h = T$.

2. Let $\mathbf{H}_1$ and $\mathbf{H}_2$ be the holohedries of the three-dimensional lattice group **G** at the points X_1 and X_2, respectively. Let τ be the translation from X_1 to X_2. Show that $(H_1)\tau^i = H_2$.

3. Let g_1, g_2, and g_3 be translations generating a primitive hexagonal lattice. What type of lattice is generated by the following:

(1) g_1, g_2, and $\frac{1}{2}(g_1 + g_2 + g_3)$?
(2) $\frac{1}{2}(g_2 + g_3)$, $\frac{1}{2}(g_3 + g_1)$, and $\frac{1}{2}(g_1 + g_2)$?
(3) g_1, g_2, $\frac{1}{2}(g_2 + g_3)$?

3.6. Crystallographic space groups

We have investigated two types of discrete groups, finite subgroups of **E** and lattice groups. In this section we study a type of group that combines these two types in a way that is of fundamental importance in the study of crystals.

Definition. A subgroup, $\mathbf{G}$, of $\mathbf{E}$ is called a *crystallographic space group* or simply a *space group* or *crystallographic group* if it is a discrete group and the subgroup, $\mathbf{G} \cap \mathbf{T}$, of all translations in $\mathbf{G}$ is a three-dimensional lattice group.

The study of space groups depends heavily on the homomorphism introduced in theorem 2.30. In that theorem we showed that for each point X and each isometry ϕ there is a unique factorization of the form $\phi = \phi_X \tau$ in which ϕ_X is an isometry leaving X fixed and τ is the translation carrying X to $(X)\phi$. The correspondence $\phi \longrightarrow \phi_X$ was shown to be a homomorphism of $\mathbf{E}$ onto $[\mathbf{E}, X]$. In corollary 2.31 we showed that this homomorphism maps a subgroup, $\mathbf{G}$, of $\mathbf{E}$ onto a subgroup of $[\mathbf{E}, X]$ called $\mathbf{G}_X$, the point group of $\mathbf{G}$, and that this homomorphism has kernel $\mathbf{G} \cap \mathbf{T}$. This homomorphism leads to the classification of space groups into crystal classes as follows.

Theorem 3.17. If $\mathbf{G}$ is a space group and X is a point, then $\mathbf{G}_X$, the point group of $\mathbf{G}$, is one of the 32 types of crystallographic point groups. Moreover, this type is independent of the chosen point X.

Proof. The group, $\mathbf{G}_X$, is a subgroup of $[\mathbf{E}, X]$ and leaves the three-dimensional lattice group $\mathbf{G} \cap \mathbf{T}$ invariant (exercise 3.6, problem 1). Hence it is one of the 32 types of crystallographic point groups. Since the various point groups for $\mathbf{G}$ are all similar, this type is independent of X.

The type of its point group determines the *crystal class* of a space group. As we mentioned at the end of section 3.4, the crystal class is obtained by ignoring the translational aspects of the symmetries of the atomic lattice of a crystal, i.e., by factoring out $\mathbf{G} \cap \mathbf{T}$.

For any discrete group, $\mathbf{G}$, it is possible to find a point X in *general position*, i.e., such that $[\mathbf{G}, X] = \{\mathbf{1}\}$. There may exist a point at the other extreme, i.e., such that $[\mathbf{G}, X] = \mathbf{G}_X$. If so, the group is called a *symmorphic group*. There are 73 isomorphism classes of symmorphic space groups. These space groups are relatively easy to describe. One starts with a crystallographic point group $\mathbf{G}$ and a three-dimensional lattice group, $\mathbf{H}$, that is left invariant by $\mathbf{G}$ and then forms the *semi-direct product* of $\mathbf{H}$ by $\mathbf{G}$, i.e., the set $\{gh | g \in \mathbf{G}, h \in \mathbf{H}\}$. Note that a product of two elements of this set is of the form

$$(g_1h_1)(g_2h_2) = (g_1g_2)(g_2^{-1}h_1g_2h_2)$$

and is therefore in the set. This set is a space group whose subgroup of translations is $\mathbf{H}$ and whose point group is $\mathbf{G}$.

In addition to the symmorphic space groups, there are other more complicated space groups. The following example illustrates some ways a crystallographic point group can be extended to a space group.

Example. The six space groups whose point groups are **4**, *a group generated by a rotation of order 4.* The smallest holohedry containing **4** is **4/mmm**; hence this group is in the tetragonal system. There are two types of lattices for this system: **P**, the primitive lattice, and **I**, the body-centered lattice. To choose specific examples of these lattices let g_1, g_2, and g_3 be the three translations sending the origin (0, 0, 0) to the points $(a, 0, 0)$, $(0, a, 0)$, and $(0, 0, c)$, $a \neq c$, in a standard cartesian coordinate system. Let **P** be the lattice group generated by g_1, g_2, and g_3 and **I** the lattice group generated by g_1, g_2, and $g_b = \frac{1}{2}(g_1 + g_2 + g_3)$. Note that **P** is a subgroup of index 2 in **I** and that the rotation ρ whose axis is the z-axis and sending the x-axis to the y-axis leaves both **P** and **I** invariant.

The two symmorphic space groups are **P4** (generated by g_1, g_3, and ρ) and **I4** (generated by g_1, g_b, and ρ). Note that we need not list g_2 as a generator in either group since $g_2 = \rho^{-1} g_1 \rho$.

The nonsymmorphic space groups arise from special combinations of screw displacements and glide reflections. Since **4** contains no opposite isometries, the corresponding space groups cannot either. Thus, we need only consider screw displacements with rotational aspect of order 2 or 4. If we let $g = (\frac{1}{4})g_3$ be the translation sending (0, 0, 0) to $(0, 0, c/4)$, then representatives of the four types of nonsymmorphic space groups in the crystal class **4** are: $\mathbf{P4_1}$ generated by ρg and g_1, $\mathbf{P4_2}$ generated by ρg^2 and g_1, $\mathbf{P4_3}$ generated by ρg^3 and g_1, and $\mathbf{I4_1}$ generated by g_1, g_b, and ρg. Note that $\mathbf{P4_1}$ contains a right-handed screw, ρg, but no left-handed one; $\mathbf{P4_3}$ contains a left-handed screw, $\rho^{-1} g$, but no right-handed one; and $\mathbf{I4_1}$ contains both right-, ρg, and left-, $\rho g g_b (\rho g)^{-4} = \rho g^{-1} g_b$, handed screws.

Although $\mathbf{P4_1}$ and $\mathbf{P4_3}$ are similar groups, they are listed separately in the International Tables and are usually considered to be distinct types of space groups. These two groups are examples of *oriented space groups*, i.e., space groups, **G**, containing some isometry ϕ such that $(\phi)\sigma^i = \sigma^{-1}\phi\sigma$ is never in **G** for any opposite isometry σ. If **G** is an oriented space group, and σ is an opposite isometry, then **G** and $\sigma^{-1}\mathbf{G}\sigma$ are called *enantiomorphic* space groups. For some dramatic physical implications of enantiomorphic crystals, see [13], pp. 28–38.

There are 219 isomorphism classes of space groups, the simplest of which contains all of the three-dimensional lattice groups. The eleven isomorphism classes containing pairs of enantiomorphic groups are split into 22 classes by taking orientation into account. These 230 types of space groups are the types of the standard classification.

A complete description of the 230 types of space groups appears in section 4.3 (pp. 73–346) of the International Tables [8]. In addition to other information, each table gives the **G**-orbit of each point, X, classified according to the point symmetry at X, i.e., according to $[\mathbf{G}, X]$. The following four paragraphs are included as a guide to reading these tables.

The crystal system for a space group, **G**, is that of its point group, $\mathbf{G}_X$, and need not be the same as that of its lattice subgroup $\mathbf{G} \cap \mathbf{T}$. This crystal system is specified at the head of the table next to the standard name of the point group. The symbol used in the tables for the space group appears both as the center heading and at the outer top corner above the Schoenflies symbol for the space group.

The coordinate system used to describe a space group is determined by the unit cell of its crystal system and is *usually not a standard cartesian system.* Relations between the lengths of the unit segments and the angles between the axes are given in Table 2.3.1 (p. 11) and in this book in section 3.5. The only ambiguity in the table arises in the trigonal system. If $\mathbf{G}_X$ is in the trigonal system, but $\mathbf{G} \cap \mathbf{T}$ is in the hexagonal system, then **G**-orbits are described relative to a hexagonal system. If $\mathbf{G}_X$ and $\mathbf{G} \cap \mathbf{T}$ are both in the trigonal system, then **G**-orbits are described both with respect to a rhombohedral (trigonal) and a hexagonal coordinate system. The preference shown the hexagonal description is due to the fact that the z-axis is perpendicular to the x, y plane in a hexagonal system but not in a rhombohedral system.

The primitive lattice group, **P**, generated by the three translations carrying the origin to the unit points on the three axes is always a subgroup of **G**. Thus, the **G**-orbit of X always consist of a union of **P**-orbits, one for each coset of **P** in **G**. Hence, in order to describe the **G**-orbit of a point X, only the coordinates of one representative point from each of the necessary **P**-orbits are given in the table. Another way of saying this is that all coordinates are reduced modulo 1. The index of **P** in $\mathbf{G} \cap \mathbf{T}$, $[\mathbf{G} \cap \mathbf{T}:\mathbf{P}]$, is 1, 2, 2, or 4 according as $\mathbf{G} \cap \mathbf{T}$ is a primitive, body-centered, end-centered, or face-centered lattice. The first symbol of the name of the space group identifies the type of lattice. If $[\mathbf{G} \cap \mathbf{T}:\mathbf{P}] > 1$, then immediately under the heading "coordinates of equivalent points" appear the coordinates of a representative point from each **P**-orbit within the $\mathbf{G} \cap \mathbf{T}$-orbit (not the **G**-orbit!) of the origin. The number of **P**-orbits required to make up the **G**-orbit of a point X is called the number of positions for X and appears in the left-hand column. The number of positions of X is $[\mathbf{G} \cap \mathbf{T}:\mathbf{P}][\mathbf{G}_X:[\mathbf{G}, X]]$ (exercise 3.6, problem 5).

Two final comments should enable the reader to proceed to the Tables and gain some insight into the structure of space groups. Additive inverses, $-x$, $-y$, etc., are often denoted $\bar{x}$, $\bar{y}$, etc. The left-hand diagram illustrates part of the **G**-orbit of a point in general position, and the right-hand diagram illustrates axes, mirrors, directions, etc., for various isometries in **G**. The notation used in these diagrams is explained on pages 48–50.

Exercise 3.6

1. Assume that ϕ and θ are in the same coset of **T** in **E**, i.e., "differ" by a translation. Show that ϕ^i and θ^i agree on translations. Use this result to show that the point group of a space group, **G**, leaves $\mathbf{G} \cap \mathbf{T}$ invariant.

2. Let **G** be the symmorphic space group **P23**, i.e., the semi-direct product of the primitive cubic lattice group **P** by the tetrahedral group, **23.** Using the unit cell of **P** as the unit cube in a standard cartesian coordinate system, and assuming a rotation of order 3 in **23** has as its axis the line $x = y = z$, find for each subgroup **H** in **23** the coordinates of a point X such that $[\mathbf{G}, X] = \mathbf{H}$. Compare your results with those in the International Tables.

3. Prove that no oriented space group contains an opposite isometry.

4. Find two more pairs of enantiomorphic space groups.

5. Let **G** be a space group and **P** the primitive lattice group of its crystal system. Show that the number of **P**-orbits in the **G**-orbit of X is $[\mathbf{G} \cap \mathbf{T}:\mathbf{P}][\mathbf{G}_X:[\mathbf{G}, X]]$.

6. Give an example of a space group, **G**, that is nonsymmorphic, but such that $[\mathbf{G}, X]$ is nontrivial for at least one point X.

3.7. Generalizations

Since the only isometries of the line are reflections and translations, it is easy to prove that there are three isomorphism classes of nontrivial, discrete groups on the line. Of these only two leave a one-dimensional lattice invariant; hence there are only two types of one-dimensional "space" groups (exercise 3.7, problem 1).

In the plane the discrete groups have been completely classified. There are two sequences, $\mathbf{C}_n$ and $\mathbf{D}_n$, of finite groups. Once again the crystallographic restriction is valid, so there are only ten similarity classes (of which exactly two ($\mathbf{C}_2 \cong \mathbf{D}_1$) combine into one isomorphism class) of crystallographic point groups in the plane. Combining these point groups with the various lattice groups which they leave invariant leads to 17 isomorphism classes of two-dimensional "space" groups, *cf.* [7], Chapter III, or [5], Chapter 4 and p. 413. The only remaining nonfinite discrete groups in the plane leave a one-dimensional lattice invariant. There are seven similarity classes of these "strip groups" (exercise 3.7, problem 2) which combine into four isomorphism classes.

We have not discussed discrete groups in space whose translations form a one-dimensional or two-dimensional lattice group. Let us call these fiber and net groups, respectively. If τ is any translation, and ρ is a rotation commuting with τ, then ρ and τ generate a fiber group. Thus, there are infinitely many isomorphism classes of fiber groups. If we allow only isometries obeying the crystallographic restriction, there are 75 types of fiber groups including eight enantiomorphic pairs, thus, *at most* 67 isomorphism classes. When we turn to net groups, the crystallographic restriction again applies, and there are only a finite number (at most 80) of isomorphism classes of net groups. For more details see the references at the end of section 4.1 in the International Tables.

Generalizing to higher dimensions, one can ask if there are only finitely many types of n-dimensional space groups, i.e., discrete groups in euclidean n-space whose translations form an n-dimensional lattice group as defined in section 5.10. David Hilbert included this question among the 23 problems he proposed in Paris at the 1900 International Congress of Mathematicians (see [9], problem 18). For comments on this problem, see [13], p. 210, or [2], sections 26 and 27. There are interesting connections between semi-simple Lie groups and discrete groups generated by the higher dimensional analogs of reflections ([4], pp. 211–212).

Exercise 3.7

1. Prove that there are only four isomorphism classes of discrete groups on the line. Which two of these contain all the one-dimensional space groups?

2. Show that there are exactly seven classes of "strip groups." For each class devise a pattern, such as . . . *N N N* . . . , whose symmetry group is in that class.

Bibliography and suggestions for further reading

[1] Birkhoff, Garrett, *Lattice Theory*, Am. Math. Soc. Colloquium Publications, Vol. 25, revised ed., Providence, R.I., 1948.

[2] Burckhardt, J. J., *Die Bewegungsgruppe der Kristallographie*, Birkhauser, Basel, 1957.

[3] Cassels, J. W. S., *An Introduction to the Geometry of Numbers*, in *Die Grundlehren der Mathematischen Wissenschaften*, Vol. 99, Springer, Berlin, 1959.

[4] Coxeter, H. S. M., *Regular Polytopes*, 2nd ed., Macmillan, New York, 1963.

[5] Coxeter, H. S. M., *Introduction to Geometry*, Wiley, New York, 1961.

[6] Curie, P., "Sur les repetitions et la symetrie," *Compt. rend.* **100,** 1393–1396, 1885.

[7] Guggenheimer, Heinrich W., *Plane Geometry and its Groups*, Holden-Day, San Francisco, 1967.

[8] Henry, N. F. M., and K. Lonsdale (eds.), *International Tables for X-Ray Crystallography*, Vol. 1, published for the International Union of Crystallography by the Kynoch Press, Birmingham, England, 1952.

[9] Hilbert, David, "Mathematical problems," *Bull. Am. Math. Soc.* **8,** 437–479 (1902). This is a translation of the original paper which appeared in the *Gottinger Nachrichten* **1900,** 253–297 and in the *Archiv der Mathematik und Physik* 3rd ser. **1,** 44–63, 213–237 (1901).

[10] Le Corbeiller, Phillipe, "Crystals and the future of physics," *Sci. Am.*, 50 ff. (Jan. 1953).

[11] Lomont, J. S., *Applications of Finite Groups*, Academic Press, New York, 1959.
[12] Schiff, L. I., "Paper representations of the non-cubic crystal classes," *Am. J. Phys.* **22,** 621–622 (1954).
[13] Weyl, Hermann, *Symmetry*, Princeton Univ. Press, Princeton, N.J., 1952.

The most important reference for this chapter is the collection of tables, [8], in which the 32 crystallographic point groups, the 14 Bravais lattices, and the 230 space groups are carefully described. Many examples to illustrate the material we have presented in chapters 1–3 can be gleaned from these tables.

Either [10] or [13] is invaluable in preserving one's perspective while studying this chapter.

A quick survey of crystallographic point groups, lattices, and space groups is in Chapter 4 of [11]. Unfortunately, the author uses the Schoenflies notation for these groups. With the aid of the International Tables, which gives the Schoenflies symbol for each space group directly below the usual name, it is not difficult to correlate the two notations.

The formal development of the theory we have surveyed in this chapter is in [2]. We have found this reliable and helpful as we struggled to correlate the various notations, incomplete proofs, and partial presentations found in other introductions to crystallography.

The solids described at the end of section 3.2 in order to represent the finite subgroups of **E** as symmetry groups are based on the ideas presented in [12]. The descriptions there may help the reader understand our presentation better; moreover, they are useful in correlating the Schoenflies and International symbols for the crystal classes.

Chapter 4

Fields and Linear Algebra: A Quick Review

In this chapter we lay the groundwork for our study of affine and projective geometries. In sections 4.1 and 4.2, we present (without proof) standard facts about fields and vector spaces, and illustrate these using finite fields. The emphasis is placed on finite fields in order to prepare the way for finite affine and projective geometries and in order to maintain the reader's interest as he reviews computational techniques that are probably familiar to him if the scalars are real numbers.

In sections 4.3–4.6, we study linear transformations and their relation to matrices. Coordinate systems are defined as mappings, and the way in which the matrix representation of a linear transformation depends on the particular coordinate system is carefully studied. As an application of this, we examine two types of linear transformations, projections and involutions. We also present an alternative proof of the crystallographic restriction.

Finally in section 4.7, we discuss the symmetries of a finite-dimensional vector space, once again with special emphasis on the interesting case in which the field of scalars is a finite field.

4.1. Fields

The reader is undoubtedly familiar with at least three fields: the field, **Q**, of rational numbers, the field, **R**, of real numbers, and the field, **C**, of complex numbers. In this section we present (without proof) techniques for constructing other fields—in particular, finite fields.

Definition. A field, **F**, is a (non-empty) set, F, with two operations, addition and multiplication, such that:

A. The set F forms an abelian group under addition. We denote the additive identity by 0.
M. The set $F - \{0\}$ forms an abelian group under multiplication. We denote the multiplicative identity by 1, and note that since $1 \in F - \{0\}, 1 \neq 0$.
D. The two distributive laws are valid, i.e., for all a, b, c in F, $a(b + c) = ab + ac$ and $(a + b)c = ac + bc$.

It is interesting that, since $0x = x0$ does not follow from axiom M, one must assume *both* distributive laws. For a discussion of this point, see [5].

Definition. If **F** and **G** are fields, then **F** is called a *subfield* of **G** or **G** is called a *field extension* of **F** if $F \subseteq G$, and the addition and multiplication in **F** agree with the addition and multiplication in **G**.

Thus, **Q** is a subfield of **R**, and both **Q** and **R** are subfields of **C**. Many other interesting subfields of **C** may be found by starting from **Q** and applying (perhaps several times over) the following technique.

Theorem 4.1. If **E** is a subfield of **F** and b is in **F**, then $\mathbf{E}(b) = \{p(b)/q(b) | p, q$ are polynomials with coefficients in **E**, and $q(b) \neq 0\}$ is also a subfield of **F**. It is the smallest subfield of **F** containing both **E** and b.

If, in the above theorem, $q(b) \neq 0$ for *all* nonzero polynomials, then we say that b is *transcendental* over **E**. In this case distinct quotients of polynomials lead to distinct elements in $\mathbf{E}(b)$, and $\mathbf{E}(b)$ is isomorphic to the field of rational forms (i.e., quotients of polynomials) with coefficients in **E**. However, if b is *algebraic* over **E**, i.e., if $q(b) = 0$ for some nonzero polynomial, then there is much duplication, for then $p(b)/q(b) = p_1(b)/q_1(b)$ whenever the polynomial $p(x)q_1(x) - p_1(x)q(x)$ has b as one of its roots. In case b is algebraic over **E**, then we define $m(x)$ to be the *minimal polynomial* of b over **E** if $m(x)$ is the monic polynomial of least degree among all the polynomials with coefficients in **E** and with b as a root. If the degree of $m(x)$ is n, we say b is algebraic *of degree n over* **E**.

Theorem 4.2. If b is algebraic of degree n over **E**, then

$$\mathbf{E}(b) = \{a_0 + a_1 b + \cdots + a_{n-1}b^{n-1} | a_i \in E, i = 0, 1, \ldots, n - 1\};$$

moreover, each element of $\mathbf{E}(b)$ has a unique representation in this form.

This theorem (proved in [1], p. 370, or [3], pp. 171–172) may seem rather surprising at first, for it seems strange that multiplicative inverses can be written in this form. The familiar example $\mathbf{C} = \mathbf{R}(i)$ illustrates the way this is done: $(a + bi)^{-1} = c + di$ with $c = a/(a^2 + b^2)$ and $d = -b/(a^2 + b^2)$. A much more complicated example is $Q(\sqrt[4]{7}) = \{a_0 + a_1\sqrt[4]{7} + a_2\sqrt{7} +$

$a_3\sqrt[4]{343}|a_i \in Q\}$ in which, for example, $1/(1 + \sqrt[4]{7}) = (u^3 - 4u^2 + 6u - 4)/6$, $u = 1 + \sqrt[4]{7}$.

It is easy to prove (exercise 4.1, problem 4) that any subfield of **C** must contain **Q** as a subfield. Hence there are no finite subfields of **C**. There are, however, finite fields, the simplest of which are the integers mod p, $\mathbf{Z}_p$, for p prime. These fields are defined as follows.

Definition. Given a positive prime integer p, let Z_p be a set with p elements $\bar{0}, \bar{1}, \bar{2}, \ldots, \overline{p-1}$. Define *addition in* $\mathbf{Z}_p$ by: $\bar{a} + \bar{b} = \bar{c}$ if c is the remainder of $a + b$ on division by p, i.e., $\bar{a} + \bar{b} = \bar{c}$ if c is $a + b$ *reduced modulo* p. Similarly, *multiplication* in $\mathbf{Z}_p$ is defined by $\bar{a} \cdot \bar{b} = \bar{c}$ if c is the remainder of ab on dividing by p. If m is a positive integer and $\bar{a} \in \mathbf{Z}_p$, then we denote by $m\bar{a}$ that element of $\mathbf{Z}_p$ obtained by adding $\bar{a}$ to itself m times in $\mathbf{Z}_p$, i.e., $m\bar{a} = \bar{b}$ if b is ma reduced modulo p. Similarly, $\bar{a}^m = \bar{b}$ if b is a^m reduced modulo p. Note that in both notations, $m\bar{a}$ and $\bar{a}^m$, the integer m is *not* reduced modulo p.

Here are some examples of computations in $\mathbf{Z}_p$. In $\mathbf{Z}_5$, $\bar{3} + \bar{4} = \bar{2}$ since $3 + 4 = 7$, which, when divided by 5, leaves a remainder of 2. Also in $\mathbf{Z}_5$, $\bar{3} \cdot \bar{4} = \bar{2}$, $10(\bar{3}) = \bar{0}$, and $\bar{3}^5 = \bar{3}$ since $3 \cdot 4 = 12$, which reduces to 2 modulo 5, $10 \cdot 3 = 30$, which reduced to 0 modulo 5, and $3^5 = 243$, which reduces to 3 modulo 5. Note, however, that in $\mathbf{Z}_7$, $\bar{3} + \bar{4} = \bar{0}$, $\bar{3} \cdot \bar{4} = \bar{5}$, $10(\bar{3}) = \bar{2}$, and $\bar{3}^5 = \bar{5}$.

Theorem 4.3. $\mathbf{Z}_p$, with the operations of addition and multiplication defined as above, is a field with zero $\bar{0}$ and multiplicative identity $\bar{1}$. For every $\bar{a} \in \mathbf{Z}_p$, $p\bar{a} = 0$ and $\bar{a}^p = \bar{a}$, in particular, $-\bar{a} = (p - 1)\bar{a}$, and if $\bar{a} \neq \bar{0}$, then $\bar{a}^{-1} = \bar{a}^{(p-2)}$.

For the details of the proof that $\mathbf{Z}_p$ is a field, we refer the reader to either [1], pp. 25–27, 40, or [3], p. 91. If one observes that "reducing modulo p" is an operation that preserves sums and products, i.e., is a ring homomorphism, then it is apparent that the associative, commutative, and distributive laws are all "inherited" by $\mathbf{Z}_p$ from their analogs in **Z**, the ring of integers. Since p divides pa in **Z**, it is clear that $p\bar{a} = 0$. The proof that $\bar{a}^p = \bar{a}$ in $\mathbf{Z}_p$ is left as an exercise (exercise 4.1, problem 1).

If **F** is a field, then all nonzero elements of **F** have the same order as far as the group **F** under addition is concerned. Furthermore, this order is either infinite, in which case we say **F** is a field of *characteristic* 0, or a finite prime, p, in which case we say **F** is of *characteristic* p. A field of characteristic 0 always contains a subfield isomorphic to **Q**, while one of characteristic p contains an isomorphic replica of $\mathbf{Z}_p$; thus, these fields are referred to as the *prime fields*. The field of rational forms, $r(x)/q(x)$, with coefficients in $\mathbf{Z}_p$

is an example of a field which is not finite even though its characteristic is finite.

The various $\mathbf{Z}_p$ are *not* the only finite fields. The facts are as follows. If $\mathbf{F}$ is a finite field, then, of course, its characteristic must be finite, say it is p, and $\mathbf{F}$ contains an isomorphic replica, $\mathbf{E}$, of $\mathbf{Z}_p$. There is an element b in $\mathbf{F}$ which is algebraic over $\mathbf{E}$ and such that $\mathbf{F} = \mathbf{E}(b)$. In view of theorem 4.2, this implies that $\#F = p^n$ if n is the degree of the minimal polynomial of b over $\mathbf{E}$. Thus, essentially the only finite fields are those of the form $\mathbf{Z}_p(b)$ with b algebraic of degree n over $\mathbf{Z}_p$. For each integer n and each positive prime p, there is a field with p^n elements, and all fields with p^n elements are isomorphic ([3], p. 316).

For small n the easiest way to describe the arithmetic in a field, $\mathbf{F}$, with p^n elements, in terms of the arithmetic in $\mathbf{Z}_p$ is as follows. Let $m(x)$ be a monic nth degree polynomial with coefficients in $\mathbf{Z}_p$ that is irreducible over $\mathbf{Z}_p$. There is a root, b, of this polynomial in $\mathbf{F}$ and thus (theorem 4.2) each element of $\mathbf{F}$ has a unique representation in the form $a_0 + a_1b + \cdots + a_{n-1}b^{n-1}$, with $a_i \in \mathbf{Z}_p$. To add or subtract two elements of $\mathbf{F}$ find their representations in terms of b and add or subtract (in $\mathbf{Z}_p$) corresponding coefficients. To multiply them find their representations in terms of b and then exploit the associative, commutative, and distributive laws in conjunction with the basic representation of b^n (in terms of lower powers of b) implied by $m(b) = 0$.

Since polynomials of degree 2 or 3 are reducible over $\mathbf{Z}_p$ if and only if they have linear factors, i.e., if and only if they have a root in the field $\mathbf{Z}_p$, it is easy to describe fields with p^2 or p^3 elements. To find irreducible quartic or quintic polynomials we also have to check for quadratic factors, and for higher degree irreducible polynomials we of course must be cautious about higher degree factors. As examples we consider the fields with 4, 8, and 9 elements.

Example 1. The field with four elements. The only quadratic polynomial that is irreducible over $\mathbf{Z}_2$ is $x^2 + x + \bar{1}$. Using 0, 1 in place of the more cumbersome $\bar{0}$, $\bar{1}$, the four elements are 0, 1, b, and $1 + b$. The addition is that of the 4-group, i.e., each element is its own inverse, and the sum of any two nonzero elements is the third. The multiplicative group of nonzero elements is cyclic.

Example 2. The field with eight elements. There are two irreducible cubic polynomials over $\mathbf{Z}_2$, $x^3 + x^2 + \bar{1}$ and $x^3 + x + \bar{1}$. Choosing $x^3 + x + \bar{1}$, and simplifying $\bar{0}$ and $\bar{1}$ as in example 1, we arrive at a field with the eight elements, 0, 1, b, $b + 1$, b^2, $b^2 + 1$, $b^2 + b$, and $b^2 + b + 1$. The most important relations in computing products are $b^3 = b + 1$ and $x + x = 0$. Thus,

$$b^2(b^2+1) = b^4 + b^2 = b(b^3) + b^2 = b(b+1) + b^2 = b^2 + b + b^2 = b\,.$$

If we let $c = b + 1$, then $\mathbf{Z}_2(b) = \mathbf{Z}_2(c)$ and $c^3 + c^2 + 1 = 0$, hinting at why we arrive at essentially the same field, regardless of which irreducible cubic polynomial we pick.

Example 3. The field with nine elements. Since $x^2 + \bar{1}$ is irreducible over $\mathbf{Z}_3$, the field with nine elements can be built up from $\mathbf{Z}_3$ in the same way that $\mathbf{C}$ is built up from $\mathbf{R}$. Here we replace $\bar{0}$, $\bar{1}$, and $\bar{2}$ by the more convenient 0, 1, and -1, and to suggest the analogy with $\mathbf{C}$ use i in place of b. The nine elements are then $0, \pm 1, \pm i, \pm 1 \pm i$. The complete addition and multiplication tables are shown in Figure 24. Note that the analog of complex conjugation is an automorphism of this field just as it is for $\mathbf{C}$.

Although we have by no means surveyed all of the ways in which fields can be constructed, we have enough examples at hand to illustrate adequately the geometric material that follows. Therefore, we shall not pursue the study of fields even though we are close to one of the most fascinating applications of symmetry in algebra, namely, Galois theory.

	0	1	-1	i	$-i$	$1+i$	$1-i$	$-1+i$	$-1-i$
0	0	1	-1	i	$-i$	$1+i$	$1-i$	$-1+i$	$-1-i$
1	1	-1	0	$1+i$	$1-i$	$-1+i$	$-1-i$	i	$-i$
-1	-1	0	1	$-1+i$	$-1-i$	i	$-i$	$1+i$	$1-i$
i	i	$1+i$	$-1+i$	$-i$	0	$1-i$	1	$-1-i$	-1
$-i$	$-i$	$1-i$	$-1-i$	0	i	1	$1+i$	-1	$-1+i$
$1+i$	$1+i$	$-1+i$	i	$1-i$	1	$-1-i$	-1	$-i$	0
$1-i$	$1-i$	$-1-i$	$-i$	1	$1+i$	-1	$-1+i$	0	i
$-1+i$	$-1+i$	i	$1+i$	$-1-i$	-1	$-i$	0	$1-i$	1
$-1-i$	$-1-i$	$-i$	$1-i$	-1	$-1+i$	0	i	1	$1+i$

Addition

	0	1	-1	i	$-i$	$1+i$	$1-i$	$-1+i$	$-1-i$
0	0	0	0	0	0	0	0	0	0
1	0	1	-1	i	$-i$	$1+i$	$1-i$	$-1+i$	$-1-i$
-1	0	-1	1	$-i$	i	$-1-i$	$-1+i$	$1-i$	$1+i$
i	0	i	$-i$	-1	1	$-1+i$	$1+i$	$-1-i$	$1-i$
$-i$	0	$-i$	i	1	-1	$1-i$	$-1-i$	$1+i$	$-1+i$
$1+i$	0	$1+i$	$-1-i$	$-1+i$	$1-i$	$-i$	-1	1	i
$1-i$	0	$1-i$	$-1+i$	$1+i$	$-1-i$	-1	i	$-i$	1
$-1+i$	0	$-1+i$	$1-i$	$-1-i$	$1+i$	1	$-i$	i	-1
$-1-i$	0	$-1-i$	$1+i$	$1-i$	$-1+i$	i	1	-1	$-i$

Multiplication

FIG. 24. Addition and multiplication tables for the nine-element field.

Exercise 4.1

1. Prove that p divides $a^p - a$ for any prime p and any integer a. Use this result to show that $\bar{a}^p = \bar{a}$ for all $\bar{a}$ in $\mathbf{Z}_p$.

2. Let $\mathbf{F}$ and $\mathbf{G}$ be two subfields in $\mathbf{C}$. What is a necessary and sufficient condition on $\mathbf{F}$ and $\mathbf{G}$ in order that $\mathbf{F} \cup \mathbf{G}$ be a subfield of $\mathbf{C}$?

3. Show how to express the multiplicative inverse of a typical element of $\mathbf{Q}(\sqrt[3]{2})$ in the form $a + b\sqrt[3]{2} + c\sqrt[3]{4}$ with a, b, and c in $\mathbf{Q}$.

4. Prove that the intersection of *any* non-empty collection of subfields of a field $\mathbf{F}$ is also a subfield of $\mathbf{F}$. What is the intersection of *all* subfields of $\mathbf{C}$?

5. Show that any field with four elements is isomorphic to the field in problem 1.

6. Find a fourth-degree polynomial that is irreducible over $\mathbf{Z}_2$. If b is a root of that polynomial in the field with 16 elements, then express b^4, b^5, b^6, b^7, and b^8 in terms of 1, b, b^2, and b^3. Use this information to find $(1 + b^2 + b^3)^3$.

7. Show that there is no subfield with four elements in a field with eight elements. The general theorem (which you may *not* use) is that (for p, q prime) the field with p^m elements has a subfield with q^n elements if and only if $p = q$ and n divides m.

4.2. Vector spaces, subspaces, and echelon form

In this section we collect the basic definitions and computational techniques concerning vector spaces, subspaces, and bases for subspaces. We assume that the reader has seen this material before, at least in the case in which the field of scalars is $\mathbf{R}$. Our main objective is to widen the reader's perspective and to be sure that the generalization to an arbitrary field (especially a finite field) is clear.

Definitions. Assume that we are given:

(1) a field, $\mathbf{F}$, called the field of *scalars*
(2) a non-empty set, V, called the set of *vectors*
(3) an operation of addition in V, called *vector addition*
(4) an operation, called *multiplication by scalars*, assigning to each α in $\mathbf{F}$ and each v in V an element αv in V.

We say then that $\mathbf{V}$ is *a vector space over* $\mathbf{F}$, provided:

A. $\mathbf{V}$ is an abelian group under vector addition.
M. For every v, w in $\mathbf{V}$ and every α, β in $\mathbf{F}$,
$\alpha(v + w) = \alpha v + \alpha w$,
$(\alpha + \beta)v = \alpha v + \beta v$,
$\alpha(\beta v) = (\alpha\beta)v$,
$1v = v$.

If W is a subset of V such that $\mathbf{W}$ is itself a vector space over $\mathbf{F}$, using the vector addition and multiplication by scalars that it inherits from $\mathbf{V}$, then we say that $\mathbf{W}$ is a *subspace* of $\mathbf{V}$.

If $v_1, v_2, \ldots, v_n \in \mathbf{V}$, then we call any vector which can be represented as $\alpha_1 v_1 + \alpha_2 v_2 + \cdots + \alpha_n v_n$, where the $\alpha_i \in \mathbf{F}$, a *linear combination* (over $\mathbf{F}$) of $v_1, v_2, \ldots, v_n$. Since we can always represent the zero vector as $0v_1 + 0v_2 + \cdots + 0v_n$ it is a linear combination of $v_1, v_2, \ldots, v_n$. We call it a *nontrivial* linear combination if we can find at least one *other* representation, i.e., if we can find at least one in which at least one α_i is not zero.

The vectors $v_1, v_2, \ldots, v_n$ are said to be *linearly dependent* if the zero vector is a nontrivial linear combination of $v_1, v_2, \ldots, v_n$. If, on the contrary, the only linear combination of $v_1, v_2, \ldots, v_n$ which is the zero vector is the trivial one (all coefficients $\alpha_i = 0$), then we say that $v_1, v_2, \ldots, v_n$ are *linearly independent.*

If **W** is a subspace of **V** (perhaps **V** itself), then we call the ordered n-tuple of vectors $w_1, w_2, \ldots, w_n$ an *ordered basis* of **W** if:

B1. The vectors $w_1, w_2, \ldots, w_n$ are linearly independent.

B2. Every vector in **W** is a linear combination of $w_1, w_2, \ldots, w_n$, i.e., $w_1, w_2, \ldots, w_n$ is a set of *generators* of **W**, or **W** is *generated by* $w_1, w_2, \ldots, w_n$.

In order to determine whether or not a subset, **W**, of a vector space **V** over **F** is a subspace it is not necessary to check all of the axioms since most of them are inherited from **V**. A minimal set of properties one must verify are as follows.

Theorem 4.4. If **V** is a vector space over **F**, then **W**, formed from a subset W of **V** and the vector addition and multiplication by scalars for **V**, is a subspace if and only if:

S1. $W \neq \varnothing$.

S2. If $w_1, w_2 \in W$, then $w_1 + w_2 \in W$.

S3. If $w \in W$, $\alpha \in F$, then $\alpha w \in W$.

These three properties are usually identified as: W must be non-empty, *closed under addition*, and *closed under multiplication by scalars*. From now on we will not be careful about the distinction between W and **W** and will think of either the *set* of vectors satisfying *S1*, *S2*, and *S3* or of that set together with the vector addition and multiplication by scalars as the subspace.

Example 1. Given a field, **F**, and a positive integer n, the vector space $\mathbf{V}_n(\mathbf{F})$ is that vector space with vectors $\langle \alpha_1, \alpha_2, \ldots, \alpha_n \rangle$, $\alpha_i \in \mathbf{F}$; vector addition defined by

$$\langle \alpha_1, \alpha_2, \ldots, \alpha_n \rangle + \langle \beta_1, \beta_2, \ldots, \beta_n \rangle = \langle \alpha_1 + \beta_1, \alpha_2 + \beta_2, \ldots, \alpha_n + \beta_n \rangle \,;$$

and multiplication by a scalar, λ, defined by

$$\lambda \langle \alpha_1, \alpha_2, \ldots, \alpha_n \rangle = \langle \lambda\alpha_1, \lambda\alpha_2, \ldots, \lambda\alpha_n \rangle \,.$$

The vectors $\langle 1, 0, \ldots, 0 \rangle, \langle 0, 1, 0, \ldots, 0 \rangle, \ldots, \langle 0, 0, \ldots, 0, 1 \rangle$ clearly form an ordered basis for $\mathbf{V}_n(\mathbf{F})$. This basis (in this order!) is called the *natural basis* of $\mathbf{V}_n(\mathbf{F})$.

This example is fundamental in our study of vector spaces, affine spaces, and projective spaces for it is the vector space of analytic geometry. We use $\mathbf{V}_n(\mathbf{F})$ whenever we perform operations with coordinates, i.e., whenever we wish to actually compute something. If we recognize that $\langle \alpha_1, \alpha_2, \ldots, \alpha_n \rangle$ is nothing but a strained (although familiar and hence comfortable) way of specifying a function from $\{1, 2, \ldots, n\}$ into $\mathbf{F}$, then the following generalization is natural.

Example 2. Let S be a set and $\mathbf{F}$ be a field. The vector space $\mathbf{V}_S(\mathbf{F})$ is that vector space whose: vectors are functions $f{:}S \to \mathbf{F}$; vector addition is defined by $(s)(f + g) = (s)f + g(s)$; and multiplication by a scalar, λ, is defined by $(s)(\lambda f) = \lambda((s)f)$. If $\#S = n$, then clearly $\mathbf{V}_S(\mathbf{F})$ and $\mathbf{V}_n(\mathbf{F})$ are isomorphic (as vector spaces!), so nothing new is gained. If, however, S is infinite, then we have an entirely new kind of vector space.

Example 3. Let $\mathbf{F}$ be a subfield of $\mathbf{G}$. If we view the elements of $\mathbf{G}$ as vectors, the addition in $\mathbf{G}$ as vector addition, and multiplication in $\mathbf{G}$ (with one factor in $\mathbf{F}$) as multiplication by scalars, then $\mathbf{G}$ is a vector space over $\mathbf{F}$. For example, $\mathbf{C}$ is a vector space over $\mathbf{R}$, and any of the following ordered pairs of complex numbers is an ordered basis of $\mathbf{C}$ over $\mathbf{R}$: $(1, i)$, $(1 + i, 1 - i)$, $((-1 + \sqrt{3}\,i)/2, (-1 - \sqrt{3}\,i)/2)$. Similarly $\mathbf{Q}(\sqrt[4]{7})$ is a vector space over $\mathbf{Q}$, and one of its ordered bases is $1, \sqrt[4]{7}, \sqrt{7}, \sqrt[4]{343}$.

Example 4. Let $\mathbf{V}$ be the set of all polynomials with coefficients in the field $\mathbf{F}$. With the usual addition of polynomials and multiplication of a polynomial by a scalar as vector addition and multiplication by a scalar, $\mathbf{V}$ is a vector space, over $\mathbf{F}$. A (very small) subspace of this space is the set, $\mathbf{W}$, of all polynomials of degree at most 3. The polynomials $1, x, x^2, x^3$ form an ordered basis for $\mathbf{W}$ as do $1, (x - 1), (x - 1)^2$, and $(x - 1)^3$.

The following theorem about subspaces and their bases is the foundation of the concept of dimension. Its proof is standard material in any linear algebra text.

Theorem 4.5. Assume that $\mathbf{V}$ is a vector space over $\mathbf{F}$ having at least one *finite* set of generators. Let $\mathbf{W}$ be a subspace of $\mathbf{V}$ (perhaps $\mathbf{V}$ itself).

(1) Any set of independent vectors in $\mathbf{W}$ can be completed to a basis of $\mathbf{W}$.
(2) From any set of generators of $\mathbf{W}$, we can extract a basis of $\mathbf{W}$.
(3) The bases of $\mathbf{W}$ all contain the same number of vectors.

Definition. The number of vectors in any basis of $\mathbf{W}$ is called the *dimension* of $\mathbf{W}$ and is denoted by $\dim(\mathbf{W})$ or $\dim_F(\mathbf{W})$ if it is necessary to keep track of the field of scalars. In the special case $\mathbf{W} = \{0\}$, we say $\dim(\mathbf{W}) = 0$ and, therefore, call $\varnothing$ a basis for $\{0\}$.

Example 5. If b is algebraic of degree n over $\mathbf{F}$, then (theorem 4.2) $1, b, b^2, \ldots, b^{n-1}$ is a basis of $\mathbf{F}(b)$ over $\mathbf{F}$; hence $\mathbf{F}(b)$ is of dimension n over $\mathbf{F}$. In particular, a finite field with p^n elements, p prime, is a vector space of dimension n over $\mathbf{Z}_p$.

Example 4 (continued). The vector space of all polynomials with coefficients in $\mathbf{F}$ is not finite-dimensional over $\mathbf{F}$ since any finite set of polynomials contains a polynomial whose degree, n, is maximal in that set, and any linear combination of polynomials of degree at most n is a polynomial of degree at most n. The subspace of *all* polynomials of degree at most n is of dimension $n + 1$ since $1, x, x^2, \ldots, x^n$ is clearly a basis.

Example 6. Let $\mathbf{V} = \mathbf{V}_F(\mathbf{F})$ be the set of all functions from F into $\mathbf{F}$. The functions that are 1 at just one point of F and zero everywhere else are linearly independent. They form a basis for $\mathbf{V}$ over $\mathbf{F}$ if and only if $\mathbf{F}$ is finite; hence $\mathbf{V}$ is finite-dimensional if and only if $\mathbf{F}$ is finite. Moreover, the dimension of $\mathbf{V}$ is just $\#F$ if $\mathbf{F}$ is finite. It is interesting to note that if $\mathbf{F}$ is finite, *any* function from F into $\mathbf{F}$ is a polynomial function ([1], pp. 56–59), and thus there are exactly n^n different polynomial functions from a finite field with n elements into itself. If the n elements of this field are $a_1, a_2, \ldots, a_n$, then two polynomials determine the same polynomial function into $\mathbf{F}$ if and only if their difference is divisible by $(x - a_1)(x - a_2) \cdots (x - a_n)$.

Definition. If $\mathbf{W}_1$ and $\mathbf{W}_2$ are subspaces of $\mathbf{V}$, then

$$\mathbf{W}_1 + \mathbf{W}_2 = \{w_1 + w_2 | w_1 \in W_1, w_2 \in W_2\}$$

is called the *sum* of $\mathbf{W}_1$ and $\mathbf{W}_2$. It is usually *not* the same as $\mathbf{W}_1 \cup \mathbf{W}_2$.

Theorem 4.6. If $\mathbf{W}_1$ and $\mathbf{W}_2$ are subspaces of $\mathbf{V}$, then both $\mathbf{W}_1 \cap \mathbf{W}_2$ and $\mathbf{W}_1 + \mathbf{W}_2$ are also. Moreover, $\mathbf{W}_1 + \mathbf{W}_2$ is the smallest subspace containing $\mathbf{W}_1 \cup \mathbf{W}_2$, and

$$\dim(\mathbf{W}_1) + \dim(\mathbf{W}_2) = \dim(\mathbf{W}_1 \cap \mathbf{W}_2) + \dim(\mathbf{W}_1 + \mathbf{W}_2)$$

if $\mathbf{W}_1 + \mathbf{W}_2$ is a finite-dimensional vector space.

Once again we refer the reader to any of the standard linear algebra texts for a proof. The basic idea is to start with a basis for $\mathbf{W}_1 \cap \mathbf{W}_2$, complete it to a basis of $\mathbf{W}_1$, start over and complete it to a basis of $\mathbf{W}_2$, and then show that all of these vectors form a basis for $\mathbf{W}_1 + \mathbf{W}_2$.

Definition. If $\mathbf{W}_1$ and $\mathbf{W}_2$ are subspaces of $\mathbf{V}$ such that $\mathbf{W}_1 \cap \mathbf{W}_2 = \{0\}$ and $\mathbf{W}_1 + \mathbf{W}_2 = \mathbf{V}$, then we say that $\mathbf{W}_1$ and $\mathbf{W}_2$ are *complementary subspaces in* $\mathbf{V}$. See exercise 4.2, problem 1 for an alternative definition.

If $\mathbf{W}_1$ is a subspace of $\mathbf{V}$, then it is always possible to find a subspace $\mathbf{W}_2$ such that $\mathbf{W}_1$ and $\mathbf{W}_2$ are complementary. One way to do this is to complete a basis of $\mathbf{W}_1$ to a basis of $\mathbf{V}$, and let $\mathbf{W}_2$ be the subspace generated by the new basis vectors.

Example 7. Let $\mathbf{V}$ be the field of four elements (example 1, section 4.1) viewed as a vector space over $\mathbf{Z}_2$. If $\mathbf{W}_1 = \{0, b\}$, then $\mathbf{W}_2 = \{0, 1\}$ is a complementary subspace, as is $\{0, 1 + b\}$. These are the only two subspaces of $\mathbf{V}$ complementary to $\mathbf{W}_1$.

Example 8. Orthogonal complements. Let (v, w) be an inner product on $\mathbf{V}_n(\mathbf{R})$, which, for our purposes, we take to be the standard dot product defined by

$$(\langle \alpha_1, \alpha_2, \ldots, \alpha_n \rangle, \langle \beta_1, \beta_2, \ldots, \beta_n \rangle) = \alpha_1\beta_1 + \cdots + \alpha_n\beta_n .$$

If $\mathbf{W}$ is a subspace, then

$$\mathbf{W}^\perp = \{v \in \mathbf{V}_n(\mathbf{R}) | (w, v) = 0 \text{ for all } w \in \mathbf{W}\}$$

is a complementary subspace ([1], p. 182, or [3], p. 157) and is called the orthogonal complement of $\mathbf{W}$. Generalizing to subspaces $\mathbf{W}$ of $\mathbf{V}_n(\mathbf{F})$ for an arbitrary field $\mathbf{F}$, we may lose the property that $\mathbf{W} \cap \mathbf{W}^\perp = \{0\}$. However, we still have $\dim(\mathbf{W}^\perp) = n - \dim(\mathbf{W})$. In other words, although the dimension of $\mathbf{W}^\perp$ is just right for it to be complementary, it sometimes fails to be so because of vectors in $\mathbf{W}$ that are orthogonal to themselves. For example, let $\mathbf{W} = \{\langle \bar{0}, \bar{0} \rangle, \langle \bar{1}, \bar{1} \rangle\}$ in $\mathbf{V}_2(\mathbf{Z}_2)$. Then $\mathbf{W}^\perp = \mathbf{W}$, so both $\mathbf{W}$ and $\mathbf{W}^\perp$ are one-dimensional but not complementary.

The result mentioned in the example above is equivalent to the fact that a subspace of $\mathbf{V}_n(\mathbf{F})$ is of dimension k if and only if it is the solution set of a system of $n - k$ independent, homogeneous, linear equations, i.e., of a homogeneous system of rank $n - k$. The most useful elementary technique to reduce a set of generators to a basis or reduce a system of linear equations to an easily solved independent set is the "echelon form" technique. We present this as our final topic in this section and emphasize by our examples that it is valid for vector spaces over arbitrary fields.

Definition. Let $\mathbf{W}$ be a subspace of $\mathbf{V}_n(\mathbf{F})$. An ordered basis $w_1, \ldots, w_p$ of $\mathbf{W}$ is called a *canonical basis* if:

(1) The initial nonzero entry of w_i is 1, $i = 1, 2, \ldots, p$.
(2) The initial nonzero entry of w_{i+1} is further to the right than that of w_i, $i = 1, 2, \ldots, p - 1$. In other words, the w_i are in *echelon form.*
(3) If the initial nonzero entry of w_i occurs in position j, then the jth component of each of the other basis vectors is zero.

Examples 9–12. 9. If $w_1 = \langle 0, 1, 0, 3, -5 \rangle$, $w_2 = \langle 0, 0, 1, -1, 4 \rangle$, then w_1, w_2 form a canonical basis of some two-dimensional subspace of $\mathbf{V}_5(\mathbf{Q})$. *10.* The vectors $w_1 = \langle 4, 0, 3\rangle$ and $w_2 = \langle 0, 1, 3\rangle$ do not form a canonical basis in $\mathbf{V}_3(\mathbf{Q})$ although $\frac{1}{4}w_1, w_2$ do. *11.* The subspace $\mathbf{W}$ of $\mathbf{V}_5(\mathbf{R})$ consisting of all vectors $\langle \alpha_1, \alpha_2, \ldots, \alpha_5\rangle$ such that

$$\alpha_1 + \alpha_2 - \alpha_3 - \alpha_4 = \alpha_1 - \alpha_2 = 0$$

has a canonical basis $w_1 = \langle 1, 1, 0, 2, 0\rangle$, $w_2 = \langle 0, 0, 1, -1, 0\rangle$, $w_3 = \langle 0, 0, 0, 0, 1\rangle$. *12.* The subspace of $\mathbf{V}_4(\mathbf{Z}_3)$ generated by $\langle \bar{1}, \bar{1}, \bar{0}, \bar{0}\rangle$, $\langle \bar{1}, \bar{1}, \bar{0}, \bar{1}\rangle$, $\langle \bar{0}, \bar{1}, \bar{1}, \bar{0}\rangle$, and $\langle \bar{2}, \bar{1}, \bar{2}, \bar{2},\rangle$ has as its canonical basis the vectors $\langle \bar{1}, \bar{0}, \bar{2}, \bar{0}\rangle$, $\langle \bar{0}, \bar{1}, \bar{1}, \bar{0}\rangle$, and $\langle \bar{0}, \bar{0}, \bar{0}, \bar{1}\rangle$. *Caution!* $\bar{2} + \bar{1} = \bar{0}$ in $\mathbf{Z}_3$.

Given a set of generators (not all the zero vector) for a subspace $\mathbf{W}$, it is always possible to perform elementary row operations (described below) on these vectors to arrive at a canonical basis for $\mathbf{W}$. The elementary row operations are:

(1) Suppress the zero vector or any vector linearly dependent on the remaining ones.
(2) Reorder the sequence of vectors.
(3) Replace any vector by a nonzero scalar multiple of itself.
(4) Replace v_i by $v_i + \lambda v_j$ if $i \neq j$ (λ arbitrary).

The sequence of row operations needed is found by working from the left and eliminating as many nonzero entries from each column as possible. Finally delete all zero vectors, reorder the vectors, and replace each by an appropriate nonzero scalar multiple of itself in order to make the initial nonzero entry $+1$. Some schematic device for keeping track of the computations is desirable. In Figure 25 we reduce the generators in example 12 above to the given canonical basis.

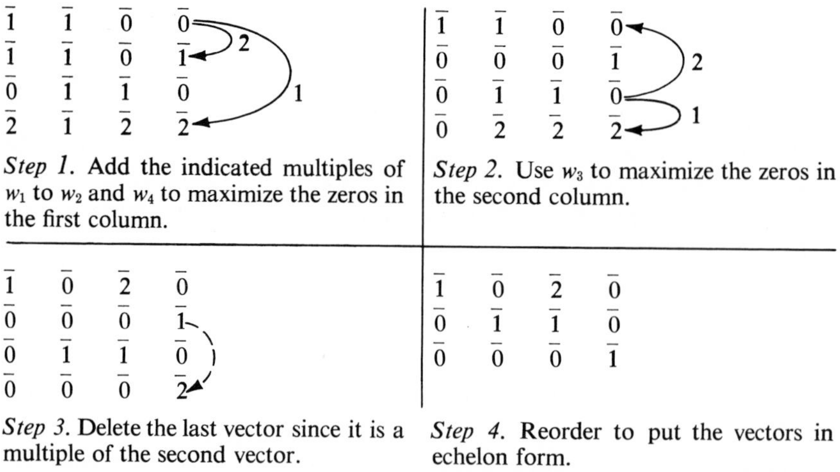

$$\begin{array}{cccc} \bar{1} & \bar{1} & \bar{0} & \bar{0} \\ \bar{1} & \bar{1} & \bar{0} & \bar{1} \\ \bar{0} & \bar{1} & \bar{1} & \bar{0} \\ \bar{2} & \bar{1} & \bar{2} & \bar{2} \end{array}$$

(arrows: 2 from w_1 to w_2; 1 from w_1 to w_4)

Step 1. Add the indicated multiples of w_1 to w_2 and w_4 to maximize the zeros in the first column.

$$\begin{array}{cccc} \bar{1} & \bar{1} & \bar{0} & \bar{0} \\ \bar{0} & \bar{0} & \bar{0} & \bar{1} \\ \bar{0} & \bar{1} & \bar{1} & \bar{0} \\ \bar{0} & \bar{2} & \bar{2} & \bar{2} \end{array}$$

(arrows: 2 from w_3 to w_1; 1 from w_3 to w_4)

Step 2. Use w_3 to maximize the zeros in the second column.

$$\begin{array}{cccc} \bar{1} & \bar{0} & \bar{2} & \bar{0} \\ \bar{0} & \bar{0} & \bar{0} & \bar{1} \\ \bar{0} & \bar{1} & \bar{1} & \bar{0} \\ \bar{0} & \bar{0} & \bar{0} & \bar{2} \end{array}$$

Step 3. Delete the last vector since it is a multiple of the second vector.

$$\begin{array}{cccc} \bar{1} & \bar{0} & \bar{2} & \bar{0} \\ \bar{0} & \bar{1} & \bar{1} & \bar{0} \\ \bar{0} & \bar{0} & \bar{0} & \bar{1} \end{array}$$

Step 4. Reorder to put the vectors in echelon form.

FIG. 25. Reducing a set of generators to a canonical basis of a subspace of $\mathbf{V}_4(Z_3)$.

The same row operations are used in an effective technique for solving systems of linear equations. The change from vectors to equations is fairly obvious so we present only one example.

Example 13. Let $\mathbf{F} = \{0, 1, b, b^2 = b + 1\}$ be the field with four elements. Consider the homogeneous system of four equations in four unknowns:

$$\begin{aligned} w + bx + y + bz &= 0 && (1)\\ w + bx + b^2z &= 0 && (2)\\ bw + b^2x + y &= 0 && (3)\\ w + bx + by + z &= 0\,. && (4) \end{aligned}$$

Since the characteristic of **F** is 2, we can subtract by adding. Using eq. (1) to eliminate w, we replace eq. (2) by [eq. (1) + eq. (2)], eq. (3) by [b times eq. (1) + eq. (3)], and eq. (4) by [eq. (1) + eq. (4)] to get the equivalent system:

$$\begin{aligned} w + bx + y + bz &= 0 && (5)\\ y + z &= 0 && (6)\\ b^2y + b^2z &= 0 && (7)\\ b^2y + b^2z &= 0\,. && (8) \end{aligned}$$

The last two equations are scalar multiples of the second so we suppress them. Finally we replace eq. (5) by [eq. (5) + eq. (6)] in order to eliminate y and arrive at the "solution":

$$\begin{aligned} w + bx + b^2z &= 0\\ y + z &= 0\,. \end{aligned}$$

Usually one interprets this solution by saying x and z are arbitrary, $y = -z$ ($=z$ in this field!), and $w = bx + b^2z$. Since **F** has only four elements, we could list explicitly the 16 solutions corresponding to the four choices for x and the four choices for z. A more efficient way to describe this solution set in terms of vectors is as follows. Choose two independent pairs of values for x and z, say, $(x, z) = (1, 0)$ or $(0, 1)$. All solutions are linear combinations of the basic solutions, $\langle b, 1, 0, 0\rangle$ and $\langle b^2, 0, 1, 1\rangle$, corresponding to these choices of x and z, i.e., *any* solution is of the form $\langle w, x, y, z\rangle = x\langle b, 1, 0, 0\rangle + z\langle b^2, 0, 1, 1\rangle$, with $x, z \in \mathbf{F}$.

Exercise 4.2

1. Prove that subspaces $\mathbf{W}_1$ and $\mathbf{W}_2$ are complementary subspaces of **V** if and only if each vector v in **V** has a unique representation in the form $v = w_1 + w_2$, $w_1 \in \mathbf{W}_1$, $w_2 \in \mathbf{W}_2$.

2. Let $\mathbf{V} = \mathbf{V}_4(\mathbf{Z}_3)$, and replace $\bar{0}, \bar{1}, \bar{2}$ for elements of $\mathbf{Z}_3$ by the less cumbersome 0, 1, -1. Let **W** be the subspace generated by $\langle -1, 1, 0, -1\rangle$, $\langle 1, -1, -1, 0\rangle$,

$\langle 1, 0, -1, 0\rangle$, and $\langle 1, 1, 0, 1\rangle$. Let $\mathbf{X}$ be the subspace generated by $\langle 1, 1, 0, 1\rangle$, $\langle 1, 0, 0, 1\rangle$, and $\langle 0, -1, -1, 1\rangle$. Find canonical bases for $\mathbf{W}$, $\mathbf{X}$, $\mathbf{W} \cap \mathbf{X}$, and $\mathbf{W} + \mathbf{X}$, and find systems of homogeneous equations describing each of these four subspaces.

3. Find necessary and sufficient conditions on the subspaces $\mathbf{W}_1$ and $\mathbf{W}_2$ of $\mathbf{V}$ for $\mathbf{W}_1 \cup \mathbf{W}_2$ to be a subspace of $\mathbf{V}$. Compare problem 2, exercise 4.1.

4. Find at least two vector spaces that are not of the form $\mathbf{V}_n(\mathbf{F})$ and were not used as examples in this section.

5. Find, if possible, four vectors in $\mathbf{V}_3(\mathbf{F})$ that generate a two-dimensional subspace, but such that *any* two of the four generators are independent:
(1) if $\mathbf{F}$ is the field of rationals
(2) if $\mathbf{F} = \mathbf{Z}_2$.

6. Let $\mathbf{V}$ be a vector space over $\mathbf{F}$, and let S be a set. Show that the set of all functions from S into $\mathbf{V}$, with addition and multiplication by scalars defined as in example 2 is a vector space over $\mathbf{F}$. Find the dimension of this space if $\#S = m$ and $\dim(\mathbf{V}) = n$ are both finite. *Hint.* Example 6 covers the case in which $\dim \mathbf{V} = 1$.

7. Let $\mathbf{V}$ be a vector space over a finite field, $\mathbf{F}$, with f elements. Show that the dimension of $\mathbf{V}$ over $\mathbf{F}$ is just $\log_f(\#\mathbf{V})$.

8. Let $\mathbf{F}$ be the nine-element field described in section 4.1. Let $\mathbf{W}$ be the subspace of $\mathbf{V}_4(\mathbf{F})$ generated by $\langle 1, i, 0, 1 + 2i\rangle$, $\langle 1, 1, 1, -i\rangle$, and $\langle 0, 2, 1, 1 - i\rangle$. Find a canonical basis for $\mathbf{W}$.

4.3. Linear transformations

In this section we review the basic definitions and properties of linear transformations, and give some examples showing that the field of scalars may be an arbitrary field and that matrices are not always essential. Then we study in detail two special types of linear transformations, projections and involutions.

Definition. Let $\mathbf{V}$ and $\mathbf{W}$ be vector spaces over the same field $\mathbf{F}$ (perhaps $\mathbf{V} = \mathbf{W}$). A function $g: \mathbf{V} \to \mathbf{W}$ with domain $\mathbf{V}$ and codomain $\mathbf{W}$ is called a *linear transformation* from $\mathbf{V}$ to $\mathbf{W}$ if it preserves all linear combinations, i.e., if for all $v_1, v_2, \ldots, v_k \in \mathbf{V}$ and all $\alpha_1, \alpha_2, \ldots, \alpha_k \in \mathbf{F}$,

$$(\alpha_1 v_1 + \cdots + \alpha_k v_k)g = \alpha_1(v_1)g + \alpha_2(v_2)g + \cdots + \alpha_k(v_k)g .$$

The *kernel* of g is the set of all $v \in \mathbf{V}$ such that $(v)g$ is the zero vector in $\mathbf{W}$. As usual, the *range* of $g = (\mathbf{V})g$ may be a proper subset of $\mathbf{W}$. The *rank* of g is $\dim(\mathbf{V})g$.

The proofs of the following three propositions are only outlined here since they may be found in any linear algebra text and are the same for an arbitrary field of scalars as for the case in which the scalars are assumed to be real numbers. In all three we assume the vector spaces **V** and **W** have the same field, **F**, of scalars.

Proposition 4.7. If g is a function mapping **V** into **W**, then g is a linear transformation if and only if:

(1) g is additive, i.e., $(v_1 + v_2)g = (v_1)g + (v_2)g$ for all $v_1, v_2 \in \mathbf{V}$.
(2) g is homogeneous, i.e., $(\alpha v)g = \alpha(v)g$ for all $v \in \mathbf{V}, \alpha \in \mathbf{F}$.

Proof (*outline*). Conditions (1) and (2) assert that certain linear combinations are preserved, and hence they must hold if g is a linear transformation. An easy induction proof shows that all linear combinations are preserved by g if (1) and (2) hold.

Proposition 4.8. If g is a linear transformation from **V** into **W**, then the kernel of g is a subspace of **V** and the range of g is a subspace of **W**. In particular, g must send the zero vector of **V** to the zero vector of **W**. If, in addition, dim **V** is finite, then dim (kernel g) + dim (range g) = dim **V**.

Proof (*outline*). Since g is additive, its kernel is a subgroup of **V**, and since g is homogeneous, this subgroup must be a subspace. Similarly, the range is a subspace of **W**. If we complete a basis of kernel g to a basis of **V**, then the images of the added basis vectors form a basis of $(\mathbf{V})g$, and thus, dim (kernel g) + dim (range g) = dim **V** if dim **V** is finite.

Proposition 4.9. Let $v_1, v_2, \ldots, v_n$ be an ordered basis for **V**, and let $w_1, w_2, \ldots, w_n$ be *any* sequence of n vectors, not necessarily distinct, in **W**. There is *exactly* one linear transformation g from **V** into **W** such that $(v_i)g = w_i$, $i = 1, 2, \ldots, n$.

Proof (*outline*). Since there are no linear relationships between basis vectors, there is nothing a linear transformation need preserve, and therefore basis vectors can be sent anywhere. On the other hand, every vector in **V** is a linear combination of the basis vectors. Hence the images of the basis vectors determine completely the linear transformation.

We meet proposition 4.9 in a different context in the next section where we use it to show that any matrix is the matrix of some linear transformation. Now we exploit it to give some examples of linear transformations without using matrices.

Example 1. Let s be a vector in **W**. There is a linear transformation, f, sending every basis vector in a particular basis of **V** to s. If $s = 0$, then $f = 0$, the zero transformation which sends *all* vectors in **V** to the zero vector in **W**. This is the only linear transformation that is constant, for if $s \neq 0$, then

$$(\alpha_1 v_1 + \cdots + \alpha_n v_n)f = (\alpha_1 + \cdots + \alpha_n)s \neq s$$

if $(\alpha_1 + \alpha_2 + \cdots + \alpha_n) \neq 1$. Note that we may or may not have $(s)f = s$. If $s \neq 0$, then $\dim(\text{range } f) = 1$, and f is a linear transformation of *rank* 1.

Example 2. Formal differentiation. Let **V** be the vector space of all polynomials with coefficients in the field **F**. The monomials, $1 = x^0, x^1, x^2, \ldots, x^n, \ldots$, form a basis of **V**, albeit an infinite one. There is no problem in generalizing proposition 4.8 to infinite bases, and, in particular, there is a unique linear transformation, d, of **V** into itself such that $(x^n)d = nx^{n-1}$. *Caution!* We interpret the exponent n as an integer but the coefficient n as the multiplicative identity of **F** added to itself n times. Thus, $x^5 + 1$ is not a constant polynomial, yet $(x^5 + 1)d = 0$ if the characteristic of **F** is 5. For applications of the formal derivative to the study of polynomials, see [1], pp. 402–403, or [3], pp. 191–193.

Definition. Let **V** and **W** be vector spaces over the same field **F**, and let g and h be linear transformations of **V** into **W**. The functions $g + h$ and αg, $\alpha \in \mathbf{F}$ are defined by $(v)(g + h) = (v)g + (v)h$ and $(v)\alpha g = \alpha(v)g$ for all $v \in \mathbf{V}$. The set of all linear transformations from **V** into **W** is denoted $L(\mathbf{V}, \mathbf{W})$. If $\mathbf{V} = \mathbf{W}$, then we also define the product of two linear transformations, g and h, in $L(\mathbf{V}, \mathbf{V})$ to be the ordinary composite, i.e., $(v)(gh) = ((v)g)h$.

Theorem 4.10. If **V** and **W** are vector spaces over the same field **F**, then $L(\mathbf{V}, \mathbf{W})$ with addition and multiplication by scalars as defined above is also a vector space over **F**. Moreover, if both $\dim \mathbf{V} = m$ and $\dim \mathbf{W} = n$ are finite, then $\dim L(\mathbf{V}, \mathbf{W}) = mn$. In case $\mathbf{V} = \mathbf{W}$, then composition of linear transformations mixes nicely with these operations, i.e., if f, g, and h are in $L(\mathbf{V}, \mathbf{V})$ and α, β are scalars:

$$f(\alpha g + \beta h) = \alpha fg + \beta fh$$
$$(\alpha f + \beta g)h = \alpha fh + \beta gh\,.$$

Once again we only outline the proof and refer the reader to [3], pp. 144–145 and 216–217 for the details. Verifying that $L(\mathbf{V}, \mathbf{W})$ is a vector space is a tedious but trivial exercise in reading definitions. One proof that $L(\mathbf{V}, \mathbf{W})$ is of dimension mn depends on proposition 4.9—namely, one chooses a basis for **V** and for **W** and shows that the mn linear transformations, each of which sends just one of the m basis vectors of **V** to one of the n basis vectors of **W** and all the rest to 0, is a basis for $L(\mathbf{V}, \mathbf{W})$. The mixed associative and distributive laws for products of scalars and linear transformations are easy to prove.

To gain some experience in the algebra of linear transformations from **V** into itself we study carefully two of the most interesting types of such transformations, involutions and projections. The prototypes of involutions are the coordinate symmetries one studies in analytic geometry, e.g., the reflection in the xy plane for 3-space is the linear transformation, g, such that $\langle x, y, z\rangle g = \langle x, y, -z\rangle$. The prototypes for projections are the mappings which collapse 3-space onto a coordinate plane or a coordinate axis, e.g., $\langle x, y, z\rangle g = \langle x, 0, z\rangle$ or $\langle x, y, z\rangle g = \langle 0, 0, z\rangle$.

Definition. Let g be a linear transformation from $\mathbf{V}$ into itself. g is called an *involution* if $g^2 = \mathbf{1}$ but $g \neq \mathbf{1}$. g is called a *projection* if $g^2 = g$.

The definition of involution agrees with our previous ideas, but the geometric implications of the definition of projection are not at all clear. These implications and the connection between projection and the idea of collapsing $\mathbf{V}$ are explained in the next theorem.

Theorem 4.11. Assume p is a linear transformation from $\mathbf{V}$ into itself. p is a projection if and only if all three of the following hold:

(1) $v \in \mathbf{V}$ implies $v = w_1 + w_2$ with $w_1 \in$ kernel p, $w_2 \in$ range p.
(2) The only vector in both kernel p and range p is the zero vector.
(3) $w_2 \in$ range p implies $(w_2)p = w_2$. Hence if $w = w_1 + w_2$ as in (1) above, then $(w)p = w_2$, i.e., p collapses $\mathbf{V}$ onto range p.

Proof. Assume p is a projection, then (3) is immediate, and (1) follows easily from the observation that $v = (v - (v)p) + (v)p$, since $(v - (v)p)p = 0$. If v is in both range p and kernel p, then by (3) $(v)p = v$, but $(v)p = 0$, so $v = 0$, which proves (2).

Conversely, assume p is a linear transformation which satisfies properties (1), (2), and (3). Then for any vector $v \in V$, $v = w_1 + w_2$ as in (1), and hence

$$((v)p)p = (w_2)p = w_2 = (v)p\,,$$

so $p^2 = p$ which completes the proof.

Note that if we combine properties (1) and (2) in the above theorem, we find that the kernel and range of a projection are complementary subspaces. If we visualize a projection as something similar to "casting a shadow," then we are led naturally to the following definition.

Definition. Let p be a projection with domain $\mathbf{V}$, range $\mathbf{W}_1$, and kernel $\mathbf{W}_2$. Then we say that p is the *projection of* $\mathbf{V}$ *onto* $\mathbf{W}_1$ *along* $\mathbf{W}_2$.

Intuitively there are many directions in which we can cast shadows on the ground. The formal analog for this would assert that there are many projections of $\mathbf{V}$ onto one of its subspaces, $\mathbf{W}$. The only obvious exceptions to this are $\mathbf{W} = \{0\}$ (casting shadows on a point) and $\mathbf{W} = \mathbf{V}$ (everything is its own shadow) corresponding to the special projections 0 and $\mathbf{1}$. The following theorem shows that even if **F** is a strange field instead of **R**, these intuitive ideas reflect the true situation.

Theorem 4.12. Let **W** be a subspace of a vector space **V** over a field **F**. For each subspace **X** complementary to **V**, there is a unique projection of **V** onto **W** along **X**. Moreover, if $0 < \dim(\mathbf{W}) < n = \dim(\mathbf{V})$, then there is always more than one projection of **V** onto **W**.

Proof. Let **W** and **X** be complementary subspaces of **V**. Then each vector $v \in \mathbf{V}$ has a unique representation in the form $v = w + x$, $w \in \mathbf{W}$, $x \in \mathbf{X}$ (problem 1, exercise 4.2). Thus, the mapping p defined by $(w + x)p = w$ for all $w \in \mathbf{W}$, $x \in \mathbf{X}$ is defined unambiguously on all of **V**. Furthermore, p is a linear transformation since if $v_1 = w_1 + x_1$ and $v_2 = w_2 + x_2$, then $(\alpha w_1 + \beta w_2) + (\alpha x_1 + \beta x_2)$ is a representation of $\alpha v_1 + \beta v_2$ in the form $w + x$, and since such a representation is unique,

$$(\alpha v_1 + \beta v_2)p = \alpha w_1 + \beta w_2 = \alpha(v_1)p + \beta(v_2)p\,.$$

Clearly the kernel and range of p are **X** and **W**, respectively. Hence by theorem 4.8, p is the projection of **V** onto **W** along **X**. There is at most one such projection since the representation of a vector in the form $w + x$ is unique.

If $0 < \dim(\mathbf{W}) = k < n = \dim \mathbf{V}$, then complete an ordered basis $w_1, w_2, \ldots, w_k$ of **W** to an ordered basis $w_1, w_2, \ldots, w_k, x_1, x_2, \ldots, x_{n-k}$ of **V**. Since $0 < k < n$, $w_1, w_2, \ldots, w_k, x_1 + w_1$ are independent and can be completed to an ordered basis of **V**. The last $n - k$ basis vectors in these two bases of **V** generate distinct subspaces, both of which are complementary to **W**. Hence there are at least two projections of **V** onto **W**.

Now we exploit the algebra of linear transformations to explore the relations between projections and involutions. The following result illustrates clearly the difficulties that fields of characteristic 2 often cause.

Theorem 4.13. If the characteristic of the field of scalars is 2 then $2p - \mathbf{1} = \mathbf{1}$ for any linear transformation, p, of **V** into itself. In all other cases the correspondence, g corresponds to p if and only if $g = 2p - \mathbf{1}$, is a one-to-one correspondence between involutions, g, and projections, p, that are not **1**.

Proof. If the characteristic is 2 then $2p = (1 + 1)p = 0$ and $-\mathbf{1} = \mathbf{1}$. Otherwise $g = (2p - \mathbf{1})$ if and only if $p = \frac{1}{2}(g + \mathbf{1})$. Easy computations show that $(2p - \mathbf{1})^2 = \mathbf{1}$ but $(2p - \mathbf{1}) \neq \mathbf{1}$ whenever p is a projection but not **1** and that $(\frac{1}{2}(g + \mathbf{1}))^2 = (\frac{1}{2})(g + \mathbf{1})$ but $(\frac{1}{2})(g + \mathbf{1}) \neq \mathbf{1}$ whenever g is an involution. Thus we have two mappings, $(p)\phi = (2p - \mathbf{1})$ and $(g)\theta = (\frac{1}{2})(g + \mathbf{1})$ that are inverses to each other and interchange nontrivial projections with involutions. Hence ϕ is a one-to-one correspondence as claimed.

Caution! If f and g are linear transformations of **V** into **V** then $(f+g)^2 = f^2 + fg + gf + g^2$ reduces to $f^2 + 2fg + g^2$ if and only if $fg = gf$. Since **1** and **−1** commute with all linear transformations from **V** into **V** this complication does not enter the computations in the proof.

Example 3. Let **V** be the field with nine elements viewed as a two-dimensional vector space over $\mathbf{Z}_3$ (example 3, section 4.1). The analog of complex conjugation is an involution of **V**, and the corresponding projection collapses **V** onto $\mathbf{Z}_3$ along the subspace $\{0, i, -i\}$. Similarly if we view **C** as a vector space over **R**, then the projection corresponding to complex conjugation is the mapping sending every complex number to its real part.

Exercise 4.3

1. Familiar involutions of $\mathbf{V}_3(\mathbf{R})$ are:

(1) the inversion with center at the origin
(2) half-turns about lines through the origin
(3) reflections in planes through the origin.

Explain how to find nonfamiliar involutions of $\mathbf{V}_3(\mathbf{R})$, i.e., involutions of $\mathbf{V}_3(\mathbf{R})$ which are not isometries.

2. Show that $p \in L(\mathbf{V}, \mathbf{V})$ is a projection if and only if $\mathbf{1} - p$ is a projection.

3. Let **V** be a vector space over a field of characteristic 2. Show that $v + v = 0$ for every $v \in \mathbf{V}$ and that $g + g = 0$ for every $g \in L(\mathbf{V}, \mathbf{V})$. Show that $p \in L(\mathbf{V}, \mathbf{V})$ is an involution if and only if $p = \mathbf{1} + q$ with $q \in L(\mathbf{V},\mathbf{V})$ and $q^2 = 0$. Also show (using proposition 4.8) that dim (kernel q) $\geq (\frac{1}{2})$dim (**V**).

4. Let p be an involution of a vector space **V** over a field of characteristic not 2. Show directly (without appealing to projections or coordinates) that $\mathbf{V}^+ = \{v|(v)p = v\}$ and $\mathbf{V}^- = \{v|(v)p = -v\}$ are complementary subspaces of **V**.

5. Show that for any one-dimensional subspace, **W**, of $\mathbf{V}_2(\mathbf{Z}_2)$ there are exactly two projections of $\mathbf{V}_2(\mathbf{Z}_2)$ onto **W**. What is the analogous result for $\mathbf{V}_2(\mathbf{Z}_3)$? for $\mathbf{V}_3(\mathbf{Z}_2)$?

6. By proposition 4.8, if g is a linear transformation from **V** into **V** and dim (**V**) $= n$ is finite, then dim (kernel g) $= k$ if and only if the rank of g is $n - k$. Use the results on projections to show that any subspace of dimension k can be realized as the *kernel* of a linear transformation of rank $n - k$. Thus, we can assert that a subspace has dimension k if and only if it is the kernel of a linear transformation of rank $n - k$.

7. Let q be a linear transformation of the vector space **V** into itself. Show that q satisfies conditions (1) and (2) of theorem 4.11 if and only if kernel q^2 = kernel q. Give an example of such a transformation which is *not* a projection. *Hint.* There are suitable q's when $\mathbf{V} = \mathbf{V}_2(\mathbf{R})$.

8. Let $f{:}V_1 \to V_2$ and $g{:}V_2 \to V_3$ be linear transformations. Show that the range of fg is contained in the range of g and the kernel of f is contained in the kernel of fg.

Using these facts and proposition 4.8, show that the rank of fg is at most the minimum of the ranks of f and g.

9. Show (by example) that it is possible for two linear transformations f and g to have the same rank, while f^2 and g^2 have different ranks.

4.4. Coordinate mappings, matrices for linear transformations

Although the vector spaces $\mathbf{V}_n(\mathbf{F})$ are not the only vector spaces of importance, they do play an important role in computations. In this section we try to identify this role and show how linear transformations and matrices are related. We illustrate the theory developed in this section with the following special example.

Example 1. Let **P** be the set of all polynomials of degree at most 3 with rational coefficients. Define addition in **P** and multiplication by a (rational) scalar to be the standard operations for polynomials. Clearly **P** is a vector space over **Q**. Moreover, it is not the same as $\mathbf{V}_n(\mathbf{Q})$ for any n, since polynomials are not n-tuples of numbers. It is clear that the derivative of a polynomial of degree at most 3 with rational coefficients is also a polynomial of degree at most 3 with rational coefficients. Furthermore, "the derivative of a sum is the sum of the derivatives," and "constants slip out from under the differentiation sign." Hence the mapping, d, which sends each polynomial in **P** to its derivative is a linear transformation of **P** into itself. This example is referred to as the **P**, d example in the rest of this section.

Although **P** is not the same as $\mathbf{V}_4(\mathbf{Q})$, it is clear that there is a natural 1–1 correspondence between **P** and $\mathbf{V}_4(\mathbf{Q})$, namely,

$$\alpha_3 x^3 + \alpha_2 x^2 + \alpha_1 x + \alpha_0 \leftrightarrow \langle \alpha_3, \alpha_2, \alpha_1, \alpha_0 \rangle .$$

This correspondence is compatible with addition and multiplication by a scalar in either **P** or $\mathbf{V}_4(\mathbf{Q})$. The associated function from **P** onto $\mathbf{V}_4(\mathbf{Q})$ is a specific example of a coordinate map.

Definition. Let $v_1, v_2, \ldots, v_n$ be an ordered basis of the vector space **V** over **F**. Each vector $v \in \mathbf{V}$ has a unique representation in the form $v = \alpha_1 v_1 + \alpha_2 v_2 + \cdots + \alpha_n v_n$. Define $v^\phi = \langle \alpha_1, \ldots, \alpha_n \rangle$. ϕ is then a function mapping **V** into $\mathbf{V}_n(\mathbf{F})$ and is called the *coordinate map of* **V** with respect to the basis $v_1, v_2, \ldots, v_n$. The n numbers, $\alpha_1, \alpha_2, \ldots, \alpha_n$, are called the *$\phi$-coordinates* of v. We shall use the exponential notation for coordinate maps in order to avoid extra parentheses.

Theorem 4.14. If **V** is a vector space of dimension n over the field **F**, and ϕ is a coordinate map of **V**, then:

(1) ϕ is a 1–1 function from **V** onto $\mathbf{V}_n(\mathbf{F})$.
(2) ϕ is a linear transformation.

Proof. Assume ϕ is the coordinate map relative to the basis $v_1, v_2, \ldots, v_n$. By definition of a basis, each $v \in \mathbf{V}$ has a *unique* representation in the form $\Sigma\alpha_i v_i$; thus, ϕ is a 1–1 function. Since any linear combination of the v_i is in $\mathbf{V}$, ϕ maps $\mathbf{V}$ onto $\mathbf{V}_n(\mathbf{F})$. Clearly,

$$\Sigma\alpha_i v_i + \Sigma\beta_i v_i = \Sigma(\alpha_i + \beta_i)v_i$$

and $\lambda\Sigma\alpha_i v_i = \Sigma\lambda\alpha_i v_i$. Thus, ϕ is a linear transformation.

Example 1 (continued). In the $\mathbf{P}$, d example, let ϕ be the coordinate map of $\mathbf{P}$ relative to the natural basis x^3, x^2, x, and 1. Another person might be interested in evaluating polynomials near $x = 2$. For him the basis $(x-2)^3$, $(x-2)^2$, $(x-2)$, and 1 and its associated coordinate map, θ, would be much more convenient. Clearly ϕ and θ are different coordinate maps, e.g., $(x-2)^\phi = \langle 0, 0, 1, -2\rangle$, but $(x-2)^\theta = \langle 0, 0, 1, 0\rangle$. A third coordinate map which would be a natural choice for a person who views polynomials as "short power series" would be the coordinate map ρ relative to the ordered basis $1, x, x^2, x^3$. $(x-2)^\rho = \langle -2, 1, 0, 0\rangle$. Consider a typical vector in $\mathbf{V}_4(\mathbf{Q})$, say, $\langle 0, 1, -1, 0\rangle$. The polynomial it represents depends on the coordinate system under discussion, e.g., $\langle 0, 1, -1, 0\rangle$ are the

(1) ϕ coordinates of $\langle 0, 1, -1, 0\rangle\phi^{-1} = x^2 - x$,
(2) θ coordinates of $\langle 0, 1, -1, 0\rangle\theta^{-1} = (x-2)^2 - (x-2)$,
(3) ρ coordinates of $\langle 0, 1, -1, 0\rangle\rho^{-1} = x - x^2$.

Note that we are shifting back and forth between the $(v)\phi$ and $(v)^\phi$ notations. The exponential notation for coordinate maps emphasizes the fact that they are special linear transformations, and also serves as a reminder that we are "passing to coordinates."

When dealing with vectors that are n-tuples, there is a natural way to represent linear transformations in terms of matrices so that the process of computing images of vectors is neatly mirrored by the standard method of multiplying matrices.

Definition. Let f be a linear transformation from $\mathbf{V}_n(\mathbf{F})$ into $\mathbf{V}_m(\mathbf{F})$. We call the matrix A with n rows and m columns the *natural matrix* of f if the ith row of A is the f-image of the ith natural basis vector of $\mathbf{V}_n(\mathbf{F})$.

Example 2. Let $f: \mathbf{V}_4(\mathbf{Q}) \to \mathbf{V}_4(\mathbf{Q})$ be the linear transformation such that $\langle w, x, y, z\rangle f = \langle 0, 3w, 2x, y\rangle$. Then the natural matrix of f is

$$\begin{pmatrix} 0 & 3 & 0 & 0 \\ 0 & 0 & 2 & 0 \\ 0 & 0 & 0 & 1 \\ 0 & 0 & 0 & 0 \end{pmatrix}$$

since $\langle 1, 0, 0, 0\rangle f = \langle 0, 3, 0, 0\rangle$, $\langle 0, 1, 0, 0\rangle f = \langle 0, 0, 2, 0\rangle, \ldots$. Note that

$$\langle w, x, y, z \rangle \begin{pmatrix} 0 & 3 & 0 & 0 \\ 0 & 0 & 2 & 0 \\ 0 & 0 & 0 & 1 \\ 0 & 0 & 0 & 0 \end{pmatrix} = \langle 0, 3w, 2x, y \rangle .$$

Theorem 4.15. Let f be a linear transformation from $\mathbf{V}_n(\mathbf{F})$ into $\mathbf{V}_m(\mathbf{F})$. Then if A is the natural matrix of f, the f-image of any n-tuple $\langle \alpha_1, \alpha_2, \ldots, \alpha_n \rangle$ is the ordinary matrix product $\langle \alpha_1, \alpha_2, \ldots, \alpha_n \rangle A$.

Proof. Let $e_i = \langle 0, 0, \ldots, 1, 0, \ldots, 0 \rangle$ be the n-tuple with 1 in the ith position and zeros elsewhere. Then the ith row of A is $(e_i)f$. However, $\langle \alpha_1, \alpha_2, \ldots, \alpha_n \rangle A$ is (α_1 times the first row of A) + (α_2 times the second) $+ \cdots + \alpha_n$(the last row) $= \alpha_1(e_1)f + \alpha_2(e_2)f + \cdots + \alpha_n(e_n)f = (\alpha_1 e_1 + \alpha_2 e_2 + \cdots + \alpha_n e_n)f = \langle \alpha_1, \alpha_2, \ldots, \alpha_n \rangle f$.

We now have a natural computational device for linear transformations involving only n-tuples. Using coordinates, we can exploit this to aid us in computing images for linear transformations of arbitrary finite-dimensional vector spaces. We restrict ourselves to cases in which $\mathbf{V} = \mathbf{W}$.

Definition. Let $f{:}\mathbf{V} \rightarrow \mathbf{V}$ be a linear transformation, and let ϕ be a coordinate mapping of $\mathbf{V}$ onto $\mathbf{V}_n(\mathbf{F})$. The *ϕ-matrix* of f, $f^{\phi'}$, is the natural matrix of $\phi^{-1}f\phi$.

Example 1 (*continued*). Consider the linear transformation d in the $\mathbf{P}$, d example. The ϕ- and θ-matrices of d happen to be the same, but the ρ-matrix is different. For example, the first rows of the ϕ-, θ-, and ρ-matrices are:

$$\langle 1, 0, 0, 0 \rangle \phi^{-1} d\phi = (x^3)d\phi = (3x^2)^\phi = \langle 0, 3, 0, 0 \rangle$$
$$\langle 1, 0, 0, 0 \rangle \theta^{-1} d\theta = ((x-2)^3)d\theta = (3(x-2)^2)^\theta = \langle 0, 3, 0, 0 \rangle$$
$$\langle 1, 0, 0, 0 \rangle \rho^{-1} d\rho = (1)d\rho = (0)^\rho = \langle 0, 0, 0, 0 \rangle .$$

We leave to the reader the task of computing the other three rows of each matrix. Both $d^{\phi'}$ and $d^{\theta'}$ are just the matrix in example 2.

We assume that the reader is familiar with the standard algebra of matrices. The set of all n by n matrices with entries from the field $\mathbf{F}$ is called $M_n(\mathbf{F})$.

Theorem 4.16. Let $\mathbf{V}$ be a vector space of dimension n over $\mathbf{F}$, and let ϕ be a coordinate map of $\mathbf{V}$ onto $\mathbf{V}_n(\mathbf{F})$. The mapping $\phi'{:}L(\mathbf{V}, \mathbf{V}) \rightarrow M_n(\mathbf{F})$ defined by $f^{\phi'} =$ the ϕ-matrix of f is an isomorphism of $L(\mathbf{V}, \mathbf{V})$ onto $M_n(\mathbf{F})$, i.e.,

(1) ϕ' is 1–1 and onto.
(2) $(fg)^{\phi'} = f^{\phi'}g^{\phi'}$.
(3) $(f + g)^{\phi'} = f^{\phi'} + g^{\phi'}$.
(4) $(\lambda f)^{\phi'} = \lambda(f^{\phi'})$.

In particular, $0^{\phi'} = 0$, the matrix whose entries are all zero. $(\mathbf{1})^{\phi'} = I$, the identity matrix. If $f^{\phi'} = A$, then f^{-1} exists if and only if A^{-1} exists; moreover, (if they exist) $(f^{-1})^{\phi'} = A^{-1}$.

We omit the proof which is tedious and consists primarily of the careful reading and interpretation of definitions. Instead we illustrate some typical applications, using again the **P**, d example.

Example 1 (conclusion). With respect to the **P**, d example, consider the linear differential operator $d^2 + 2d$. Since the ϕ-matrix of d is

$$A = \begin{pmatrix} 0 & 3 & 0 & 0 \\ 0 & 0 & 2 & 0 \\ 0 & 0 & 0 & 1 \\ 0 & 0 & 0 & 0 \end{pmatrix},$$

the ϕ-matrix of $d^2 + 2d$ is

$$A^2 + 2A = \begin{pmatrix} 0 & 6 & 6 & 0 \\ 0 & 0 & 4 & 2 \\ 0 & 0 & 0 & 2 \\ 0 & 0 & 0 & 0 \end{pmatrix}.$$

Clearly neither A^{-1} nor $(A^2 + 2A)^{-1}$ exists; hence neither d^{-1} nor $(d^2 + 2d)^{-1}$ exists in $L(\mathbf{P}, \mathbf{P})$, i.e., one cannot find *in* **P** an anti-derivative (integral) for each polynomial in **P**, nor can one solve *in* **P** the differential equation $y'' + y' = q$ for each polynomial q in **P**. Clearly the fourth derivative of any polynomial in **P** is 0; hence $d^4 = 0$. We can exploit this fact to compute B^{-1} if $B = A + I$:

$$0 = d^4 = [(\mathbf{1} + d) - \mathbf{1}]^4 = (\mathbf{1} + d)^4 - 4(\mathbf{1} + d)^3 + 6(\mathbf{1} + d)^2 - 4(\mathbf{1} + d) + \mathbf{1}\,.$$

Hence by theorem 4.16,

$$B^4 - 4B^3 + 6B^2 - 4B + I = 0\,,$$

and, therefore, $B^{-1} = 4I - 6B + 4B^2 - B^3$.

Exercise 4.4

1. Let g be a linear transformation of $\mathbf{V}_4(\mathbf{Q})$ into itself, and let v be a vector in $\mathbf{V}_4(\mathbf{Q})$ such that v, $(v)g$, $(v)g^2$, $(v)g^3$ are linearly independent. Let θ be the coordinate map relative to this basis of $\mathbf{V}_4(\mathbf{R})$. Find the θ-matrix of g if $g^4 + 3g^3 - 2g + \mathbf{1} = 0$.

2. Let f be the linear transformation sending each polynomial $p(x)$ in **P(P** as in example 1) to the polynomial $p(x - 2)$. Find the ϕ, θ, and ρ matrices of f if ϕ, θ, and ρ are the coordinate maps of example 1.

3. In view of theorems 4.16 and 4.10, it is clear that $M_n(\mathbf{F})$ is a vector space over $\mathbf{F}$. What is its dimension? Let $\mathbf{W}$ be the subspace of $M_n(\mathbf{F})$ consisting of all symmetric matrices. What is dim $(\mathbf{W})$? Find a basis if $n = 3$ and $\mathbf{F} = \mathbf{Q}$.

4. The matrix B in $M_4(\mathbf{Q})$ with *all* entries 1 satisfies a quadratic equation, i.e., there are rational numbers a and b such that $B^2 = aB + bI$. Find this equation, and use it to find A^{-1} if

$$A = B + I = \begin{pmatrix} 2 & 1 & 1 & 1 \\ 1 & 2 & 1 & 1 \\ 1 & 1 & 2 & 1 \\ 1 & 1 & 1 & 2 \end{pmatrix}.$$

5. Use proposition 4.9 to show that any matrix, say with m rows and n columns and all entries in the field $\mathbf{F}$, is the matrix of some linear transformation, here from $\mathbf{V}_m(\mathbf{F})$ into $\mathbf{V}_n(\mathbf{F})$.

4.5. Similar matrices and commutative diagrams

In this section we develop the machinery for relating the matrices of a linear transformation with respect to different coordinate systems. The schematic device introduced is used in many other areas of mathematics, so we define it in some generality.

Assume that we have several sets and several functions whose domains and codomains are among the given sets. A diagram involving these sets as vertices and arrows from the domain to the codomain of each function is said to be a *commutative diagram* if it is immaterial which of the available paths we follow along the arrows in going from any one set to another.

Example 1. Given sets A, B, and C, the assertion that the diagram

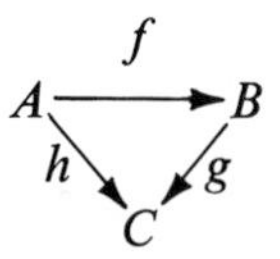

$$\begin{array}{ccc} A & \xrightarrow{f} & B \\ & {\scriptstyle h}\searrow \quad \swarrow{\scriptstyle g} & \\ & C & \end{array}$$

is commutative is the assertion that f, g, and h are functions with the indicated domains and codomains, and that for all $a \in A$, $(a)fg = (a)h$. Similarly,

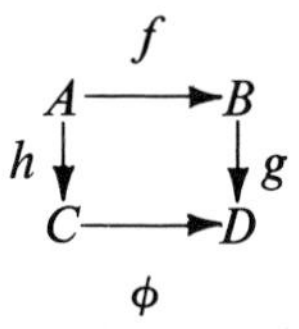

$$\begin{array}{ccc} A & \xrightarrow{f} & B \\ {\scriptstyle h}\downarrow & & \downarrow{\scriptstyle g} \\ C & \xrightarrow[\phi]{} & D \end{array}$$

is commutative if and only if for all $a \in A$, $(a)fg = (a)h\phi$.

If one of the functions involved in a commutative diagram is 1–1 and onto, then the diagram obtained by reversing this arrow is also commutative, e.g., if f is 1–1 and onto in both of the diagrams above, then the diagrams

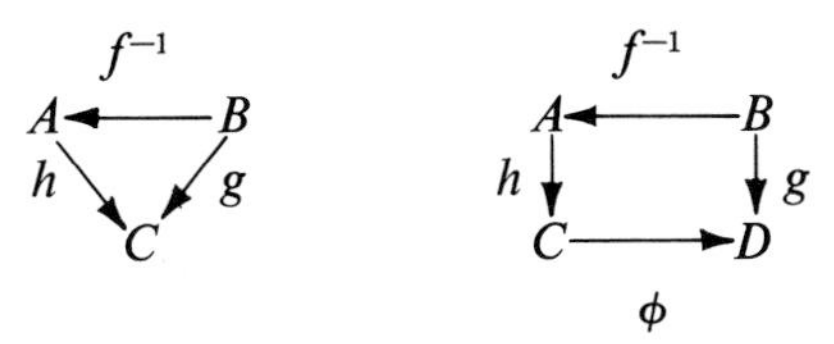

are also commutative, i.e., for all $b \in B$, $(b)f^{-1}h = (b)g$ in the first situation, and $(b)f^{-1}h\phi = (b)g$ in the second.

Now assume that we have a vector space **V** over a field **F** and a coordinate map, ϕ, of **V** onto $\mathbf{V}_n(\mathbf{F})$. For any linear transformation $f:\mathbf{V} \to \mathbf{V}$, the diagram

$$\begin{array}{ccc} V & \xrightarrow{\ f\ } & V \\ \phi\downarrow & & \downarrow\phi \\ V_n(F) & \xrightarrow[\phi^{-1}f\phi]{} & V_n(F) \end{array}$$

is commutative. On the lower or computational level, we are not really concerned with $\phi^{-1}f\phi$, but rather with its natural matrix. *Therefore, on the lower level in all diagrams we identify linear transformations of* $\mathbf{V}_n(\mathbf{F})$ *and their natural matrices.* Thus, either the name of a matrix or the name of a linear transformation of $\mathbf{V}_n(\mathbf{F})$ may appear on the lower level.

Assume that θ is another coordinate map for the vector space **V**. Then there are two natural ways to complete

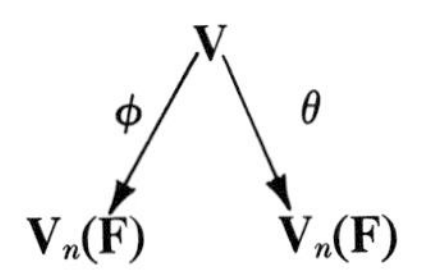

to a commutative diagram. These are:

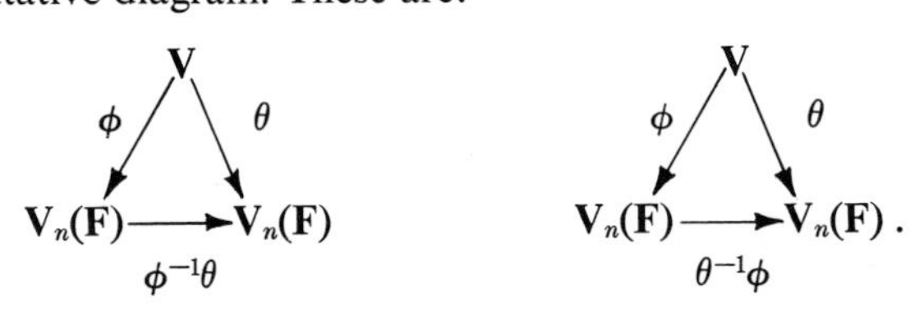

Since $\phi^{-1}\theta$ is 1–1 and onto, and since $(\phi^{-1}\theta)^{-1} = \theta^{-1}\phi$, the commutativity of the second diagram is equivalent to the commutativity of the first. The natural matrix for $\phi^{-1}\theta$ is the matrix for changing from ϕ-coordinates to θ-coordinates, as might be suspected if one views ϕ^{-1} as "recovering the vectors from their ϕ-coordinates" and θ as "imposing the θ-coordinates."

Example 2. Returning to the **P**, d example of section 4.4, consider the two coordinate maps: ϕ relative to the basis x^3, x^2, x, and 1; and θ relative to $(x-2)^3$, $(x-2)^2$, $(x-2)$, and 1. The natural matrix for $\phi^{-1}\theta$ is determined by

$$\langle 1, 0, 0, 0\rangle\phi^{-1}\theta = (x^3)^\theta = \cdots \text{(supply computations!)} \cdots = \langle 1, 6, 12, 8\rangle,$$
$$\langle 0, 1, 0, 0\rangle\phi^{-1}\theta = (x^2)^\theta = ((x-2)^2 + 4(x-2) + 4)^\theta = \langle 0, 1, 4, 4\rangle,$$
$$\langle 0, 0, 1, 0\rangle\phi^{-1}\theta = \langle 0, 0, 1, 2\rangle,$$

and

$$\langle 0, 0, 0, 1\rangle\phi^{-1}\theta = \langle 0, 0, 0, 1\rangle .$$

Thus, the matrix is

$$\begin{pmatrix} 1 & 6 & 12 & 8 \\ 0 & 1 & 4 & 4 \\ 0 & 0 & 1 & 2 \\ 0 & 0 & 0 & 1 \end{pmatrix}.$$

A similar computation shows that the natural matrix of $\theta^{-1}\phi$ is

$$\begin{pmatrix} 1 & -6 & 12 & -8 \\ 0 & 1 & -4 & 4 \\ 0 & 0 & 1 & -2 \\ 0 & 0 & 0 & 1 \end{pmatrix}.$$

The ϕ-coordinates of $9x^3 + 3x^2 - x + 5$ are $\langle 9, 3, -1, 5\rangle$ and the θ-coordinates are

$$\langle 9, 3, -1, 5\rangle\phi^{-1}\theta = \langle 9, 3, -1, 5\rangle \begin{pmatrix} 1 & 6 & 12 & 8 \\ 0 & 1 & 4 & 4 \\ 0 & 0 & 1 & 2 \\ 0 & 0 & 0 & 1 \end{pmatrix} = \langle 9, 57, 119, 87\rangle ,$$

which the reader may verify by expanding $9(x-2)^3 + 57(x-2)^2 + 119(x-2) + 87$. Clearly $(\phi^{-1}\theta)(\theta^{-1}\phi) = \mathbf{1}$. Thus, the product of the two matrices above should be I. Verify!

Theorem 4.17. Let **V** be a vector space over the field **F**, and let ϕ be a coordinate map of **V** onto $\mathbf{V}_n(\mathbf{F})$. The correspondence matching θ and the natural matrix of $\phi^{-1}\theta$ is a 1–1 correspondence between the set of coordinate maps of **V** and the n-square, invertible matrices B with entries in **F**. Moreover, if B is the natural matrix of $\phi^{-1}\theta$, then both of the following diagrams are commutative.

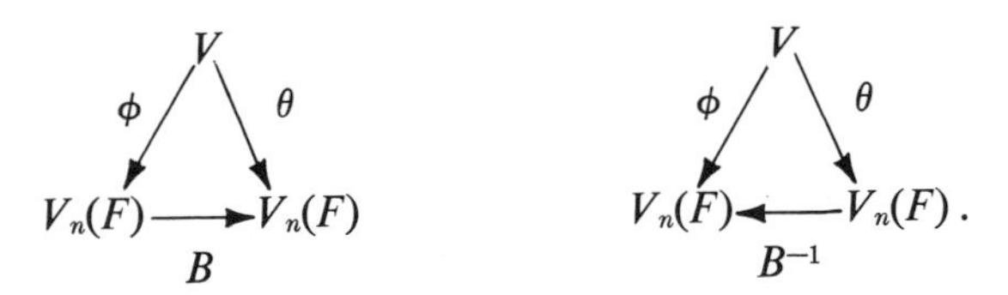

Proof. Let θ be a coordinate map of **V** onto $\mathbf{V}_n(\mathbf{F})$; then $\phi^{-1}\theta$ and $\theta^{-1}\phi$ are linear transformations of $\mathbf{V}_n(\mathbf{F})$ onto itself. These linear transformations are inverses of each other. Hence their natural matrices are inverses of each other, and thus the natural matrix of $\phi^{-1}\theta$ is invertible. Conversely if B is an invertible matrix, then let it be the natural matrix of the linear transformation $b:\mathbf{V}_n(\mathbf{F}) \to \mathbf{V}_n(\mathbf{F})$. Since B is invertible, b is 1–1 and onto. Let $\theta = \phi b$, and then θ is a 1–1 linear transformation of **V** onto $\mathbf{V}_n(\mathbf{F})$. We leave it to the reader to show that any such linear transformation is a coordinate map of **V**. This completes the proof that the correspondence between coordinate maps θ and invertible matrices B is "onto." We show that it is one-to-one in two steps. First, $\phi^{-1}\theta_1 = \phi^{-1}\theta_2$ if and only if $\theta_1 = \phi\phi^{-1}\theta_1 = \phi\phi^{-1}\theta_2 = \theta_2$, and, second, the correspondence between a linear transformation [of $\mathbf{V}_n(\mathbf{F})$ into itself] and its natural matrix is 1–1. If we identify $\phi^{-1}\theta$ and its natural matrix B, then it is trivial to prove that the diagrams commute.

With these concepts available, it becomes a matter of "pasting diagrams" to keep track of the matrices of a linear transformation with respect to different coordinate systems. The method is illustrated in the following example.

Example 3. Returning to the **P**, d example again, recall that θ is the coordinate map of **P** onto $\mathbf{V}_4(\mathbf{Q})$ with respect to the basis $(x-2)^3$, $(x-2)^2$, $(x-2)$, and 1, and that ρ is the coordinate map with respect to the basis 1, x, x^2, x^3. The reader (hopefully) found that the ρ-matrix of d is

$$B = \begin{pmatrix} 0 & 0 & 0 & 0 \\ 1 & 0 & 0 & 0 \\ 0 & 2 & 0 & 0 \\ 0 & 0 & 3 & 0 \end{pmatrix}.$$

Thus, the diagram

$$\begin{array}{ccc} \mathbf{P} & \xrightarrow{\ d\ } & \mathbf{P} \\ \rho\downarrow & & \downarrow\rho \\ \mathbf{V}_4(\mathbf{Q}) & \xrightarrow[B]{} & \mathbf{V}_4(\mathbf{Q}) \end{array}$$

commutes. To convert this matrix to the θ-matrix of d we need to find a similar diagram with θ's down the sides. This can be done by "pasting" two triangles on the sides, i.e.,

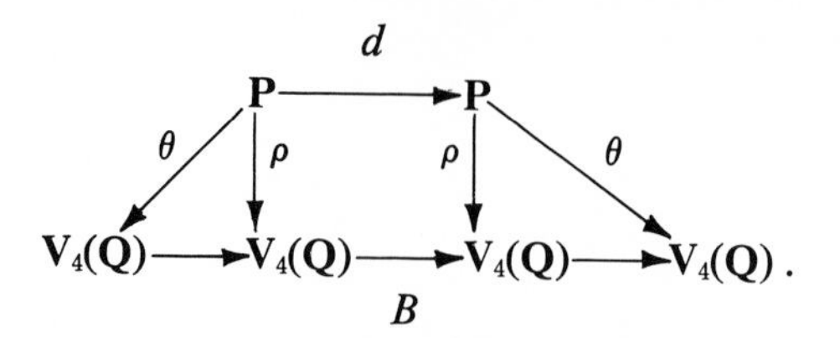

The only question then is what matrices to use to fill in the two blanks on the bottom of the diagram. The obvious solution is to choose those matrices which make the large diagram commutative. The complete diagram is as follows.

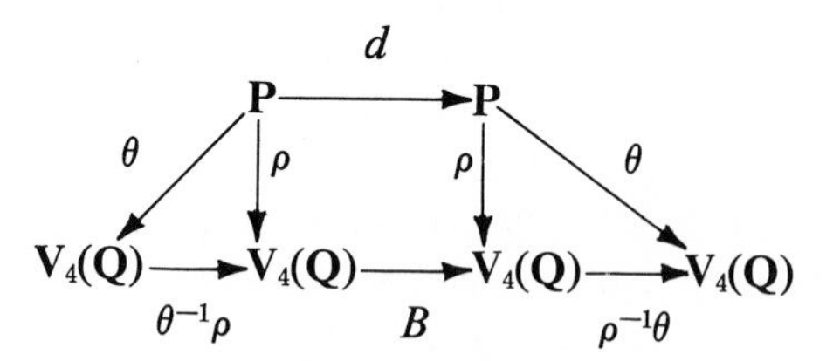

Following the direct path from $\mathbf{V}_4(\mathbf{Q})$ to $\mathbf{V}_4(\mathbf{Q})$ across the bottom, we find that the θ-matrix for d is $(\theta^{-1}\rho)B(\rho^{-1}\theta)$. A simple computation shows that

$$\theta^{-1}\rho = \begin{pmatrix} -8 & 12 & -6 & 1 \\ 4 & -4 & 1 & 0 \\ -2 & 1 & 0 & 0 \\ 1 & 0 & 0 & 0 \end{pmatrix}$$

and that

$$\rho^{-1}\theta = \begin{pmatrix} 0 & 0 & 0 & 1 \\ 0 & 0 & 1 & 2 \\ 0 & 1 & 4 & 4 \\ 1 & 6 & 12 & 8 \end{pmatrix}$$

The reader should check that

$$\begin{pmatrix} -8 & 12 & -6 & 1 \\ 4 & -4 & 1 & 0 \\ -2 & 1 & 0 & 0 \\ 1 & 0 & 0 & 0 \end{pmatrix} \begin{pmatrix} 0 & 0 & 0 & 0 \\ 1 & 0 & 0 & 0 \\ 0 & 2 & 0 & 0 \\ 0 & 0 & 3 & 0 \end{pmatrix} \begin{pmatrix} 0 & 0 & 0 & 1 \\ 0 & 0 & 1 & 2 \\ 0 & 1 & 4 & 4 \\ 1 & 6 & 12 & 8 \end{pmatrix} = \begin{pmatrix} 0 & 3 & 0 & 0 \\ 0 & 0 & 2 & 0 \\ 0 & 0 & 0 & 1 \\ 0 & 0 & 0 & 0 \end{pmatrix}$$

equals the θ-matrix for d as found in section 4.4.

Theorem 4.18. Let $\mathbf{V}$ be a vector space over $\mathbf{F}$, ϕ a coordinate map of $\mathbf{V}$ onto $\mathbf{V}_n(\mathbf{F})$, f a linear transformation of $\mathbf{V}$ into $\mathbf{V}$, and A the ϕ-matrix of f. For each invertible matrix B, there is a coordinate map θ such that the θ-

matrix of f is $B^{-1}AB$. Conversely, if ρ is any coordinate map of $\mathbf{V}$ onto $\mathbf{V}_n(\mathbf{F})$, then the ρ-matrix of f is of the form $C^{-1}AC$ for some invertible matrix C.

Proof. Given an invertible matrix B, then by theorem 4.17 there is a coordinate map θ such that $B = \phi^{-1}\theta$, i.e., such that B is the matrix for changing ϕ-coordinates to θ-coordinates. Then

$$B^{-1}AB = (\theta^{-1}\phi)(\phi^{-1}f\phi)(\phi^{-1}\theta) = \theta^{-1}f\theta \,,$$

and hence $B^{-1}AB$ is the θ-matrix of f.

Again by theorem 4.17, if ρ is a coordinate map, then there is an invertible matrix C such that $\phi^{-1}\rho = C$. Thus,

$$f^{\rho'} = (\text{the } \rho\text{-matrix of } f) = \rho^{-1}f\rho = \rho^{-1}\phi\phi^{-1}f\phi\phi^{-1}\rho = C^{-1}AC \,,$$

which completes the proof.

Definition. Let A and B be n-square matrices with entries in the field $\mathbf{F}$. A and B are *similar* if and only if there is an invertible n-square matrix, C, with entries in $\mathbf{F}$ such that $C^{-1}AC = B$.

With this definition we can paraphrase theorem 4.18 as follows: Although distinct matrices never represent the same linear transformation in one coordinate system they may in distinct coordinate systems, and in fact they will if and only if they are similar. This makes it obvious that similarity is an equivalence relation among the n-square matrices. The equivalence class of αI for any $\alpha \in \mathbf{F}$ contains only the matrix αI (exercise 4.5, problem 1). In section 4.7 we see that these are the only one-element equivalence classes for similarity.

Corollary 4.19. If B is an invertible n-square matrix with entries in the field $\mathbf{F}$, i.e., if B is invertible and $B \in M_n(\mathbf{F})$, then for any matrices A and C in $M_n(\mathbf{F})$ and any $\alpha \in \mathbf{F}$:

(1) $B^{-1}(AC)B = (B^{-1}AB)(B^{-1}CB)$
(2) $B^{-1}(A + C)B = B^{-1}AB + B^{-1}CB$
(3) $B^{-1}(\alpha A)B = \alpha(B^{-1}AB)$.

Proof. These three properties are easy to derive from the familiar associative and distributive laws for matrix algebra, so we omit a formal proof. We call attention to the geometric interpretation however. Let f and g be linear transformations of $\mathbf{V}$ into $\mathbf{V}$ whose ϕ-matrices are A and C, respectively. Then the ϕ-matrices of fg, $f + g$, and αf are AC, $A + C$, and αC, respectively. If B is the matrix for changing ϕ-coordinates to θ-coordinates then $B^{-1}AB$ and $B^{-1}CB$ are the θ-matrices of f and g. Thus the three equations above simply assert that the θ-matrices for fg, $f + g$, and αf can be computed either by performing the appropriate algebraic operation on the ϕ-matrices and then changing coordinates or by first changing coordinates and then performing the algebraic operations on the θ-matrices. In other words, changing coordinates mixes nicely with the basic algebraic operations on linear transformations or on matrices.

Exercise 4.5

1. Prove that if $B \in M_n(\mathbf{F})$ is similar to $\alpha I \in M_n(\mathbf{F})$, $\alpha \in \mathbf{F}$, then $B = \alpha I$.

2. Let $f \in L(\mathbf{V}, \mathbf{V})$, and assume for every vector, v, there is a corresponding scalar, α, such that $(v)f = \alpha v$. Show that $f = \lambda\mathbf{1}$ for some $\lambda \in \mathbf{F}$.

3. Assume A and B are both in $M_n(\mathbf{F})$.

(1) Show that if A is nonsingular, then AB is similar to BA.

(2) Using 2 by 2 matrices, find two pairs of matrices A and B demonstrating that if A is singular, then AB may or may not be similar to BA.

4. Let

$$A = \begin{pmatrix} 3 & 0 & 0 \\ 0 & 0 & 0 \\ 0 & 0 & 0 \end{pmatrix}, \quad B = \begin{pmatrix} 1 & 1 & 1 \\ 1 & 1 & 1 \\ 1 & 1 & 1 \end{pmatrix}, \quad \text{and } C = \begin{pmatrix} 1 & 1 & 1 \\ 0 & 0 & 0 \\ 0 & 0 & 0 \end{pmatrix}.$$

(1) Show that $AC = CB$.

(2) Are the equations $AC = CB$ and $C^{-1}AC = B$ equivalent?

(3) Find a matrix $D \in M_3(\mathbf{Q})$ such that $D^{-1}AD = B$.

5. Generalize the preceding problem, and show that the n-square matrix, all of whose entries are 1, is similar to an n-square matrix with only one nonzero entry. (Assume that the field of scalars is the rationals.)

6. (continuation of problem 3, exercise 4.3). Assume that $q \in L(\mathbf{V}, \mathbf{V})$, $q^2 = 0$, and that $\mathbf{V}$ is a vector space of dimension n over a field of characteristic 2. Show that if the rank of $q = k$, then there is a coordinate map ϕ such that the ϕ-matrix of q has the form

$$\left(\begin{array}{c|c} 0_1 & 0_2 \\ \hline I & 0_3 \end{array}\right)$$

where 0_1, 0_2, and 0_3 have all entries zero, and I is a k by k identity matrix.

7. Let B be a matrix in $M_2(\mathbf{R})$ such that $B^2 + I = 0$ (for example, $B = \begin{pmatrix} -1 & -2 \\ 1 & 1 \end{pmatrix}$ satisfies this equation). Show that B is similar to the matrix $\begin{pmatrix} 0 & +1 \\ -1 & 0 \end{pmatrix}$, but do this geometrically, i.e., not by solving (brute force) for A such that $A^{-1}BA = \begin{pmatrix} 0 & 1 \\ -1 & 0 \end{pmatrix}$, but rather by looking for an appropriate basis. *Hint.* Your first basis vector can be *any* $v \neq 0$.

4.6. *Applications of matrix similarity*

In this section we present several results which demonstrate the advantages of considering matrices other than the natural matrix for a linear transformation. In other words, we show that it is often advantageous to change bases or coordinates. First we recall two elementary concepts concerning matrices.

The *trace* of a matrix A is the sum of the entries on the main diagonal of A. It is a simple exercise in "chasing subscripts" to prove that trace $(AB) =$ trace (BA). From this it follows immediately that trace $(B^{-1}AB) =$ trace $(ABB^{-1}) =$ trace (A). Hence similar matrices have the same trace.

We assume the reader is familiar with the *determinant* of a matrix and with the theorem that $\det(AB) = (\det A)(\det B)$. The concept of a determinant and its multiplicative property is valid for any field $\mathbf{F}$ even if the characteristic of $\mathbf{F}$ is 2 ([3], pp. 279–287). From this multiplicative property, it is clear that an invertible matrix has a nonzero determinant. We omit the proof of the converse, i.e., we assume that a matrix is invertible if and only if its determinant is not zero. The *characteristic polynomial* of a matrix A is the polynomial

$$\det(xI - A) = x^n - (\text{trace } A)x^{n-1} + \cdots + (-1)^n \det(A).$$

The characteristic polynomials of similar matrices are the same. The roots of this polynomial are called characteristic roots of A, but let us define them in a way that is more geometric, and then show that this algebraic definition is equivalent.

Definition. Let $\mathbf{V}$ be a vector space over the field $\mathbf{F}$, and let f be a linear transformation of $\mathbf{V}$ into itself. A number $\lambda \in \mathbf{F}$ is called a *characteristic value* of f if there is a nonzero vector $v \in \mathbf{V}$ such that $(v)f = \lambda v$. If λ is a characteristic value of f, then we call $\{v | (v)f = \lambda v\}$ the *characteristic subspace belonging to* λ. Any *nonzero* vector, v, such that $(v)f = \lambda v$ is called a *characteristic vector* belonging to λ.

Since $\{v \in \mathbf{V} | (v)f = \lambda v\} = \{v \in \mathbf{V} | (v)(f - \lambda\mathbf{1}) = 0\}$ is the kernel of $f - \lambda\mathbf{1}$, the characteristic subspace belonging to λ is in fact a subspace of $\mathbf{V}$. This also shows that λ is a characteristic value if and only if the kernel of $f - \lambda\mathbf{1}$ is not $\{0\}$. We pursue this further in the following theorem but first remark that in the literature characteristic values have a number of aliases, among which are: latent values, eigenvalues, proper roots, and spectral values.

Theorem 4.20. Let A be the ϕ-matrix of the linear transformation f. A number $\lambda \in \mathbf{F}$ is a characteristic value of f if and only if λ is a root of the characteristic polynomial of A.

Proof. If λ is a characteristic value, then $\lambda\mathbf{1} - f$ maps at least one nonzero vector to the zero vector, and hence $\lambda\mathbf{1} - f$ is not 1–1. Therefore, it cannot have an inverse. Hence its ϕ-matrix, $\lambda I - A$, is not invertible. But this implies $\det(\lambda I - A) = 0$, i.e., that λ is a root of the characteristic polynomial. Now assume $\det(\lambda I - A) = 0$. Then $(\lambda I - A)$ is not invertible; hence the kernel of $(\lambda\mathbf{1} - f)$ is not $\{0\}$, i.e., λ is a characteristic value of f.

Example 1. Let $\mathbf{V} = \mathbf{V}_3(\mathbf{R})$, and let p be the projection of $\mathbf{V}_3(\mathbf{R})$ onto the plane $x + y + z = 0$ with kernel the line $x = y = z$. The natural matrix for p is

$$\begin{pmatrix} \frac{2}{3} & -\frac{1}{3} & -\frac{1}{3} \\ -\frac{1}{3} & \frac{2}{3} & -\frac{1}{3} \\ -\frac{1}{3} & -\frac{1}{3} & \frac{2}{3} \end{pmatrix} \quad \begin{aligned} \text{since } \langle 1, 0, 0\rangle &= \frac{1}{3}\langle 2, -1, -1\rangle + \frac{1}{3}\langle 1, 1, 1\rangle, \\ \langle 0, 1, 0\rangle &= \frac{1}{3}\langle -1, 2, -1\rangle + \frac{1}{3}\langle 1, 1, 1\rangle, \\ \text{and } \langle 0, 0, 1\rangle &= \frac{1}{3}\langle -1, -1, 2\rangle + \frac{1}{3}\langle 1, 1, 1\rangle. \end{aligned}$$

However, there is a basis for $\mathbf{V}_3(\mathbf{R})$ consisting entirely of characteristic vectors of p, and the matrix for p with respect to the associated coordinate map is much easier to use. Every vector in the plane $x + y + z = 0$ is sent to itself by p; moreover, these are the only fixed vectors for p (review theorem 4.11). Thus, 1 is a characteristic value, and this plane is its characteristic subspace. Similarly, the number 0 is a characteristic value, and the line $x = y = z$ is its characteristic subspace. Now choose a basis for $\mathbf{V}_3(\mathbf{R})$ such that the first two basis vectors are in the plane $x + y + z = 0$, and the third lies on the line $x = y = z$, e.g., $\langle 0, 1, -1\rangle$, $\langle 1, 0, -1\rangle$, and $\langle 1, 1, 1\rangle$. Let θ be the coordinate map of $\mathbf{V}_3(\mathbf{R})$ onto $\mathbf{V}_3(\mathbf{R})$ with respect to this map, then the θ-matrix of p is

$$\begin{pmatrix} 1 & 0 & 0 \\ 0 & 1 & 0 \\ 0 & 0 & 0 \end{pmatrix}.$$

Note that $q = \mathbf{1} - p$ is also a projection, namely, the projection of $\mathbf{V}_3(\mathbf{R})$ onto the line $x = y = z$ along the plane $x + y + z = 0$. The involution $t = 2q - \mathbf{1}$, associated with q in theorem 4.13, has θ-matrix

$$\begin{pmatrix} -1 & 0 & 0 \\ 0 & -1 & 0 \\ 0 & 0 & 1 \end{pmatrix}.$$

and hence it maps every vector in the plane $x + y + z = 0$ to its negative and leaves every vector on the line $x = y = z$ fixed. Since the line and the plane are perpendicular, this means that t is the half-turn with axis "$x = y = z$." The involution $2p - \mathbf{1} = -t$ is the reflection with mirror "$x + y + z = 0$."

The following theorem shows that the example above is typical for projections, regardless of the dimension of **V**, and regardless of the field **F**. Because of the special nature of involutions when the characteristic of **F** is 2 (problem 3, exercise 4.3) we avoid "characteristic 2" for involutions.

Theorem 4.21. Let p be a projection of the vector space **V** onto the subspace **W** of **V**. If dimension $\mathbf{V} = n$ and dimension $\mathbf{W} = k$, then we can choose a coordinate map, ϕ, of **V** such that the ϕ-matrix of p is a diagonal matrix with k ones followed by $n - k$ zeros on the diagonal, i.e.,

$$\begin{array}{r} k \text{ rows} \\ \\ n-k \text{ rows} \end{array}\left(\begin{array}{cccccc|c} 1 & 0 & 0 & 0 & \cdots & 0 & \\ 0 & 1 & 0 & 0 & \cdots & 0 & \\ 0 & 0 & 1 & 0 & \cdots & 0 & \\ 0 & 0 & 0 & 1 & & & \text{all zeros} \\ \cdot & & & & \cdot & \cdot & \\ \cdot & & & & \cdot & \cdot & \\ \cdot & & & & \cdot & \cdot & \\ 0 & & & & & 1 & \\ \hline \text{all zeros} & & & & & & \text{all zeros} \end{array}\right)$$

The only characteristic values of p are 1 and 0. If $k = 0$, then $p = 0$, and the only characteristic value is 0. If $k = n$, then $p = \mathbf{1}$, and the only characteristic value is 1.

If the characteristic of **F** is *not* 2, then any involution is of the form $t = 2p - \mathbf{1}$. Hence a coordinate map, ϕ, can be chosen such that the ϕ-matrix of t is a diagonal matrix with k ones followed by $n - k$ minus ones down the diagonal. The only characteristic values of an involution are $+1$ and -1.

Proof. It is an easy consequence of theorem 4.11 that a coordinate map ϕ exists such that the ϕ-matrix of p (and hence of t) is in the desired form. The ϕ-matrix of p has characteristic polynomial $(x - 1)^k x^{n-k}$ whose only roots are 1 and 0. The characteristic polynomial for the ϕ-matrix of t is $(x - 1)^k(x + 1)^{n-k}$. Hence t has only $+1$ and -1 for characteristic values.

The "nicest" matrix for a given linear transformation is a diagonal matrix. From the definition of characteristic vectors, it is clear that a linear transformation f has a diagonal matrix if and only if there is a basis of **V** consisting entirely of characteristic vectors. This is not always the case and, in particular, is hardly ever true for rotations of space (or the plane).

Theorem 4.22. Let ρ be a rotation of space. ρ is a linear transformation if and only if ρ leaves the origin fixed. If ρ is linear, then there is a coordinate map, θ, such that the θ-matrix of ρ is

$$\begin{pmatrix} 1 & 0 & 0 \\ 0 & \cos\alpha & \sin\alpha \\ 0 & -\sin\alpha & \cos\alpha \end{pmatrix}.$$

Furthermore, ρ has a diagonal matrix if and only if ρ is **1** or a half-turn about a line through the origin.

Proof. Any linear transformation leaves the origin fixed. Hence if ρ is to be linear, it must satisfy $(0)\rho = 0$. We omit the tedious (but trivial) proof that a rotation (or more generally an isometry) leaving the origin fixed is a linear transformation.

Now assume ρ is a rotation with axis M through an angle of radian measure $\alpha (0 \leq \alpha \leq \pi)$, and assume that M contains the origin so that ρ is linear. Choose as basis vectors v_1 lying along M, v_2 and v_3 perpendicular to M, such that v_1, v_2, and v_3 are mutually perpendicular unit vectors and ρ carries v_2 toward v_3. Clearly $(v_1)\rho = v_1$, and (just as in elementary analytic geometry)

$$(v_2)\rho = (\cos\alpha)v_2 + (\sin\alpha)v_3,\ (v_3)\rho = (-\sin\alpha)v_2 + (\cos\alpha)v_3\,.$$

Let θ be the coordinate map with respect to the basis v_1, v_2, v_3, and then the θ-matrix of ρ is

$$\begin{pmatrix} 1 & 0 & 0 \\ 0 & \cos\alpha & \sin\alpha \\ 0 & -\sin\alpha & \cos\alpha \end{pmatrix}.$$

The characteristic polynomial of this matrix, $(x - 1)(x^2 - (2\cos\alpha)x + 1)$, has 1 as its only root (in **R**) unless $\cos\alpha = \pm 1$. If $\cos\alpha = \pm 1$, then ρ is **1** or a half-turn. If $|\cos\alpha| < 1$, then M is the characteristic subspace of the only characteristic value. Hence there is no basis of $\mathbf{V}_3(\mathbf{R})$ consisting entirely of characteristic vectors, i.e., ρ cannot have a diagonal matrix.

Corollary 4.23 (the crystallographic restriction). Let ρ be a rotation leaving a lattice invariant. Then the angle of rotation has radian measure 0, $\pi/3$, $\pi/2$, $2\pi/3$, or π. In other words, ρ is the identity or is of order 6, 4, 3, or 2.

Proof. By theorem 4.22 there is a coordinate map, θ, such that the θ-matrix of ρ has trace $1 + 2\cos\alpha$, where α is the radian measure of the angle of rotation. Let w_1, w_2, and w_3 be basis vectors for the lattice left invariant by ρ, and let ϕ be the coordinate map of $\mathbf{V}_3(\mathbf{R})$ with respect to this basis. Since ρ leaves the lattice invariant, it must send w_1, w_2, and w_3 to vectors which are *integral* linear combinations of w_1, w_2, and w_3, and thus, the ϕ-matrix of ρ has only integers for entries. The trace of this matrix is an integer, *but* the trace is also $1 + 2\cos\alpha$. The only solutions, for $0 \leq \alpha \leq \pi$, of $1 + 2\cos\alpha =$ (an integer) are the values of α listed in the theorem.

Exercise 4.6

1. Let $A, B \in M_n(\mathbf{R})$. Show that $AB - BA = I$ is impossible. What change must be made if A and B are in $M_n(\mathbf{F})$, $\mathbf{F}$ an arbitrary field? *Hint.* Consider traces on both sides.

2. Assume $g \in L(\mathbf{V}, \mathbf{V})$ and that *every* nonzero vector of $\mathbf{V}$ is a characteristic vector for g. Show that g is a "scalar," i.e., that $g = \alpha\mathbf{1}$ for some $\alpha \in \mathbf{F}$. *Hint.* Compare the stretching factors of v, w, and $v + w$ when v and w are linearly independent. It is trivial to show that v and βv are stretched by the same factor. Compare this problem with problem 2 of exercise 4.5.

3. Assume A is nonsingular with characteristic roots $\lambda_1, \lambda_2, \ldots, \lambda_n$. Show that the characteristic roots of A^{-1} are $1/\lambda_1, \ldots, 1/\lambda_n$. Find characteristic roots of $A^2 + A + I$.

4. Let λ be a characteristic value of g, $g \in L(\mathbf{V}, \mathbf{V})$.

(1) Show that if $p(x)$ is any polynomial whose coefficients are in the field of scalars for $\mathbf{V}$, then $p(\lambda)$ is a characteristic value of $p(g)$.

(2) Using the result above and that of problem 3, exercise 4.3, show that the only characteristic value of an involution of a vector space over a field of characteristic 2 is $+1$.

(3) Assume $f \in L(\mathbf{V}, \mathbf{V})$ has only one characteristic value, λ. Show that f has a diagonal matrix in one coordinate system if and only if $f = \lambda\mathbf{1}$.

5. One of the major theorems about matrices is the Cayley–Hamilton theorem which asserts that any matrix satisfies its characteristic equation, i.e., if

$$\det(xI - A) = x^n + a_{n-1}x^{n-1} + \cdots + a_0 ,$$

then

$$A^n + a_{n-1}A^{n-1} + \cdots + a_1A + a_0I = 0 .$$

Use this theorem to show that if $\det A \neq 0$, then A^{-1} exists and can be expressed as a polynomial in A.

6. Let

$$A = \begin{pmatrix} 5 & -1 \\ -1 & 5 \end{pmatrix} .$$

(1) Find the characteristic roots of A and their corresponding characteristic subspaces.

(2) Use characteristic vectors of A to find a coordinate map ϕ such that changing to ϕ-coordinates "puts A in diagonal form."

(3) Using the preceding (and the technique of changing coordinates and then changing back), find A^{15}.

7. Let ϕ be an isomorphism of $\mathbf{S}_3$ into $GL_3(\mathbf{R})$, the group of nonsingular 3 by 3 real matrices. Recall that $\mathbf{S}_3$ is the group of all permutations of $\{1, 2, 3\}$. Discuss the possible candidates for the characteristic polynomials of the six matrices in the range of ϕ. Find one such isomorphism, ϕ.

8. Prove that an isometry, σ, of $\mathbf{V}_3(\mathbf{R})$ which leaves the origin fixed has a diagonal matrix if and only if it is an involution.

4.7. Symmetries of $V_n(\mathbf{F})$, $GL_n(\mathbf{F})$

In this section we study the (algebraic) symmetries of $\mathbf{V}_n(\mathbf{F})$. In particular, we find the center of this group and the size of the group if $\mathbf{F}$ is a finite field.

Definition. A *nonsingular* linear transformation of $\mathbf{V}_n(\mathbf{F})$ is a linear transformation, g, of $\mathbf{V}_n(\mathbf{F})$ into itself that is 1–1 and onto, i.e., it is a linear transformation such that g^{-1} exists. The set of all matrices which are natural matrices of nonsingular linear transformations is denoted by $GL_n(\mathbf{F})$. In other words

$$GL_n(\mathbf{F}) = \{B \in M_n(\mathbf{F}) | B^{-1} \text{ exists}\} .$$

$GL_n(\mathbf{F})$ is called the *general linear group* over $\mathbf{F}$, or the *n-square general linear group* if it is vital to keep track of the size of the matrices.

Recall that in section 1.2 we defined a symmetry operation of X to be a permutation of X that preserves the structure of X. If we view the "structure" of $\mathbf{V}_n(\mathbf{F})$ as the addition and multiplication by scalars defined in $\mathbf{V}_n(\mathbf{F})$, then the symmetry operations of $\mathbf{V}_n(\mathbf{F})$ are just the nonsingular linear transformations of $\mathbf{V}_n(\mathbf{F})$. Since the structure preserved is algebraic, we view these symmetry operations as the "algebraic" symmetry operations of $\mathbf{V}_n(\mathbf{F})$. Later (section 6.7) we focus on the geometric structure of $\mathbf{V}_n(\mathbf{F})$, i.e., the subspaces of $\mathbf{V}_n(\mathbf{F})$, and investigate the symmetries of the set of subspaces of $\mathbf{V}_n(\mathbf{F})$.

In the definition above, we defined $GL_n(\mathbf{F})$ in the classical way as a set of matrices. As we have seen in section 4.6, it is often convenient to be able to shift from one matrix to a similar matrix in order to simplify computations. Stated a different way, it is often convenient to change coordinates. The following theorem shows that $GL_n(\mathbf{F})$ is "essentially" the group of algebraic symmetries of any n-dimensional vector space over $\mathbf{F}$.

Theorem 4.24. Let $\mathbf{V}$ be a vector space of dimension n over $\mathbf{F}$. Let $\mathbf{G}$ be the group consisting of those permutations of $\mathbf{V}$ that are also linear transformations. For each coordinate map θ of $\mathbf{V}$ onto $\mathbf{V}_n(\mathbf{F})$, there is an isomorphism, θ', of $\mathbf{G}$ onto $GL_n(\mathbf{F})$ defined by $g^{\theta'} = B$ if and only if $B = \theta^{-1}g\theta$, i.e., if and only if B is the θ-matrix of g.

Proof. Theorem 4.16 states (among other things) that the correspondence above [extended to $L(\mathbf{V}, \mathbf{V})$] is a 1–1 correspondence between $L(\mathbf{V}, \mathbf{V})$ and $M_n(\mathbf{F})$ which preserves products. Moreover, it states that g^{-1} exists if and only if B^{-1} exists. From these statements it follows easily that the restriction of the correspondence between $L(\mathbf{V}, \mathbf{V})$ and $M_n(\mathbf{F})$ to $\mathbf{G}$ is an isomorphism between $\mathbf{G}$ and $GL_n(\mathbf{F})$.

Note, in particular, that the theorem above implies that $GL_n(\mathbf{F})$ is a group, justifying our use of the term general linear *group*. A change in coordinates amounts to an inner automorphism of $GL_n(\mathbf{F})$ as we now show.

Theorem 4.25. Let $\mathbf{V}$ be an n-dimensional vector space over $\mathbf{F}$, and let ϕ and θ be two coordinate maps for $\mathbf{V}$. Consider the mapping $b: GL_n(\mathbf{F}) \to GL_n(\mathbf{F})$ defined by $A^b = C$ if and only if A is the ϕ-matrix and C is the θ-matrix of the same $g \in L(\mathbf{V}, \mathbf{V})$. This mapping, b, is an inner-automorphism of $GL_n(\mathbf{F})$, i.e., there is a matrix B such that $A^b = C$ if and only if $B^{-1}AB = C$.

This theorem follows easily from theorem 4.18.

Let $\mathbf{F}$ be a finite field with f elements. For each $v \in \mathbf{V}_n(\mathbf{F})$, there are then f possible choices for each component of v, i.e., there are f^n possible vectors in $\mathbf{V}_n(\mathbf{F})$. Since a coordinate map is a 1–1 correspondence, we have the following.

Proposition 4.26. If $\mathbf{V}$ is an n-dimensional vector space over a finite field $\mathbf{F}$ with f elements, then $\#\mathbf{V} = f^n$.

As an application of this theorem, we count the number of elements in $GL_n(\mathbf{F})$ and show, for example, that if $\mathbf{F}$ is the field with two elements, then there are $7 \cdot 6 \cdot 4 = 168$ nonsingular 3 by 3 matrices with entries in this field.

Theorem 4.27. If $\mathbf{F}$ is a finite field with f elements, then there are f^{n^2} matrices in $M_n(\mathbf{F})$ of which $(f^n - 1)(f^n - f)(f^n - f^2) \cdots (f^n - f^{n-1})$ are in $GL_n(\mathbf{F})$.

Proof. Let $v_1, v_2, \ldots, v_n$ be an ordered basis for $\mathbf{V}_n(\mathbf{F})$. By proposition 4.9 there is exactly one linear transformation, g, of $\mathbf{V}_n(\mathbf{F})$ for each *ordered* set of vectors $w_1, w_2, \ldots, w_n$ such that $(v_i)g = w_i$, $i = 1, 2, \ldots, n$. Since there are f^n vectors in $\mathbf{V}_n(\mathbf{F})$, this implies that there are $(f^n)^n = f^{n^2}$ matrices in $M_n(\mathbf{F})$. The linear transformation g is nonsingular if and only if the vectors $w_1, \ldots,$ w_n also form an ordered basis of $\mathbf{V}_n(\mathbf{F})$. If we let $S(w_1, \ldots, w_i)$ be the subspace generated by $w_1, \ldots, w_i$, then we note that $w_1, \ldots, w_n$ form an ordered basis if and only if $w_1 \neq 0$, $w_2 \notin S(w_1)$, $w_3 \notin S(w_1, w_2), \ldots, w_n \notin S(w_1, w_2, \ldots, w_{n-1})$. Thus, there are (for a nonsingular g) $f^n - 1$ choices for w_1, $f^n - f$ choices for w_2, $f^n - f^2$ choices for $w_3, \ldots,$ and $f^n - f^{n-1}$ choices for w_n. Hence there are $(f^n - 1)(f^n - f) \cdots (f^n - f^{n-1})$ matrices in $GL_n(\mathbf{F})$.

We now turn to the problem of determining the center of $GL_n(\mathbf{F})$. Recall that the center of a group is the set of all elements in the group which commute with every element in the group. From the several methods available, we have chosen one which seems to be the most geometric in that it avoids lengthy computations with matrices. It relies on a study of involutions, and, therefore, there is a slight difference between the cases for characteristic 2 and those for other fields. The characteristic 2 case is left as a problem (exercise 4.7, problem 2) with hints to indicate the broad outline of the proof. First we prove a lemma which is applicable, regardless of the characteristic of **F**.

Lemma 4.28. Let g be an involution in $L(\mathbf{V}, \mathbf{V})$, and let $\mathbf{W} = \{v \in \mathbf{V} | (v)g = v\}$. If $h \in L(\mathbf{V}, \mathbf{V})$ commutes with g, then $\mathbf{W}^h \subseteq \mathbf{W}$, and if, furthermore, h is invertible, then $\mathbf{W}^h = \mathbf{W}$.

Proof. If $w \in \mathbf{W}$, then $w^h = w^{gh} = w^{hg}$, so $w^h \in \mathbf{W}$. If h is invertible, then $\mathbf{W}^h = \mathbf{W}$ by proposition 1.9.

Theorem 4.29. The center of $GL_n(\mathbf{F})$ is the set of scalar matrices in $GL_n(\mathbf{F})$. In other words, the only matrices in $GL_n(\mathbf{F})$ that commute with every matrix in $GL_n(\mathbf{F})$ are the matrices λI, $\lambda \in \mathbf{F}$, $\lambda \neq 0$.

Proof. If $n = 1$, the theorem is trivial.

Assume $n > 1$ and that the characteristic of **F** is not 2. Let **V** be an n-dimensional vector space over **F**, and let **G** be the group of nonsingular linear transformations of **V**. Since **G** and $GL_n(\mathbf{F})$ are isomorphic, we need only show that the center of $\mathbf{G} = \{\lambda\mathbf{1} | \lambda \in \mathbf{F}, \lambda \neq 0\}$. The inclusion, "$\supseteq$," is obvious. Assume g is in the center of **G**. We must show $g = \lambda\mathbf{1}$, $\lambda \neq 0$, to complete the proof.

Given $v \neq 0$ in **V**, choose a projection, p, collapsing **V** onto $S(v)$ = the subspace generated by v. Let $h = 2p - \mathbf{1}$; then h is an involution in **G** such that $S(v) = \{w \in \mathbf{V} | w^h = w\}$. Since $h \in \mathbf{G}$, g commutes with h, and hence by lemma 4.28 $S(v)g = S(v)$. In particular, there is a λ in **F** such that $(v)g = \lambda v$. This shows that every nonzero vector in **V** is a characteristic vector for g. To complete the proof we need only show that the "stretching factor," λ, is the same for all $v \in \mathbf{V}$. This follows easily from the linearity of g (problem 2, exercise 4.6). Note that $\lambda \neq 0$ since g^{-1} exists.

The proof for the case in which the characteristic of **F** is 2, except for the construction of the required involutions, is similar, so we leave it as a problem (exercise 4.7, problem 2).

Exercise 4.7

1. For every matrix $A \in M_n(\mathbf{F})$, let A^t denote the transpose of A. It is well known (you may assume it without proof) that $(AB)^t = B^tA^t$. Prove that if $n > 1$, then there do not exist matrices B and C such that $BAC = A^t$ for *every* $A \in M_n(F)$.

Hint. Try $A = I$ first, and then derive a contradiction by looking at $(AB)^t$ in two ways. You need to use theorem 4.25 at one point and corollary 4.19 at another.

2. Assume **F** is a field of characteristic 2 and that $n > 1$. Follow the steps below, and show that the center of $GL_n(\mathbf{F})$ is the set of scalar matrices in $GL_n(\mathbf{F})$. First review problem 3 in exercise 4.3.

(1) If $p, h \in L(\mathbf{V}, \mathbf{V})$ and h commutes with p, then h commutes with $p + \mathbf{1} = q$.
(2) For each nonzero vector s in $\mathbf{V}$, there is a $q \in L(\mathbf{V}, \mathbf{V})$ such that $q^2 = 0$ and range q = the one-dimensional subspace generated by s. Use proposition 4.9.
(3) If h commutes with q, then (range q)$h \subseteq$ range q, and if h is invertible, then (range q)h = range q.
(4) Complete the proof in a manner analogous to the proof of theorem 4.29.

Bibliography and suggestions for further reading

[1] Birkhoff, Garrett, and Saunders MacLane, *A Survey of Modern Algebra*, 3rd ed., Macmillan, New York, 1965.
[2] Dieudonné, Jean, "On the automorphisms of the classical groups," *Memoirs Am. Math. Soc., No. 2*, 1951. Reprinted by Univ. Microfilms, Ann Arbor, Mich., 1964.
[3] Herstein, I. N., *Topics in Algebra*, Blaisdell, Boston, 1964.
[4] Levine, Jack, and H. M. Nahikian, "On the construction of involutory matrices," *Am. Math. Monthly* **69,** 267–271 (1962).
[5] Lightstone, A. H., "A remark concerning the definition of a field," *Math. Magazine* **37,** 12–13 (1964).
[6] Mitchell, Barry, *Theory of Categories*, Academic Press, New York, 1965.

We have selected [1] and [3] from the many available texts on algebra to refer to for the missing or incomplete proofs in this chapter. Needless to say, these proofs may also be found in many other books.

To see what an important role involutions play in the study of the classical groups see [2]. Many of our ideas concerning involutions came from a study of sections 3 and 11 of this monograph. A treatment of involutions that depends more on matrices, but covers even the characteristic 2 case, is in [4].

A much more abstract view of commutative diagrams and their uses in dealing with homomorphisms of various types of mathematical systems is one of the main topics in [6]. This branch of mathematics is an example of some bona fide "new math."

Chapter 5

Affine Spaces

In Chapters 2 and 3 (see the proof of theorem 3.10), we observed that the translations of euclidean space form a three-dimensional vector space over the field of real numbers. In this chapter we present the theory of affine spaces over arbitrary fields as a generalization of that observation, i.e., we define an affine space to be a space in which there are translations that behave the way translations "ought to behave." The axioms that we use are almost the same as those in reference [9].

Using the theory of vector subspaces as reviewed in the preceding chapter, we discuss affine subspaces, their dimension, and the distinction between parallel subspaces and skew subspaces (sections 5.2 and 5.3). All of this is accomplished without introducing coordinates; nevertheless, it is important to understand the connection between arbitrary n-dimensional affine spaces over a field **F** and the standard affine space of n-tuples taken from **F.** In section 5.4 we develop this connection and use it, i.e., use coordinates, to show that the theory of affine subspaces can be viewed as the theory of non-homogeneous linear equations.

As with any geometry, we are primarily interested in the symmetries of affine geometry. We saw in Chapter 2 that the type of symmetry of euclidean space depends on the structure one views as basic for euclidean geometry. We found two different groups, **E** and **S,** according to whether we viewed distances or relative distances as the basic structure to be preserved. Likewise in affine spaces, we find three main groups of symmetries: dilatations if we preserve subspaces and direction, affine transformations if we preserve the underlying vector operations for translations (vectors!), or collineations (semi-affine transformations) if we merely insist on preserving subset relations between subspaces. In sections 5.5–5.8 we investigate all three types of symmetries.

Finally in the last three sections, we fill the gaps left in Chapter 3 and examine volume, lattices, and collineations in real affine spaces.

5.1. Axioms for affine spaces

Hermann Weyl was one of the early proponents of the idea that vectors and translations are essentially the same thing, [11], p. 45, or [3], p. 112. He started from the notion of vector space and presented axioms for affine spaces in terms of vectors (*cf.* [8], chapter 9). We go one step further and modify his axioms to insist that translations *are* vectors.

Definition. Let **F** be a field, A be a non-empty set, and **T** be a subgroup of the group, $\mathbf{S}_A$, of permutations of A. We say A is an *affine space over* **F** *with translation group* **T** if and only if the following two axioms are satisfied.

Transitivity axiom. **T** is exactly transitive on A, i.e., given $P, Q \in A$, there is exactly one $t \in \mathbf{T}$ such that $(P)t = Q$. We denote this t by $\overrightarrow{PQ}$.

Vector axiom. **T** is a vector space over **F** such that addition of vectors is the usual composition of mappings, i.e., for all $P \in A$ and all $t_1, t_2 \in \mathbf{T}$, $(P)(t_1 + t_2) = ((P)t_1)t_2$.

We refer to elements of A as *points*, elements of **T** as *translations*, and elements of **F** as *scalars*. Often, for brevity, we use the phrase "**F**-*affine space*" in place of "affine space over **F**." In particular, we speak of a *rational*, *real*, or *complex affine space* when $\mathbf{F} = \mathbf{Q}$, $\mathbf{R}$, or $\mathbf{C}$, respectively.

The unique translation, $\overrightarrow{PQ}$, carrying the point P to the point Q is called the *vector from P to Q*. Since we view translations as vectors, we *use the additive notation in the group* **T**. **T** is an abelian group so this does not violate our restriction on the use of the additive notation. *Caution!* In the context of **T**, we denote the identity map for A by 0 instead of **1** since we use the additive notation in **T**; however, when we study other symmetries of A, we may revert to **1**.

A hidden, but essential, feature of the vector axiom is that a multiplication of translations (vectors) by scalars is defined, and that this mixes well (the associative and distributive laws!) with the composition of translations. In particular, $(-1)t$ is the inverse of t, and $0t$ is the identity mapping.

Any vector space can be used to form an affine space in a natural way. The method to accomplish this is described carefully in the following theorem.

Theorem 5.1. Let **V** be a vector space over a field **F**. For each $v \in \mathbf{V}$ define a function $v^*: V \to V$ by $(w)v^* = w + v$. Let $V^* = \{v^* | v \in V\}$, and define $\alpha(v^*) = (\alpha v)^*$ for all $\alpha \in \mathbf{F}$, $v^* \in V^*$. Then V is an affine space over **F** with translation group $\mathbf{V}^*$.

Proof. For all $x, w, v \in \mathbf{V}$, $w + v = x + v$ if and only if $w = x$; moreover, $w = (w - v) + v$. From this it is clear that v^* is a permutation of V. Moreover, $(x)(v + w)^* = ((x)v^*)w^*$ since

$$x + (v + w) = (x + v) + w\,.$$

Thus, the mapping $*:\mathbf{V} \to \mathbf{V}^*$ is a (group) homomorphism of $\mathbf{V}$ onto $\mathbf{V}^*$. Since $x + w = x + v$ if and only if $w = v$, we find that $\mathbf{V}^*$ is exactly transitive on V and that $*:\mathbf{V} \to \mathbf{V}^*$ is one-to-one. Thus, $\mathbf{V}^*$ must be a subgroup of $\mathbf{S}_V$ which satisfies the transitivity axiom.

If we decree that $\alpha(v^*) = (\alpha v)^*$ for all $\alpha \in \mathbf{F}$, $v^* \in \mathbf{V}^*$, then the *group* isomorphism, $*$, between $\mathbf{V}$ and $\mathbf{V}^*$ is immediately converted to a *vector space* isomorphism between $\mathbf{V}$ and $\mathbf{V}^*$. Thus, $\mathbf{V}^*$ satisfies the vector axiom, and the proof is complete.

The reader should not jump to the conclusion that this is the only way to convert a vector space into an affine space (exercise 5.1, problem 2). We only claim that it is the most natural way, i.e., the *canonical* way.

When we specialize the construction above to $\mathbf{V}_n(\mathbf{F})$, it is convenient to drop the "*" and yet have a notational device with which to distinguish points from vectors. For this reason we adopt the conventions specified in the following definition.

Definition. The *standard n-dimensional affine space* over the field $\mathbf{F}$, $A_n(\mathbf{F})$, consists of: the set of points,

$$F^n = \{(\alpha_1, \alpha_2, \ldots, \alpha_n) | \alpha_i \in \mathbf{F}\}\,,$$

and the translations (vectors),

$$\mathbf{V}_n(\mathbf{F}) = \{\langle \alpha_1, \alpha_2, \ldots, \alpha_n \rangle | \alpha_i \in \mathbf{F}\}\,,$$

such that if $P = (\alpha_1, \ldots, \alpha_n)$ and $t = \langle \beta_1, \beta_2, \ldots, \beta_n \rangle$, then

$$(P)t = (\alpha_1 + \beta_1, \alpha_2 + \beta_2, \ldots, \alpha_n + \beta_n)\,.$$

Clearly this differs only in notation from the construction given in the theorem above. Hence $A_n(\mathbf{F})$ *is* an affine space over $\mathbf{F}$. Note that in this notation $(\alpha_1, \ldots, \alpha_n)$ is a point, and $\langle \alpha_1, \ldots, \alpha_n \rangle$ is the vector (translation) sending $(0, 0, \ldots, 0)$ to $(\alpha_1, \alpha_2, \ldots, \alpha_n)$. As usual, $(0, 0, \ldots, 0)$ is called the *origin*, and $\langle \alpha_1, \ldots, \alpha_n \rangle$ is called the *position vector* of $(\alpha_1, \ldots, \alpha_n)$.

Example 1. The standard two-dimensional affine space over the field, $\mathbf{Z}_2$, with two elements is $A_2(\mathbf{Z}_2)$. Its points are $(\bar{0}, \bar{0})$, $(\bar{0}, \bar{1})$, $(\bar{1}, \bar{0})$, and $(\bar{1}, \bar{1})$. The position vector, $\langle \bar{1}, \bar{0} \rangle$, of the point $(\bar{1}, \bar{0})$ carries:

$$(\bar{0}, \bar{0}) \text{ to } (\bar{0} + \bar{1}, \bar{0} + \bar{0}) = (\bar{1}, \bar{0})$$
$$(\bar{0}, \bar{1}) \text{ to } (\bar{0} + \bar{1}, \bar{1} + \bar{0}) = (\bar{1}, \bar{1})$$

$$(\bar{1}, \bar{0}) \text{ to } (\bar{1} + \bar{1}, \bar{0} + \bar{0}) = (\bar{0}, \bar{0})$$
$$(\bar{1}, \bar{1}) \text{ to } (\bar{1} + \bar{1}, \bar{1} + \bar{0}) = (\bar{0}, \bar{1}) .$$

Note, in particular, that it carries $(\bar{1}, \bar{0})$ right back to $(\bar{0}, \bar{0})$; hence $\langle\bar{1}, \bar{0}\rangle$ is its own inverse. Clearly in all affine spaces over fields of characteristic 2, all translations are their own inverses, i.e., are involutions. This is a peculiar state of affairs, and thus we often have to avoid the characteristic 2 case or treat it by special methods.

Exercise 5.1

1. Show that any one-element set can be used as the set of points for an affine space over any field **F**.

2. Let $\alpha \to \bar{\alpha}$ be an automorphism of the field **F**, and modify the definitions in theorem 5.1 by decreeing that $\alpha(v^*) = (\bar{\alpha}v)^*$, instead of $(\alpha v)^*$. Show that $\mathbf{V}^*$ is still a vector space over **F**, and hence V is still an affine space over **F** with translation group $\mathbf{V}^*$.

3. Use $\mathbf{V}_2(\mathbf{C})$ to illustrate some differences between the affine space of theorem 5.1 and that of problem 2 above.

4. Assume that A is an affine space over a field **F** and that A has at least two points. Show that all nontrivial translations of A have the same order and that this order is the characteristic of **F** unless the characteristic of **F** is zero, in which case the order is infinite.

5.2. *Affine subspaces*

Throughout this section let A be an affine space over the field **F** with translation group **T**. We consider the affine subspaces of A, define dimension, and relate the concept of dimension to the maximum number of "independent" points in the subspace.

Definition. Let B be a non-empty subset of A. We say that B is an *affine subspace* of A if and only if $\{\overrightarrow{PQ} | P, Q \in B\} = \mathbf{W}$ is a (vector) subspace of **T** such that whenever $P \in B$ and $\overrightarrow{PQ} \in \mathbf{W}$, $Q \in B$. The subspace **W** is called the *directing space* of B.

Theorem 5.2. If P is any point in A, and **W** is any subspace of **T**, then $(P)\mathbf{W} = \{Q | \overrightarrow{PQ} \in \mathbf{W}\}$ is an affine subspace of A. Conversely, if B is an affine subspace with directing space **W**, and if $P \in B$, then $B = (P)\mathbf{W}$. If we make the trivial change from **W** to $\mathbf{W}' = \{w' | w' \text{ is the restriction to } B \text{ of } w \in \mathbf{W}\}$, then B is an affine space over **F** with translation group $\mathbf{W}'$.

Proof. Clearly $B = (P)\mathbf{W}$ if $P \in B$ and B is an affine subspace with directing space **W**.

To prove the converse, assume $P \in A$ and that $\mathbf{W}$ is a (vector) subspace of $\mathbf{T}$. Since $0 \in \mathbf{W}$ and hence $P \in (P)\mathbf{W}$, it is clear that $\mathbf{W} \subseteq \{\overrightarrow{QR} | Q, R \in (P)\mathbf{W}\}$. If $Q, R \in (P)\mathbf{W}$, then $\overrightarrow{QR} = \overrightarrow{PR} - \overrightarrow{PQ}$ is in $\mathbf{W}$, and thus, $\{\overrightarrow{QR} | Q, R \in (P)\mathbf{W}\} = \mathbf{W}$ is a subspace of $\mathbf{T}$. If $Q \in (P)\mathbf{W}$ and $\overrightarrow{QR} \in \mathbf{W}$, then $\overrightarrow{PR} = \overrightarrow{PQ} + \overrightarrow{QR} \in \mathbf{W}$ so $(P)\mathbf{W}$ is an affine subspace.

Now consider $\mathbf{W}'$, consisting of the restrictions to $B = (P)\mathbf{W}$ of the translations in $\mathbf{W}$. In view of the transitivity axiom, the correspondence $w \leftrightarrow w'$ is a one-to-one correspondence between $\mathbf{W}$ and $\mathbf{W}'$. Since each translation in $\mathbf{W}$ permutes the points in B, and $\mathbf{W}$ is exactly transitive on B, $\mathbf{W}'$ is a subgroup of $\mathbf{S}_B$ and is exactly transitive on B. If we decree that addition and multiplication of vectors by scalars for $\mathbf{W}'$ agrees with the addition and multiplication by scalars for $\mathbf{W}$, then clearly $\mathbf{W}'$ is a vector space over $\mathbf{F}$ isomorphic to $\mathbf{W}$. Thus, B is an affine space over $\mathbf{F}$ with translation group $\mathbf{W}'$.

Definition. If P is a point, $\mathbf{W}$ is a (vector) subspace of $\mathbf{T}$, and $B = (P)\mathbf{W}$, then we refer to B as the affine subspace *through P in the direction of* $\mathbf{W}$.

Example 1. One of the affine spaces most familiar to the reader is the real affine space, $A_3(\mathbf{R})$, which can be thought of as ordinary euclidean space stripped of the notions of distance and perpendicularity. The affine subspaces of $A_3(\mathbf{R})$ are just points, lines, planes, or $A_3(\mathbf{R})$ itself. For example, the plane with equation $x + y - z = 2$ is the affine subspace $(P)\mathbf{W}$ with $P = (2, 1, 1)$ and $\mathbf{W}$ the subspace of $\mathbf{V}_3(\mathbf{R})$ generated by $\langle 1, 0, 1\rangle$ and $\langle -1, 2, 1\rangle$. Clearly the choice of P is not unique nor is the choice of the generators of $\mathbf{W}$. A subspace of the plane $(P)\mathbf{W}$ is the line $(P)\mathbf{X}$ where $\mathbf{X}$ is the subspace of $V_3(\mathbf{R})$ generated by $\langle 0, 1, 1\rangle$. This line can be described as the line whose parametric equations are

$$\begin{aligned} x &= 2 \\ y &= 1 + t \\ z &= 1 + t\,. \end{aligned}$$

Note that

$$\langle 0, 1, 1\rangle = \tfrac{1}{2}(\langle 1, 0, 1\rangle + \langle -1, 2, 1\rangle) \in \mathbf{W}\,,$$

so $\mathbf{X} \subseteq \mathbf{W}$. The subspace $(P)\mathbf{X}$ is disjoint from $(Q)\mathbf{W}$, $(Q)\mathbf{X}$, and $(Q)\mathbf{Y}$ if $Q = (1, 1, 2)$ and $\mathbf{Y}$ is the subspace of $V_3(\mathbf{R})$ generated by $\langle -1, 1, 0\rangle$. Note that $\overrightarrow{PQ} = \langle -1, 0, 1\rangle$ is not in $\mathbf{X} + \mathbf{Y}$, i.e., is not a linear combination of $\langle 0, 1, 1\rangle$ and $\langle -1, 1, 0\rangle$.

The following theorem generalizes these observations and provides simple criteria for recognizing when one affine subspace is contained in another or when two subspaces are disjoint.

Theorem 5.3. Let $(P)\mathbf{V}$ and $(Q)\mathbf{W}$ be affine subspaces of A. Then $(Q)\mathbf{W} \subseteq (P)\mathbf{V}$ if and only if $\overrightarrow{PQ} \in \mathbf{V}$ and $\mathbf{W} \subseteq \mathbf{V}$. The two subspaces, $(P)\mathbf{V}$ and $(Q)\mathbf{W}$, are disjoint if and only if $\overrightarrow{PQ} \notin \mathbf{V} + \mathbf{W}$.

Proof. If $(Q)\mathbf{W} \subseteq (P)\mathbf{V}$, then $Q \in (P)\mathbf{V}$, so $\overrightarrow{PQ} \in \mathbf{V}$, and

$$\mathbf{W} = \{\overrightarrow{RS} \mid R, S \in (Q)\mathbf{W}\} \subseteq \{\overrightarrow{RS} \mid R, S \in (P)\mathbf{V}\} = \mathbf{V}.$$

Conversely, if $\overrightarrow{PQ} \in \mathbf{V}$ and $\mathbf{W} \subseteq \mathbf{V}$, then $\overrightarrow{PR} = \overrightarrow{PQ} + \overrightarrow{QR} \in \mathbf{V}$ whenever $\overrightarrow{QR} \in \mathbf{W}$, i.e., $R \in (P)\mathbf{V}$ whenever $R \in (Q)\mathbf{W}$.

To prove the criterion for disjointness we consider the equivalent statement: $\overrightarrow{PQ} \in \mathbf{V} + \mathbf{W}$ if and only if $(P)\mathbf{V} \cap (Q)\mathbf{W} \neq \varnothing$. If $R \in (P)\mathbf{V} \cap (Q)\mathbf{W}$, then $\overrightarrow{PQ} = \overrightarrow{PR} - \overrightarrow{QR}$ is in $\mathbf{V} + \mathbf{W}$. Conversely, if $\overrightarrow{PQ} = v + w$ is in $\mathbf{V} + \mathbf{W}$, then $R = (P)v = (Q)(-w)$ is in $(P)\mathbf{V} \cap (Q)\mathbf{W}$.

Affine subspaces which are not disjoint intersect in an affine subspace. The manner in which they intersect is described in the following theorem.

Theorem 5.4. Assume P is in each one of a (finite or infinite) collection of affine subspaces. Let $\mathbf{U}$ be the intersection of their directing spaces. Then $(P)\mathbf{U}$ is the intersection of the affine subspaces. In particular, if $P \in (Q)\mathbf{V} \cap (R)\mathbf{W}$, then $(Q)\mathbf{V} \cap (R)\mathbf{W} = (P)(\mathbf{V} \cap \mathbf{W})$.

Proof. Let P and $\mathbf{U}$ be as described in the theorem, let $u \in \mathbf{U}$, and let $(Q)\mathbf{V}$ be any one of the affine subspaces in the collection. If $t = \overrightarrow{QP}$, then $(P)u = (Q)(t + u)$ is in $(Q)\mathbf{V}$ since $t \in \mathbf{V}$ $(P \in (Q)\mathbf{V})$ and $u \in \mathbf{V}$ $(\mathbf{U} \subseteq \mathbf{V})$. Thus, $(P)\mathbf{U} \subseteq (Q)\mathbf{V}$, and hence $(P)\mathbf{U} \subseteq$ (the intersection of the collection of affine subspaces).

Conversely, if S is in the intersection of the affine subspaces, then $\overrightarrow{PS}$ is in the directing space of each affine subspace in the collection, i.e., $\overrightarrow{PS}$ is in $\mathbf{U}$, and hence $S \in (P)\mathbf{U}$.

We now employ this result to justify our formal definition of the "smallest" affine subspace containing a given set of points.

Definition. Let B be a non-empty set of points in the affine space A. The intersection of all affine subspaces of A containing B (it is a subspace by theorem 5.4) is called *the subspace generated by* B and is denoted by $[B]$. Points in B are called *generators* of $[B]$. In the special case in which $B = \{P, Q\}$ contains exactly 2 points, we replace $[\{P, Q\}]$ by $P + Q$.

Example 2. If P and Q are distinct points, then the smallest affine subspace containing both points is the line joining P and Q. If $\mathbf{V}$ is the (vector) subspace generated by $\overrightarrow{PQ}$, then $P + Q = (P)\mathbf{V} = (Q)\mathbf{V}$.

Theorem 5.5. If $(P)\mathbf{V}$ is an affine subspace, and $\mathbf{V}$ is finite-dimensional, then $\dim(\mathbf{V}) = d$ if and only if $(P)\mathbf{V}$ can be generated by $d + 1$ points but not by any set containing fewer points.

Proof. Since (theorem 5.3) $(Q)\mathbf{W} \subseteq (P)\mathbf{V}$ if and only if $Q \in (P)\mathbf{V}$ and $\mathbf{W} \subseteq \mathbf{V}$, it follows that if $(Q)\mathbf{W} \subseteq (P)\mathbf{V}$, then $\dim \mathbf{W} \leq \dim \mathbf{V}$ with equality only if $(Q)\mathbf{W} = (P)\mathbf{V}$.

Now assume B is an affine subspace generated by $k + 1$ points, $P, Q_1, Q_2, \ldots, Q_k$. Let $\mathbf{V}$ be the vector subspace generated by $\overrightarrow{PQ_1}, \overrightarrow{PQ_2}, \ldots, \overrightarrow{PQ_k}$. All $k + 1$ points are in $(P)\mathbf{V}$; hence $B \subseteq (P)\mathbf{V}$. Since $\dim \mathbf{V} \leq k$, we see by the remarks above that the dimension of the directing space of B is less than or equal to k.

To complete the proof we need to show that if $(P)\mathbf{V}$ is an affine subspace such that $\dim \mathbf{V} = d$, then $(P)\mathbf{V}$ can, in fact, be generated by $d + 1$ points. Given such a subspace, let $v_1, v_2, \ldots, v_d$ be a basis of $\mathbf{V}$ and let $Q_i = (P)v_i$, i.e., $\overrightarrow{PQ_i} = v_i$, $i = 1, 2, \ldots, d$. Let $(P)\mathbf{W}$ be the subspace generated by P, $Q_1, Q_2, \ldots, Q_d$. Since the $Q_i \in (P)\mathbf{W}$, each of the $\overrightarrow{PQ_i} \in \mathbf{W}$, and hence $V \subseteq W$. But this implies that $(P)\mathbf{V} \subseteq (P)\mathbf{W}$. Since $(P)\mathbf{W}$ is the smallest affine subspace containing $P, Q_1, Q_2, \ldots, Q_n$, we also have $(P)\mathbf{W} \subseteq (P)\mathbf{V}$. Thus, $(P)\mathbf{V} = (P)\mathbf{W}$ is generated by $d + 1$ points.

Definition. The *dimension* of an affine subspace is simply the dimension of its directing space. In view of the result above, if this dimension is finite, it is one less than the number of points in a minimal set of generators. Such a set of points is said to be an *independent* set of points. Note that affine subspaces of dimension zero are simply points. We call affine subspaces of dimensions 1, 2, or $(\dim A) - 1$, *lines*, *planes*, or *hyperplanes* in A, respectively. Lines or planes are hyperplanes if $\dim A = 2$ or 3.

A word of caution is in order here. A set of points need not be independent, i.e., $d + 1$ points may generate a subspace whose dimension is less than d. For example, 10 points in a plane generate a line or a plane, not a nine-dimensional subspace. It could happen that of these 10 points no three are collinear. In such a case, we express the intuitive idea that there are no "special" relations between these points by saying they are in *general position* in the plane. We extend this idea as follows.

Definition. Let B be a set of points, perhaps very many points, in an n-dimensional affine space. These points are in *general position in this space* if no k of them lie in a subspace of dimension $k - 2$, $k = 2, 3, \ldots, n + 1$. These requirements are redundant for most sets. In particular, if $\#B \geq n + 1$ then it is enough to assume that no $n + 1$ points in B lie in a hyperplane.

As with vector subspaces, the union, $B_1 \cup B_2$, of two affine subspaces,

B_1 and B_2, is usually not an affine subspace. There is, however, a smallest subspace, namely, $[B_1 \cup B_2]$, among the subspaces containing both B_1 and B_2.

Theorem 5.6. If $B_1 = (P)\mathbf{V}$ and $B_2 = (Q)\mathbf{W}$ are two affine subspaces, and if $\mathbf{X}$ is the subspace generated by $\overrightarrow{PQ}$, then:

(1) $[B_1 \cup B_2] = (P)(\mathbf{V} + \mathbf{W} + \mathbf{X})$

(2) $B_1 \cap B_2 \neq \varnothing$ if and only if $[B_1 \cup B_2] = (P)(\mathbf{V} + \mathbf{W})$.

Proof. Let $\mathbf{Y}$ be the directing space of $[B_1 \cup B_2]$. Since $(P)\mathbf{V} \subseteq (P)(\mathbf{V} + \mathbf{W} + \mathbf{X})$ and

$$(Q)\mathbf{W} \subseteq (P)(\mathbf{W} + \mathbf{X}) \subseteq (P)(\mathbf{V} + \mathbf{W} + \mathbf{X}),$$

it follows that

$$(P)\mathbf{Y} = [B_1 \cup B_2] \subseteq (P)(\mathbf{V} + \mathbf{W} + \mathbf{X}).$$

By theorem 5.3 this implies that $\mathbf{Y} \subseteq \mathbf{V} + \mathbf{W} + \mathbf{X}$. Both P and Q are in $[B_1 \cup B_2]$. Hence $\overrightarrow{PQ} \in \mathbf{Y}$, i.e., $\mathbf{X} \subseteq \mathbf{Y}$. Moreover, $B_1 = (P)\mathbf{V} \subseteq (P)\mathbf{Y}$ so $\mathbf{V} \subseteq \mathbf{Y}$. Similarly, $\mathbf{W} \subseteq \mathbf{Y}$. Since $\mathbf{V} + \mathbf{W} + \mathbf{X}$ is the smallest vector subspace containing $\mathbf{V}$, $\mathbf{W}$, and $\mathbf{X}$, it follows that $\mathbf{V} + \mathbf{W} + \mathbf{X} \subseteq \mathbf{Y}$, completing the proof that $\mathbf{V} + \mathbf{W} + \mathbf{X} = \mathbf{Y}$, and hence that $[B_1 \cup B_2] = (P)(\mathbf{V} + \mathbf{W} + \mathbf{X})$.

Part (2) follows immediately from part (1) and theorem 5.3.

Theorem 5.7. Assume B and C are affine subspaces *with at least one point in common*. Then if $\dim A = n$ is finite:

(1) $0 \leq \dim (B \cap C) \leq \min \{\dim B, \dim C\}$

(2) $n \geq \dim [B \cup C] \geq \max \{\dim B, \dim C\}$

(3) $\dim B + \dim C = \dim (B \cap C) + \dim [B \cup C]$.

Proof. Since $B \cap C \neq \varnothing$, the directing space of $[B \cup C]$ is the sum of the directing spaces of B and C by theorem 5.6. Moreover, by theorem 5.4 the directing space of $B \cap C$ is the intersection of the directing spaces of B and C. Thus, the three statements follow from the analogous results for vector subspaces (see section 4.2).

We discuss the relationships between B, C, and $[B \cup C]$ in case $B \cap C = \varnothing$ in the next section. The theorem above catalogs completely the ways in which B, C, and $[B \cup C]$ can be related if $B \cap C \neq \varnothing$.

Example 3. Let B and C be planes in a four-dimensional affine space. If $B \cap C = \varnothing$, then we cannot say anything (yet!) about $[B \cup C]$. But if $B \cap C \neq \varnothing$, then we have the following possibilities:

	$B \cap C$	$[B \cup C]$
Case 1.	point	the entire 4-space
Case 2.	line	a 3-space
Case 3.	plane	plane ($B = C$).

Example 4. Let B be a 3-space and C a 4-space in a six-dimensional affine space. If $B \cap C \neq \varnothing$, then $B \cap C$ must be either a line, a plane, or else B itself. Corresponding to these three possibilities, $[B \cup C]$ may be the entire six-dimensional space, a 5-space, or the 4-space C. Note that $B \cap C$ cannot be a single point since $[B \cup C]$ cannot be seven-dimensional.

Exercise 5.2

1. Show that two affine subspaces $(P)\mathbf{V}$ and $(Q)\mathbf{W}$ are equal if and only if $P \in (Q)\mathbf{W}$ and $\mathbf{V} = \mathbf{W}$.

2. Show that in $A_3(\mathbf{R})$ the affine subspace generated by (1, 0, 0), (0, 1, 0), and (0, 0, 1) is the plane whose equation is $x + y + z = 1$.

3. Give an example of a line and two planes in real affine 3-space such that the union of the line and the first plane *is* a subspace, but the union of the line with the second plane is not a subspace.

4. Clearly if B_1 and B_2 are affine subspaces such that one is contained in the other, then $B_1 \cup B_2$ is an affine subspace. The converse is *almost*, but not quite, true. Show that if $\mathbf{F} = \mathbf{Z}_2$, then one can find a counter-example, but if $\mathbf{F}$ has more than two elements, then the converse holds.

5. Show that any 0-dimensional affine space has just one point.

6. Assume H is a hyperplane in the affine space A and that $(P)\mathbf{V}$ is a subspace of A. Show that $H \cap (P)\mathbf{V} = \varnothing$ only if $\mathbf{V}$ is contained in the directing space of H. Show that the converse fails if and only if $(P)\mathbf{V} \subseteq H$.

7. In $A_3(\mathbf{Z}_2)$ find two affine subspaces B_1 and B_2 such that:
(1) $B_1 \not\subseteq B_2$
(2) $B_1 \cap B_2 \neq \varnothing$
(3) $\dim B_1 < \dim B_2$.

8. Show that a set of $d + 1$ points in a d-dimensional affine space is in general position if and only if it is an independent set of points.

9. List all of the affine subspaces of $A_2(\mathbf{Z}_2)$ and $A_2(\mathbf{Z}_3)$. Show that there are 28 lines and 14 planes in $A_3(\mathbf{Z}_2)$.

10. Show that if P and Q are distinct and are both in the affine subspace B, then $P + Q \subseteq B$.

11. Show that any affine subspace of dimension d in an n-dimensional affine space can be realized as the intersection of $n - d$ hyperplanes, but not as the intersection of $n - d - 1$ hyperplanes.

12. Find conditions, if possible, on j, k, and n such that if A is an n-dimensional affine space with two subspaces B_1 and B_2 of dimensions j and k, respectively, and such that $B_1 \cap B_2 \neq \varnothing$, then it is *impossible* for:
(1) $B_1 \cap B_2$ to be a single point

(2) $[B_1 \cup B_2]$ to be a hyperplane
(3) $B_1 \cap B_2$ to be a line.

13. Assume B and C are both three-dimensional affine subspaces in a five-dimensional affine space. Discuss the possibilities for $B \cap C$ and $[B \cup C]$. Show that each possibility can actually occur by providing an example in $A_5(\mathbf{R})$.

5.3. Parallel and skew subspaces

In euclidean space we are familiar with parallel lines, lines parallel to planes, and parallel planes. We have adopted the convention that any line or plane is parallel to itself. If we agree to extend this convention and consider any line to be parallel to any plane within which it is contained, then the following seems to be the natural definition of parallelism in arbitrary affine spaces.

Definition. Let $(P)\mathbf{V}$ and $(Q)\mathbf{W}$ be affine subspaces of an affine space A. $(P)\mathbf{V}$ and $(Q)\mathbf{W}$ are *parallel* if and only if $\mathbf{V} \subseteq \mathbf{W}$ *or* $\mathbf{W} \subseteq \mathbf{V}$. We refer to the cases in which $(P)\mathbf{V} \subseteq (Q)\mathbf{W}$ or $(Q)\mathbf{W} \subseteq (P)\mathbf{V}$ as *trivial* cases of parallelism (*cf.* theorem 5.3).

Note that among the trivial parallel relations are: any point is parallel to any affine subspace containing it, and any subspace is parallel to the entire affine space.

Theorem 5.8 (the parallel axiom). Let P be a point and $(Q)\mathbf{W}$ be an affine subspace. There is exactly one affine subspace, B, such that $P \in B$, B and $(Q)\mathbf{W}$ are parallel, and $\dim B = \dim (Q)\mathbf{W}$.

Proof. $(P)\mathbf{W}$ is an affine subspace satisfying all three conditions, so there is at least one such subspace. If $(P)\mathbf{U}$ were another such subspace, then we would have either $U \subseteq W$ or $W \subseteq U$. But the equality of dimensions then guarantees that $\mathbf{U} = \mathbf{W}$, and hence $(P)\mathbf{U} = (P)\mathbf{W}$.

Our experience in euclidean space would lead us to expect that, aside from the trivial cases, parallel subspaces are disjoint. Moreover, in the case of subspaces of equal dimension, parallel subspaces are simply translates of each other. The following two theorems show that these ideas carry over to arbitrary affine spaces.

Theorem 5.9. If $(P)\mathbf{V}$ and $(Q)\mathbf{W}$ are distinct but parallel, then exactly one of $(P)\mathbf{V} \subseteq (Q)\mathbf{W}$, $(Q)\mathbf{W} \subseteq (P)\mathbf{V}$, or $(P)\mathbf{V} \cap (Q)\mathbf{W} = \varnothing$ holds.

Proof. Assume $(P)\mathbf{V} \nsubseteq (Q)\mathbf{W}$, $(Q)\mathbf{W} \nsubseteq (P)\mathbf{V}$, and $(P)\mathbf{V} \cap (Q)\mathbf{W} \neq \varnothing$. Then we can choose points S_1, S_2, and R such that $S_1 \in (P)\mathbf{V}$ but not to $(Q)\mathbf{W}$, $S_2 \in (Q)\mathbf{W}$ but not to $(P)\mathbf{V}$, and $R \in (P)\mathbf{V} \cap (Q)\mathbf{W}$. However, then $\overrightarrow{RS_1} \in \mathbf{V}$

but not to $\mathbf{W}$, $\overrightarrow{RS_2} \in \mathbf{W}$ but not to $\mathbf{V}$, and hence neither $\mathbf{V} \subseteq \mathbf{W}$ nor $\mathbf{W} \subseteq \mathbf{V}$, i.e., $(P)\mathbf{V}$ and $(Q)\mathbf{W}$ are not parallel. This shows that at least one of the three alternatives must hold if $(P)\mathbf{V}$ and $(Q)\mathbf{W}$ are parallel. If, in addition, they are distinct, then, since neither is empty, it is clear that at most one of the three alternatives can hold.

Theorem 5.10. Assume B and C are affine subspaces such that $\dim B \leq \dim C$. Then B and C are parallel if and only if there is a translation, t, such that $(B)t \subseteq C$. In particular, subspaces of the same dimension are parallel if and only if they are translates of each other.

Proof. Clearly $\overrightarrow{(R)t\,(S)t} = \overrightarrow{RS}$ for any translation t; hence the directing spaces of B and $(B)t$ are the same. Since $(B)t \subseteq C$ is a trivial case of parallelism, it follows that B and C are parallel whenever $(B)t \subseteq C$ for some translation t.

To prove the converse when $\dim B \leq \dim C$ assume that $\mathbf{V}$ and $\mathbf{W}$ are the directing spaces of B and C, respectively, and that $\mathbf{V} \subseteq \mathbf{W}$. Choose points $P \in B$ and $Q \in C$, and let $t = \overrightarrow{PQ}$. Then

$$(B)t = ((P)\mathbf{V})t = (P)(\mathbf{V} + t) = (P)(t + \mathbf{V}) = (Q)\mathbf{V}\,.$$

Since $Q \in C$ and $\mathbf{V} \subseteq \mathbf{W}$, $(Q)\mathbf{V} \subseteq C$ by theorem 5.3. Thus, B has a translate $(B)t = (Q)\mathbf{V}$ contained in C.

Since $((B)t)(-t) = B$, the theorem above can be paraphrased as follows: The only affine subspaces parallel to C are translates of subspaces contained in or containing C. Note that translates of subspaces of C are disjoint from C if and only if the translation used is not in the directing space of C.

Generalizing from the standard terminology for lines in euclidean space, we define skew subspaces as follows.

Definition. Subspaces B and C are called *skew subspaces* if and only if $B \cap C = \varnothing$, but B and C are not parallel.

Theorem 5.11. Assume A is an n-dimensional affine space. There are skew subspaces, $(P)\mathbf{V}$ and $(Q)\mathbf{W}$, of dimensions j and k, respectively, if and only if $1 \leq j \leq n - 2$ and $1 \leq k \leq n - 2$.

Proof. If $(P)\mathbf{V}$ and $(Q)\mathbf{W}$ are disjoint subspaces, then by theorem 5.3 $t = \overrightarrow{PQ} \notin \mathbf{V} + \mathbf{W}$. If neither $\mathbf{V} \subseteq \mathbf{W}$ nor $\mathbf{W} \subseteq \mathbf{V}$, then we can choose vectors v and w such that $v \in \mathbf{V}$ but not to $\mathbf{W}$, and $w \in \mathbf{W}$ but not to $\mathbf{V}$. Since $v \neq 0$ and $v \in \mathbf{V}$, $1 \leq \dim \mathbf{V}$. Moreover, $\dim \mathbf{V} < \dim(\mathbf{V} + \mathbf{W})$ since $w \in \mathbf{V} + \mathbf{W}$ but not to $\mathbf{V}$. Finally, $\dim(\mathbf{V} + \mathbf{W}) < n$ since $t \notin \mathbf{V} + \mathbf{W}$. Thus, $1 \leq \dim \mathbf{V} \leq n - 2$. A similar argument shows that $1 \leq \dim \mathbf{W} \leq n - 2$.

To prove the converse let $t_1, t_2, \ldots, t_n$ be a basis of the n-dimensional space of translations. Let $\mathbf{V}$ be the vector subspace generated by $t_1, \ldots, t_j$, and let **W** be the vector subspace generated by $t_{n-1}, t_{n-2}, \ldots, t_{n-k}$. If j and k satisfy the inequalities $1 \leq j \leq n-2$, $1 \leq k \leq n-2$, then $t_n \notin \mathbf{V} + \mathbf{W}$, $t_{n-1} \in \mathbf{W}$ but not to **V**, and $t_1 \in \mathbf{V}$ but not to **W**. Hence if we choose points P and Q such that $\overrightarrow{PQ} = t_n$, the subspaces $(P)\mathbf{V}$ and $(Q)\mathbf{W}$ are skew subspaces of dimensions j and k, respectively.

It is clear from this theorem that the only possible skew subspaces in a three-dimensional affine space are lines. This agrees with our intuition about euclidean space. In affine 4-space, however, the possible pairs of skew subspaces are two lines, a line and a plane, or even two planes. The last possibility illustrates the curious fact that in affine spaces of dimension greater than 3, one can have skew subspaces which contain parallel lines (exercise 5.3, problem 2).

Exercise 5.3

1. Show that two hyperplanes are parallel if and only if they are either equal or disjoint.

2. Show that two skew subspaces, $(P)\mathbf{V}$ and $(Q)\mathbf{W}$, contain a pair of parallel d-dimensional subspaces if $\dim(\mathbf{V} \cap \mathbf{W}) = d$. Using this, prove that two k-dimensional skew subspaces in a $2k$-dimensional affine space must contain parallel lines if $k > 1$.

3. Catalog the possible pairs of skew subspaces in a five-dimensional affine space. Give an example of each possible pair in $A_5(\mathbf{R})$. *Hint.* Look ahead at theorem 5.12, and note that solution sets of systems of equations are disjoint if and only if the combined system is inconsistent.

4. Distinct lines in euclidean space are parallel if and only if they are disjoint yet contained in the same plane. Prove the following generalization: If B and C are affine subspaces of dimension j and k, respectively, and if neither $B \subseteq C$ nor $C \subseteq B$, then B and C are parallel if and only if $B \cap C = \varnothing$ and $\dim [B \cup C] = 1 + \max(j, k)$.

5. Find a pair of skew planes in $A_4(\mathbf{Z}_2)$.

5.4. *Affine coordinates*

Let A be an n-dimensional affine space over the field **F** with translation group **T**. Let $v_1, v_2, \ldots, v_n$ be an ordered basis for **T**, and let ϕ be the corresponding coordinate map of **T** onto $\mathbf{V}_n(\mathbf{F})$, i.e., $v^\phi = \langle \alpha_1, \ldots, \alpha_n \rangle$ if and only if $v = \alpha_1 v_1 + \cdots + \alpha_n v_n$. Choose a point P and call it the *origin*. We define $\phi^*: A \to A_n(\mathbf{F})$ by $(Q)\phi^* = (\alpha_1, \ldots, \alpha_n)$ if and only if $(\overrightarrow{PQ})^\phi = \langle \alpha_1, \ldots, \alpha_n \rangle$. Since ϕ is one to one from **T** onto $\mathbf{V}_n(\mathbf{F})$, it follows from the transitivity axiom that ϕ^* is one–to–one from A onto $A_n(\mathbf{F})$. This one-to-one correspondence, ϕ^*, is referred to as the *coordinate map* for A with *origin* P and *unit vectors* v_i. The points Q_i such that $\overrightarrow{PQ_i} = v_i$ are called the *unit points*. Clearly $P, Q_1, \ldots, Q_n$ are $n + 1$ independent points, and for any set of $n + 1$ independent points, $R, S_1, \ldots, S_n$, there is a coordinate map for A with origin R and unit points $S_1, \ldots, S_n$. Thus, to choose a coordinate system we must choose an origin and the n unit points. Given a coordinate map for A, we can define, exactly as in elementary analytic geometry, the graph of a system of linear equations in n unknowns to be the set of points in A whose coordinates satisfy the equations.

Theorem 5.12. Assume A is an n-dimensional affine space. The graph in A of a system of linear equations in n unknowns is either $\varnothing$ or an affine subspace of A. Conversely, any affine subspace of A is the graph of at least one system of linear equations. The system of equations is homogeneous if and only if the origin is in the subspace.

Proof. If a system of linear equations is consistent, i.e., if its graph is non-empty, then we can choose $P \in A$ such that the coordinates of P, $(P)\phi^*$, satisfy the equations. Let $\mathbf{V}^\phi \subseteq \mathbf{V}_n(\mathbf{F})$ be the solution set of the associated homogeneous system. Then $(P)\mathbf{V}$ is the graph of the system in A since $(Q)\phi^*$ is a solution if and only if $\overrightarrow{PQ} \in \mathbf{V}$.

Conversely, given a subspace $(P)\mathbf{V}$, choose a linear transformation in $L(\mathbf{T}, \mathbf{T})$ whose kernel is $\mathbf{V}$ (*cf.* exercise 4.3, problem 6), and let M be the ϕ-matrix of the linear transformation. The ϕ-coordinates of P, $(P)\phi^*$ form an ordered n-tuple $(\alpha_1, \alpha_2, \ldots, \alpha_n)$. The system of equations (in matrix form)

$$(x_1, x_2, \ldots, x_n)M = (\alpha_1, \alpha_2, \ldots, \alpha_n)$$

is a system whose graph is $(P)\mathbf{V}$ (exercise 5.4, problem 1) as desired. If $(\alpha_1, \ldots, \alpha_n) = (0, 0, \ldots, 0)$, i.e., if the system is homogeneous, then the position vector for P is in $\mathbf{V}$, and hence the origin is in $(P)\mathbf{V}$. Conversely, if the origin is in $(P)\mathbf{V}$, then, since P is also in $(P)\mathbf{V}$, the position vector of P is in $\mathbf{V}$, so $(P)\phi^*M = (0, 0, \ldots, 0)$.

Exercise 5.4

1. Complete the proof of theorem 5.12.

2. Find a system of nonhomogeneous equations whose graph in $A_4(\mathbf{R})$ is the subspace generated by (1, 1, 1, 1), (2, 1, 0, 0), and (1, 2, 3, 4).

3. Let B be a subset of an affine space A of dimension n. Show that B is a hyperplane if and only if it is the graph of a single linear equation.

4. Prove that a k-dimensional subspace of an n-dimensional affine space is the intersection of exactly n-k hyperplanes.

5.5. *Affine symmetries I: Dilatations*

In this and the next two sections, we consider the three most important types of symmetries of an affine geometry. The first, and most restrictive, type are the dilatations that we met in Chapter 2 when studying similarity transformations. This is the natural type of symmetry to study if one views each affine geometry as a set whose structure is subspaces and directions of subspaces. In other words, these are the symmetries which map each subspace onto another subspace with the same direction. It is not necessary to require this for subspaces of all dimensions, and in our definition of dilatations below, we assume as little as possible. Note that this type of symmetry has no significance for affine spaces of dimension 0 or 1.

The second type of affine symmetry, to be considered in the next section, consists of those permutations of points for which the induced permutation of the translations is linear. This type of symmetry is the natural type to consider if one views the algebraic structure of the translations as an integral feature of the geometry. Historically this has proved to be the most important group of symmetries so it is known as the affine group.

The third type of affine symmetry, to be considered in section 5.8, consists of the semi-affine transformations, i.e., the permutations of points that map subspaces into subspaces of the same dimension. This is the natural type of symmetry if one views the structure of affine geometry to be the subspaces and their "size." One of the most interesting results in geometry is the fact that this type of symmetry is almost, and in some very important cases is, as restrictive as the preceding type. Essentially the only added feature for semi-affine transformations as opposed to affine transformations is that one allows automorphisms of the field as well as of the vector space involved. Results of this nature are usually gathered under the topic, "The Fundamental Theorem of Geometry."

Let us now turn to the first and most restrictive type, the dilatations. Many of the results of Chapter 2 on dilatations in euclidean space carry over. In fact, the reader should use those results, in section 2.6, as a guide for his intuition.

Definition. Assume A is an affine space of dimension at least 2. A permutation, σ, of A is called a *dilatation* if and only if for all pairs of points P and Q in A, $P + Q \mathbin{/\!/} (P)\sigma + (Q)\sigma$. Clearly the set of all dilatations of A forms a group. We call this group $\mathbf{D}$, or, if necessary, $\mathbf{D}_A$.

This definition for an arbitrary affine space of dimension at least 2 is the same as that given in section 2.6 for euclidean space. In that case we showed that the only dilatations were the translations and central dilatations. We now generalize this result to an arbitrary affine space. First we must generalize the definition of central dilatations and verify that they and translations are actually dilatations.

Definition. Given a point P and a nonzero scalar μ we define the *central dilatation*, p_μ, as follows. For any point Q there is (by the transitivity axiom) a unique point R such that $\overrightarrow{PR} = \mu\overrightarrow{PQ}$. Let $(Q)p_\mu = R$. Note that, as for inversions in Chapter 2, we use a capital letter as the name of a point and the corresponding lowercase letter in the name of the dilatation. In fact, if $A = A_3(\mathbf{R})$, p_{-1} is the inversion, p, with center P.

Theorem 5.13. All translations of A and all central dilatations are dilatations of A.

Proof. Any translation is a permutation which, by theorem 5.10, carries any subspace, especially lines, onto a parallel subspace, hence it is a dilatation.

It is easy to verify that $p_\mu p_{(1/\mu)} = p_{(1/\mu)}p_\mu = \mathbf{1}$, hence p_μ is a permutation of A by theorem 1.1. Given two points, Q and R, let $Q_1 = (Q)p_\mu$ and $R_1 = (R)p_\mu$. Then $\overrightarrow{Q_1R_1} = \overrightarrow{Q_1P} + \overrightarrow{PR_1} = \mu(\overrightarrow{QP}) + \mu(\overrightarrow{PR}) = \mu(\overrightarrow{QP} + \overrightarrow{PR}) = \mu(\overrightarrow{QR})$; hence $Q_1 + R_1$ is parallel to $Q + R$, for both lines have the (vector) subspace generated by $\overrightarrow{QR}$ as their directing space. Thus p_μ is a dilatation.

Theorem 5.14. A dilatation is uniquely determined by its action on two points.

Proof. The proof is exactly the same as the proof of proposition 2.20 since (theorem 5.8) the parallel axiom for lines is valid in an arbitrary affine space.

Theorem 5.15. A dilatation with at least two fixed points is the identity. With exactly one fixed point, P, it is a central dilatation, p_μ, and with none it is a (nontrivial) translation. In particular, any dilatation is a translation or a central dilatation.

Proof. A dilatation with two fixed points agrees with the identity on those two points. Hence it must be the identity by theorem 5.14.

Assume σ is a dilatation with one fixed point, P. Choose a point $R \neq P$, and let $S = (R)\sigma$. Since σ is a dilatation, $P + R \| P + S$, and hence P, R, and S are collinear. Thus, there is a scalar μ such that $\overrightarrow{PS} = \mu\overrightarrow{PR}$. This scalar, μ, is nonzero since σ is one-to-one, i.e., $S = (R)\sigma \neq P = (P)\sigma$. But then P_μ and σ agree at P and R, so $\sigma = p_\mu$.

Now assume σ is a dilatation with no fixed points. Choose a point P. Since σ has no fixed points, $P \neq (P)\sigma$. Since A has dimension at least 2, we can choose Q such that Q is not on $P + (P)\sigma$. Since $P + Q \| (P)\sigma + (Q)\sigma$, the four points, P, Q, $(P)\sigma$, and $(Q)\sigma$, are contained in the same plane. Hence the lines

$P + (P)\sigma$ and $Q + (Q)\sigma$ cannot be skew lines, i.e., they are either parallel or else intersect at point R. A short argument shows that if they intersect at R, then R must be a fixed point of σ (exercise 5.5, problem 1). Since ϕ has no fixed point, we must have $P + (P)\sigma \| Q + (Q)\sigma$. To keep the notation manageable let $(P)\sigma = P_1$ and $(Q)\sigma = Q_1$. Then we have: $\overrightarrow{QQ_1} = \lambda\overrightarrow{PP_1}$ since $P + P_1 \| Q + Q_1$, and $\overrightarrow{P_1Q_1} = \mu\overrightarrow{PQ}$ since $P_1 + Q_1 \| P + Q$. Thus,

$$\overrightarrow{PQ_1} = \overrightarrow{PP_1} + \overrightarrow{P_1Q_1} = \overrightarrow{PP_1} + \mu\overrightarrow{PQ},$$

and

$$\overrightarrow{PQ_1} = \overrightarrow{PQ} + \overrightarrow{QQ_1} = \overrightarrow{PQ} + \lambda\overrightarrow{PP_1}.$$

By subtracting we get

$$(1 - \lambda)\overrightarrow{PP_1} + (\mu - 1)\overrightarrow{PQ} = 0.$$

However, P, Q, and P_1 are noncollinear; hence $\overrightarrow{PP_1}$ and $\overrightarrow{PQ}$ are linearly independent vectors, and thus, $\lambda = \mu = 1$. In particular, $\overrightarrow{QQ_1} = \overrightarrow{PP_1}$; hence the translation carrying P to P_1 also carries Q to Q_1, or, in other words, $\overrightarrow{PP_1}$ and σ agree at P and Q. Thus, $\sigma = \overrightarrow{PP_1}$ is a translation.

It follows immediately from the theorem above that the product of a central dilatation and a translation is always a central dilatation or a translation. For the coordinate representation of a dilatation, it is useful to know that any dilatation can be factored into a product of this type using a preassigned point P as the fixed point of the central dilatation.

Theorem 5.16. Let σ be a dilatation and P be a point. There is a nonzero scalar, μ, and a translation, t, such that $\sigma = p_\mu t$.

Proof. Let $t = \overrightarrow{P(P)\sigma}$. Then σt^{-1} is a dilatation leaving P fixed, i.e., $\sigma t^{-1} = p_\mu$ with $\mu = 1$ or not, according as σt^{-1} has another fixed point or not. Thus, $\sigma = p_\mu t$ as desired.

We apply this result to determine the coordinate form of a dilatation as follows. If $\dim A = n$, then let ϕ^* be a coordinate map of A onto $A_n(\mathrm{F})$ with origin at P. By the theorem above, we can factor a dilatation, σ, in the form $\sigma = p_\mu t$. If the point $(P)\sigma = (P)t$ has ϕ^*-coordinates $(\alpha_1, \ldots, \alpha_n)$, then for any point Q with $(Q)\phi^* = (x_1, x_2, \ldots, x_n)$:

$$(Q)\sigma\phi^* = (x_1, x_2, \ldots, x_n)\begin{pmatrix} \mu & & & & 0 \\ & \mu & & & \\ & & \cdot & & \\ & & & \cdot & \\ & & & & \cdot \\ 0 & & & & \mu \end{pmatrix} + (\alpha_1, \ldots, \alpha_n).$$

Note that this representation is independent of the unit points of the coordinate system. Problems 5–7 in exercise 5.5 show that the scalar μ, called the *dilatation ratio*, is also independent of the choice of the origin and that the mapping sending each dilatation to its dilatation ratio is a homomorphism.

Exercise 5.5

1. Assume that α is a dilatation and that P, Q, and R are points such that $(P + (P)\alpha) \cap (Q + (Q)\alpha) = \{R\}$. Show that R must be a fixed point of α.

2. Let t be a translation and p_μ a central dilatation with $\mu \neq 1$. Show that $p_\mu t = q_\mu$ for some point Q. Be sure to explain how to find Q.

3. Let p_λ and q_μ be central dilatations. Discuss the product $p_\lambda q_\mu$ explaining when it is a translation and when it is a central dilatation r_ν. Be sure to explain how to find R and ν when $p_\lambda q_\mu = r_\nu$.

4. Let P, Q, P_1, and Q_1 be four points in an affine space (of dimension at least 2) such that $P \neq Q$, $P_1 \neq Q_1$, and $P + Q \parallel P_1 + Q_1$. Show that there is exactly one dilatation carrying P to P_1 and Q to Q_1. *Hint*. Cover first the case in which $P + Q \neq P_1 + Q_1$.

5. Let α be a dilatation. For any two points, P and Q, $P + Q \parallel (P)\alpha + (Q)\alpha$, and hence $\overrightarrow{(P)\alpha(Q)\alpha} = \mu\overrightarrow{PQ}$. Show that μ is independent of P and Q, and show that μ is the dilatation ratio, i.e., show that if we factor α in the form $r_\lambda t$ for any point R then $\lambda = \mu$.

6. Show that the dilatation ratio and the action on one point determine a dilatation uniquely.

7. Show that the mapping sending each dilatation to its dilatation ratio is a homomorphism of **D** onto **F***, the multiplicative group of nonzero scalars. Show that **T** is the kernel of this homomorphism and hence is a normal subgroup of **D**. This result is used in the next section.

5.6. *Affine symmetries II: Affine transformations*

It is easier to state the definition of affine transformations if we first recall some elementary facts about inner automorphisms. Recall that in Chapter 1 we showed that any permutation, ϕ, of a set, A, induces in a natural way a permutation, ϕ^i, of $\mathbf{S}_A$ called the inner automorphism of $\mathbf{S}_A$ induced by ϕ. Namely, ϕ^i sends any permutation σ to "σ shifted by ϕ," i.e., $(\sigma)\phi^i = \phi^{-1}\sigma\phi$. We have seen how $(\sigma)\phi^i$ often behaves like "σ shifted by ϕ," e.g., half-turns, inversions, and reflections when shifted by similarities always behaved as expected.

In affine geometry we have distinguished a certain subgroup of permutations of the space A, namely, the group **T** of translations. Thus, when we consider symmetries of an affine space we are only interested in permutations of A which shift translations to translations. In other words, we only consider permutations whose induced inner automorphisms leave **T** invariant. A method of determining the action of ϕ^i on **T** for such a permutation is given in the next theorem.

Theorem 5.17. Let ϕ be a permutation of the affine space A such that the induced inner automorphism, ϕ^i, shifts the translation t to a translation. If $t = \overrightarrow{PQ}$, then $(t)\phi^i = \overrightarrow{(P)\phi(Q)\phi}$.

Proof. Since we know that $(t)\phi^i$ is a translation, and since a translation is uniquely determined (transitivity axiom) by its action on one point, we need only check that $(t)\phi^i$ carries $(P)\phi$ to $(Q)\phi$. We proved this in Chapter 1, proposition 1.9.

Corollary 5.18. If σ is a dilatation with dilatation ratio μ, then σ^i sends the translation t to the translation μt.

Proof. By theorem 5.15, σ is either a translation or a central dilatation. If σ is a translation, then the dilatation ratio of σ is 1, and $(t)\sigma^i = -\sigma + t + \sigma = t$, as desired. If $\sigma = p_\mu$, then the dilatation ratio is μ, and we need to show $p_{(1/\mu)}tp_\mu = \mu t$. Instead of doing this directly, we use theorem 5.17 as follows. Since $p_{(1/\mu)}tp_\mu$ is a dilatation with dilatation ratio 1, σ^i does shift t to a translation. Choose Q such that $t = \overrightarrow{PQ}$, i.e., let $Q = (P)t$. Then

$$\overrightarrow{(P)\sigma(Q)\sigma} = \overrightarrow{P(Q)\sigma} = \mu\overrightarrow{PQ}$$

since $p_\mu = \sigma$ leaves P fixed, and the image of Q under p_μ is defined so that the last equation is valid. Thus, by theorem 5.17, $(t)\sigma^i = \mu t$, thus completing the proof.

One way of interpreting this theorem is to say that dilatations stretch all translations by a fixed amount, the dilatation ratio. It is easy to verify the converse (exercise 5.6, problem 4), Hence we can say that dilatations are precisely those symmetries which "dilate" translations. Thus, the term "dilatation" is a reasonable one.

When studying similarities, we first considered them in terms of stretching distances by a fixed factor, and then we showed that this was equivalent to "preserving relative distances." We now make the analogous transition from dilatations, which stretch vectors by a fixed ratio, to affine transformations, which preserve "relative lengths of parallel vectors." The analogy between the euclidean and affine cases ends here, for in the affine case the group of symmetries is enlarged considerably.

Definition. Let A be an affine n-space over the field **F** with translation group **T**. An *affine transformation* of A is a permutation, σ, of A such that:

(1) Whenever $t \in \mathbf{T}$, $(t)\sigma^i \in \mathbf{T}$.

(2) For any $t \in \mathbf{T}$ and any $\alpha \in \mathbf{F}$, $(\alpha t)\sigma^i = \alpha(t)\sigma^i$.

The group of all affine transformations of A is denoted by $\mathfrak{A}$, or if necessary $\mathfrak{A}_A$.

By corollary 5.18 any dilatation is an affine transformation. Using theorem 5.17, one can reinterpret the second requirement for affine transformations as follows: If P, Q, R, and S are points such that $\overrightarrow{RS} = \alpha\overrightarrow{PQ}$, then the same is true for their image points. In other words, affine transformations must preserve "relative distances on parallel lines." A word of caution! Even in euclidean space this is much less stringent a requirement than preserving all relative distances (exercise 5.6, problem 2).

Proposition 5.19. If σ is an affine transformation of A, then σ^i (restricted to $\mathbf{T}$) is a nonsingular linear transformation of $\mathbf{T}$.

Proof. Requirement (1) on affine transformations asserts that σ^i maps $\mathbf{T}$ into $\mathbf{T}$. By requirement (2), σ^i is homogeneous. For any translations, t and s, $t + s$ is the composite of t and s. Since σ^i is an inner automorphism of $\mathbf{S}_A$,

$$(t + s)\sigma^i = (t)\sigma^i + (s)\sigma^i,$$

and thus, σ^i is additive. This completes the proof that σ^i is a linear transformation. By theorem 5.17, σ^i does not send any nonzero translation to the zero translation (exercise 5.6, problem 1); hence σ^i is nonsingular.

Theorem 5.20. Affine transformations send affine subspaces to affine subspaces of the same dimension.

Proof. Let $(P)\mathbf{V}$ be an affine subspace, and assume the affine transformation, σ, sends P to Q. Since σ^i on $\mathbf{T}$ is nonsingular, it sends the vector subspace $\mathbf{V}$ to a subspace $\mathbf{W}$ of the same dimension. Assume the point R in $(P)\mathbf{V}$ is sent by σ to the point S. The translation $\overrightarrow{PR}$ is in $\mathbf{V}$, and hence $(\overrightarrow{PR})\sigma^i = \overrightarrow{QS}$ is in $\mathbf{W}$. However, $\overrightarrow{QS} \in \mathbf{W}$ implies $S \in (Q)\mathbf{W}$, and so σ maps $(P)\mathbf{V}$ into $(Q)\mathbf{W}$. The same argument shows that σ^{-1} sends $(Q)\mathbf{W}$ into $(P)\mathbf{V}$, i.e., that σ maps $(P)\mathbf{V}$ *onto* $(Q)\mathbf{W}$.

To complete this section we examine the affine transformations of order 2, i.e., the affine involutions. These are analogous to the involutions in euclidean space as long as the field $\mathbf{F}$ is not of characteristic 2. We first need some preliminary results concerning characteristic and midpoints.

Definition. An affine space, A, is said to be of *characteristic* p if and only if p is the smallest positive integer such that adding any nonzero translation to itself p times gives the zero translation (the identity map of A!). If no nontrivial translation has finite order then we say the characteristic of A is zero.

Proposition 5.21. Let A be an affine space over the field $\mathbf{F}$. Then A is of characteristic p if and only if $\mathbf{F}$ is of characteristic p.

Proof. Let t be a nonzero translation, then

$$\underbrace{t + t + \cdots + t}_{p \text{ times}} = (\underbrace{1 + 1 + \cdots + 1}_{p \text{ times}})t\,.$$

Since $t \neq 0$, $\alpha t = 0$ if and only if $\alpha = 0$. The theorem follows immediately since the characteristic of $\mathbf{F}$ is zero if the scalar 1 does not have finite order and otherwise is the smallest positive integer such that adding 1 to itself p times gives the scalar 0.

Definition. Let P, Q, and R be points in an affine space A. The point Q is the *midpoint* of P and R if $\overrightarrow{PQ} = \overrightarrow{QR}$.

Proposition 5.22. If A is not of characteristic 2, midpoints exist and are unique for every pair of points. Moreover, the midpoint of P and R is also the midpoint of R and P. If $P = R$, then $P = R =$ the midpoint of P and R.

Proof. Given P and R, let $t = (1/2)\overrightarrow{PR}$, and let $Q = (P)t$. Then $(Q)t = (P)2t = R$; hence $\overrightarrow{PQ} = \overrightarrow{QR} = t$. Thus, P and R have at least one midpoint, which is both P and R if $P = R$. If S is also a midpoint, then $2\overrightarrow{PS} = \overrightarrow{PS} + \overrightarrow{SR} = \overrightarrow{PR}$, and so $\overrightarrow{PS} = \overrightarrow{PQ}$. This implies (the transitivity axiom!) that $Q = S$, so midpoints are unique. Finally, if Q is the midpoint of P and R, then by taking negatives of $\overrightarrow{PQ}$ and $\overrightarrow{QR}$ we find that $\overrightarrow{RQ} = \overrightarrow{QP}$, so that Q is also the midpoint of R and P.

Proposition 5.23. An affine transformation preserves midpoints, again assuming that the characteristic is not 2.

Proof. Let Q be the midpoint of P and R, and let σ be an affine transformation. Since $\overrightarrow{PQ} = \overrightarrow{QR}$, $(\overrightarrow{PQ})\sigma^i = (\overrightarrow{QR})\sigma^i$. But, in view of theorem 5.17, this is equivalent to: $(Q)\sigma$ is the midpoint of $(P)\sigma$ and $(R)\sigma$.

Theorem 5.24. Let σ be an affine involution of an affine space, A, such that the characteristic of A is not 2. Then there are affine subspaces B^+ and B^- such that:

(1) B^+ and B^- are "complementary" in the sense that $B^+ \cap B^- = \{P\}$ for some point P and $[B^+ \cup B^-] = A$.
(2) B^+ is the set of fixed points of σ.
(3) For every point Q, the midpoint of Q and $(Q)\sigma$ is a fixed point and hence is in B^+.
(4) For every point Q not in B^+, $Q + (Q)\sigma$ is parallel to B^-.

Proof. Let Q be any point in A, and let R be the midpoint of Q and $(Q)\sigma$. Since σ preserves midpoints, $(R)\sigma$ is the midpoint of $(Q)\sigma$ and $(Q)\sigma\sigma = Q$. Thus, R and $(R)\sigma$ are both midpoints of Q and $(Q)\sigma$, so $R = (R)\sigma$ is a fixed point of σ. In particular, σ has at least one fixed point.

Since σ is an (affine) involution, σ^i (restricted to $\mathbf{T}$) is also an (linear) involution; hence there are (vector) subspaces $\mathbf{V}^+ = \{t \in \mathbf{T} | (t)\sigma^i = t\}$ and $\mathbf{V}^- = \{t \in \mathbf{T} | (t)\sigma^i = -t\}$. $\mathbf{V}^+$ and $\mathbf{V}^-$ are complementary subspaces of $\mathbf{T}$ (exercise 4.3, problem 4).

Now let P be any fixed point of σ, and let $B^+ = (P)\mathbf{V}^+$ and $B^- = (P)\mathbf{V}^-$. Then $B^+ \cap B^- = (P)(\mathbf{V}^+ \cap \mathbf{V}^-) = \{P\}$, and $[B^+ \cup B^-] = (P)(\mathbf{V}^+ + \mathbf{V}^-) = A$, so we have found our desired complementary spaces.

Since $(\overrightarrow{PR})\sigma^i = \overrightarrow{(P)\sigma(R)\sigma}$ and $(P)\sigma = P$, the point R is a fixed point for σ if and only if $\overrightarrow{PR}$ is a fixed vector for σ^i. V^+ is the set of fixed vectors for σ^i. Hence $(P)\mathbf{V}^+ = B^+$ is the set of fixed points for σ. This completes the proof of parts (1), (2), and (3).

Given any point Q such that $Q \notin B^+$, i.e., such that $Q \neq (Q)\sigma$, we have

$$(\overrightarrow{Q(Q)\sigma})\sigma^i = \overrightarrow{(Q)\sigma(Q)\sigma\sigma} = \overrightarrow{(Q)\sigma Q} = -(\overrightarrow{Q(Q)\sigma}).$$

Thus, $\overrightarrow{Q(Q)\sigma}$ is in $\mathbf{V}^- = \{t \in \mathbf{T} | (t)\sigma^i = -t\}$. But this means that $Q + (Q)\sigma$ is parallel to B^- and proves part (4).

Definition. If σ is an affine involution, and if B^+ and B^- are as in theorem 5.24, then B^+ is called the *mirror* of σ, and B^- is called the *direction* of σ. The involution σ is referred to as the *reflection with mirror* B^+ and *direction* B^-. The mirror is unique and consists of all the fixed points of σ, but the direction is not unique and may be selected from any of the translates of B^-.

We should now prove the converse of the theorem 5.24, i.e., show that given complementary affine subspaces, B^+ and B^-, conditions (2), (3), and (4) uniquely determine a permutation which is, in fact, an affine involution. This can be done (Try it!), but the details are long and tedious. It is more efficient to develop a general technique of converting linear transformations of $\mathbf{T}$ into affine transformations and then use our results about the existence of linear involutions. This we do in the next section as we explore the connections between affine transformations and linear transformations of $\mathbf{T}$.

Exercise 5.6

1. Assume σ is an affine transformation. Show that $(t)\sigma^i \neq 0$ if $t \neq 0$ in $\mathbf{T}$.

2. Prove that any similarity transformation of euclidean space is also an affine transformation. Describe one or two affine transformations (no proofs required!) that are not similarities.

3. Recall the elementary method of dividing the segment $[PQ]$ into n equal parts. This is usually done as follows: Choose a line L through P, $L \neq P + Q$, and a line M through Q such that $L \mathbin{/\!/} M$. Lay off n congruent segments on L starting from P and n segments on M (congruent to those on L) starting from Q on the opposite side

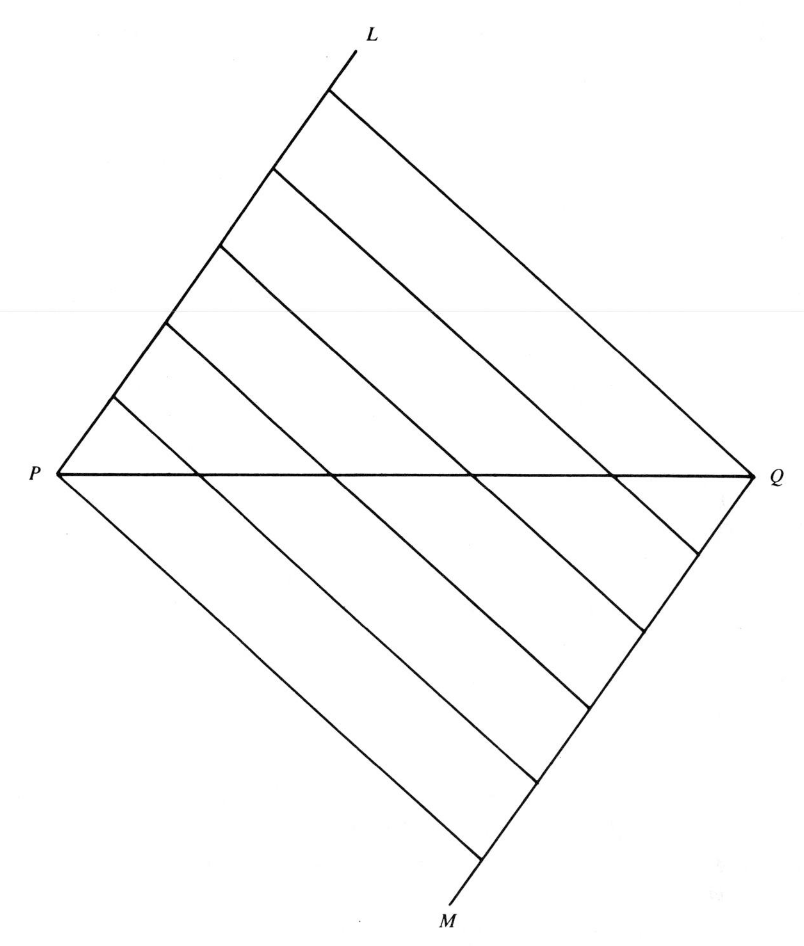

FIG. 26. Dividing the segment $[PQ]$ into n equal parts.

of $P+Q$, as in Figure 26. Connect corresponding points on L and M to get the division points in $[PQ]$. Generalize this construction to an arbitrary affine space whose characteristic does not divide n. Be sure to explain how to formalize the concepts of "congruent segments on L and M" and "opposite sides of $P+Q$."

4. Let σ be a permutation of an affine space A such that for some scalar μ, $(t)\sigma^i = \mu t$ for all translations t. Show that σ is a dilatation.

5. Let P, Q, R, and S be four independent points in an affine space of characteristic not 2. There are three ways to partition $\{P, Q, R, S\}$ into disjoint two-element sets. These determine the three pairs of opposite edges of the "tetrahedron" P, Q, R, S. Show that the lines joining midpoints of opposite edges meet in a point. Are there any additional restrictions on the characteristic?

6. Define "affine homomorphism." If $f\colon A \to A$ is an affine homomorphism, what properties should it have to be considered an affine projection? *Hint.* Note the analogies between affine and linear involutions, and attempt to mimic them for projections.

Term-paper topics

(1) Discuss involutions of affine spaces of characteristic 2. Review the exercises in Chapter 4 concerning involutions of vector spaces over fields of characteristic 2.
(2) Discuss the geometric significance of the unimodular affine group. Let A be an n-dimensional space over the field $\mathbf{F}$ with translation group $\mathbf{T}$. Let det be a determinant map of $\mathbf{T}$ into $\mathbf{F}$ (*cf.* [7], chapter 11, or [6], chapter 13, section 4). For any affine transformation, ϕ, let

$$\det \phi = \det((t_1)\phi^i, (t_2)\phi^i, \ldots, (t_n)\phi^i)/\det(t_1, t_2, \ldots, t_n)$$

for any set of independent translations $t_1, \ldots, t_n$. Show that $\det \phi$ is independent of the choice of the t_i and is a homomorphism of the affine group of A into $\mathbf{F}$. The *unimodular affine* group, $\mathcal{S}\mathcal{A}$, is the kernel of this homomorphism. Show that it is generated by the transvections (shears) of A. A transvection is an affine transformation, ϕ, whose fixed points form a hyperplane, H, such that $P + (P)\phi$ is parallel to H for any point P with $P \neq (P)\phi$.

5.7. The analytic representation of affine transformations

In the last section we showed that any affine transformation is closely related to a nonsingular linear transformation (proposition 5.19). In euclidean space we often think of a linear transformation as leaving the origin fixed and moving the other points (viewed as tips of their position vectors) in a way consistent with the way their position vectors are transformed. In the following theorem we show that this can be done in any affine space.

Theorem 5.25. Let P be a point in the affine n-space A whose translation group is $\mathbf{T}$. If f is a nonsingular linear transformation of $\mathbf{T}$, then there is a unique affine transformation, ϕ, such that $(P)\phi = P$ and $f = \phi^i$ restricted to $\mathbf{T}$.

Proof. Assume ϕ and σ are affine transformations leaving P fixed and such that $\phi^i = \sigma^i$ on $\mathbf{T}$. Then for any point Q, $\overrightarrow{P(Q)\phi} = (\overrightarrow{PQ})\phi^i = (\overrightarrow{PQ})\sigma^i = \overrightarrow{P(Q)\sigma}$ (theorem 5.17), and hence $(Q)\phi = (Q)\sigma$ by the transitivity axiom. This shows that there can be at most one affine transformation leaving a given point fixed and inducing a given linear transformation on the translations.

The uniqueness proof shows how we must define ϕ in order to establish existence, namely, $(Q)\phi = R$ if and only if $(\overrightarrow{PQ})f = \overrightarrow{PR}$. Since f is nonsingular (1–1 and onto $\mathbf{T}$), it follows from the transitivity axiom that ϕ is unambiguously defined and is also 1–1 and onto A. Clearly $(P)\phi = P$. Next we show that ϕ^i and f agree on $\mathbf{T}$. To this end let $t \in \mathbf{T}$, and assume $(t)f = s$. To show that $(t)\phi^i$ is also s we must verify that $(Q)\phi^{-1}t\phi = (Q)s$ for any point Q. To simplify notation let $R = (Q)\phi^{-1}$, i.e., choose R such that $(\overrightarrow{PR})f = \overrightarrow{PQ}$. Then

$$(\overrightarrow{P(R)t})f = (\overrightarrow{PR} + t)f = \overrightarrow{PQ} + s = \overrightarrow{P(Q)s}\,,$$

which shows that $(R)t\phi = (Q)s$, i.e., $(Q)\phi^{-1}t\phi = (Q)s$ as desired.

Since ϕ^i and f agree on $\mathbf{T}$, it follows that ϕ^i leaves $\mathbf{T}$ invariant and is homogeneous on $\mathbf{T}$. Hence ϕ is an *affine* transformation, and the proof is complete.

Definition. If f is a nonsingular linear transformation of $\mathbf{T}$, then the *affine form of f leaving P fixed* is the affine transformation, ϕ, such that $(P)\phi = P$ and $\phi^i = f$.

Clearly not all affine transformations are affine forms of linear transformations, e.g., nontrivial translations, which have no fixed points. The following theorem shows that (for a given origin) any affine transformation can be represented uniquely as a translation following the affine form of a linear transformation.

Theorem 5.26. Let P be a point and ϕ an affine transformation. If $t = \overrightarrow{P(P)\phi}$, and σ is the affine form of ϕ^i leaving P fixed, then $\phi = \sigma t$ is the unique way of factoring ϕ such that the first factor is an affine transformation leaving P fixed, and the second factor is a translation.

Proof. Given two such factorizations, the translations would send P to the same point, $(P)\phi$; hence they would be equal by the transitivity axiom. This (by cancellation) would imply that the first factors were equal. Thus, there is at most one such factorization.

Let ϕ, σ, and t be as in the hypotheses of the theorem. Let Q be any point, and let $(Q)\sigma = R$, i.e., $(\overrightarrow{PQ})\phi^i = \overrightarrow{PR}$. Then

$$\begin{aligned}
\overrightarrow{R(Q)\phi} &= \overrightarrow{RQ} + \overrightarrow{Q(Q)\phi}\\
&= \overrightarrow{RQ} + (\overrightarrow{QP} + \overrightarrow{P(P)\phi} + \overrightarrow{(P)\phi(Q)\phi})\\
&= \overrightarrow{RQ} + (\overrightarrow{QP} + t + (\overrightarrow{PQ})\phi^i)\\
&= \overrightarrow{RQ} + (\overrightarrow{QP} + t + \overrightarrow{PR})\\
&= \overrightarrow{RQ} + (t + \overrightarrow{QR})\\
&= t\,.
\end{aligned}$$

However, $\overrightarrow{R(Q)\phi} = t$ if and only if $(Q)\phi = (R)t$. Thus, we have shown that ϕ and σt agree at all points Q, i.e., that $\phi = \sigma t$.

Corollary 5.27. An affine transformation is the affine form of a linear transformation leaving P fixed if and only if it leaves P fixed.

Proof. ϕ leaves P fixed if and only if $\overrightarrow{P(P)\phi}$ is the identity, but $\phi = \sigma$ in the above factorization if and only if t is the identity.

Combining theorems 5.25 and 5.26, we see that the general method of constructing an affine transformation is to choose a linear transformation and a translation and combine them judiciously. For example, we can now prove the converse to theorem 5.24, for if $(P)\mathbf{V}$ and $(P)\mathbf{W}$ are complementary subspaces in A, then $\mathbf{V}$ and $\mathbf{W}$ are complementary in $\mathbf{T}$. Let f be the projection of $\mathbf{T}$ onto $\mathbf{W}$ with kernel $\mathbf{V}$ (theorem 4.12), and let $g = \mathbf{1} - 2f$. If the field underlying our affine space is not of characteristic 2, then g is the involution of $\mathbf{T}$ leaving every vector in $\mathbf{V}$ fixed and sending every vector in $\mathbf{W}$ to its negative. The affine form of g leaving P fixed is the affine involution with mirror $(P)\mathbf{V}$ and direction $(P)\mathbf{W}$.

Now let γ^* be a coordinate map of A onto $A_n(\mathbf{F})$ with origin at P. Given an affine transformation ϕ, we factor ϕ as in theorem 5.26 to get $\phi = \sigma t$. Let B be the γ-matrix of $\phi^i = \sigma^i$, and let $\langle \alpha_1, \ldots, \alpha_n \rangle$ be the γ-coordinates of t. If $(x_1, \ldots, x_n)$ are the γ^*-coordinates of a point Q in A, then $(x_1, \ldots, x_n)B + (\alpha_1, \ldots, \alpha_n)$ are the γ^*-coordinates of $(Q)\phi$.

Example 1. Let our affine space be the standard rational 4-space, $A_4(\mathbf{Q})$, and let ϕ be the affine reflection whose mirror, $(P)\mathbf{V}$, is the solution set of the equations $x_1 + x_2 = 1$, $x_1 - x_3 = -1$ and whose direction is any plane left invariant by both of the translations $\langle 0, 1, 0, 1 \rangle$ and $\langle 0, 0, 1, 1 \rangle$. Let γ^* be the natural coordinate map of $A_4(\mathbf{Q})$, i.e., origin at $(0, 0, 0, 0)$ and unit points at $(1, 0, 0, 0)$, $(0, 1, 0, 0)$, $(0, 0, 1, 0)$, and $(0, 0, 0, 1)$, respectively. A rather long computation (to be explained below) leads to the result that if (x_1, x_2, x_3, x_4) are the γ^*-coordinates of Q, then the γ^*-coordinates of $(Q)\phi$ are

$$(x_1, x_2, x_3, x_4)\begin{pmatrix} 1 & -2 & 2 & 0 \\ 0 & -1 & 0 & -2 \\ 0 & 0 & -1 & -2 \\ 0 & 0 & 0 & 1 \end{pmatrix} + (0, 2, 2, 4) =$$
$$(x_1, -2x_1 - x_2 + 2, 2x_1 - x_3 + 2, -2x_2 - 2x_3 + x_4 + 4)\,.$$

Before trying to follow the derivation of these equations, the reader should choose a few points in $(P)\mathbf{V}$ and see that they are left fixed. Also choose a few points not in $(P)\mathbf{V}$, and note that $Q + (Q)\phi$ is parallel to $(P)\mathbf{W}$, and the midpoint of Q and $(Q)\phi$ is in $(P)\mathbf{V}$.

To find the appropriate matrix and vector involved in the equation above it is easiest to first find the analytic expression for ϕ^i in a coordinate system that is more appropriate for the problem and then change coordinates. Note that the vector space $\mathbf{V}$ (i.e., the direction of the mirror) has a basis $\langle 1, -1, 1, 0 \rangle$ $\langle 0, 0, 0, 1 \rangle$. We were already given a basis, $\langle 0, 1, 0, 1 \rangle$ $\langle 0, 0, 1, 1 \rangle$, for $\mathbf{W}$. These four vectors (in this order) form a basis for $\mathbf{V}_4(\mathbf{Q})$. Let β be the coordinate map of $\mathbf{V}_4(\mathbf{Q})$ relative to this basis, then the β-matrix of ϕ^i is

$$B = \begin{pmatrix} 1 & 0 & 0 & 0 \\ 0 & 1 & 0 & 0 \\ 0 & 0 & -1 & 0 \\ 0 & 0 & 0 & -1 \end{pmatrix}$$

since ϕ^i leaves the first two β-basis vectors fixed and sends the last two to their negatives. Chasing the diagram

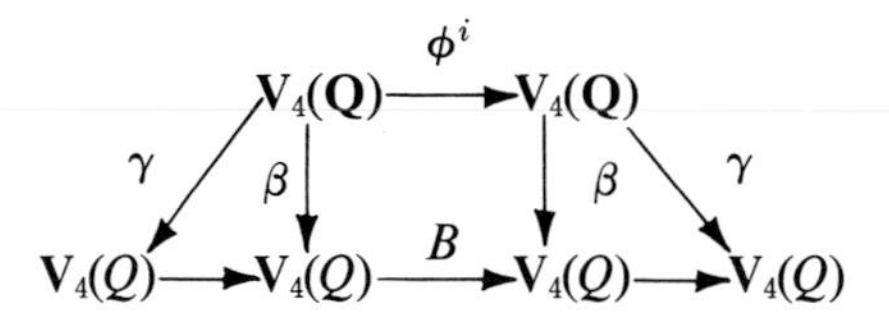

shows that $C = (\gamma^{-1}\beta)B(\beta^{-1}\gamma)$ is the γ-matrix for ϕ^i. It is easy to find the natural matrix for $\beta^{-1}\gamma$ and then invert it to get $\gamma^{-1}\beta$. The results are

$$(\gamma^{-1}\beta)B(\beta^{-1}\gamma) = \begin{pmatrix} 1 & 0 & 1 & -1 \\ 0 & -1 & 1 & 0 \\ 0 & -1 & 0 & 1 \\ 0 & 1 & 0 & 0 \end{pmatrix} \begin{pmatrix} 1 & 0 & 0 & 0 \\ 0 & 1 & 0 & 0 \\ 0 & 0 & -1 & 0 \\ 0 & 0 & 0 & -1 \end{pmatrix} \begin{pmatrix} 1 & -1 & 1 & 0 \\ 0 & 0 & 0 & 1 \\ 0 & 1 & 0 & 1 \\ 0 & 0 & 1 & 1 \end{pmatrix}$$

$$= \begin{pmatrix} 1 & -2 & 2 & 0 \\ 0 & -1 & 0 & -2 \\ 0 & 0 & -1 & -2 \\ 0 & 0 & 0 & 1 \end{pmatrix} = C.$$

Finally, to find the vector t^γ such that $(Q)\phi\gamma^* = ((Q)\gamma^*)C + t^\gamma$ for all points Q we choose a point, P, in $(P)\mathbf{V}$ and solve the equation $(P)\gamma^* = ((P)\gamma^*)C + t^\gamma$. One choice of P is $(1, 0, 2, 0)$, which leads to the equation (with $t^\gamma = \langle \alpha_1, \alpha_2, \alpha_3, \alpha_4 \rangle$).

$(1, 0, 2, 0)C + t^\gamma = (1, -2, 0, -4) + (\alpha_1, \alpha_2, \alpha_3, \alpha_4) = (1, 0, 2, 0)$. Note that in this last part we are using the fact that for P in $(P)\mathbf{V}$ we must have $(P)\phi = P$.

As a final application of the connection between affine and linear transformations, we present the following result concerning the transitivity of the group of affine transformations.

Theorem 5.28. Let $P_0, P_1, \ldots, P_n$ and $Q_0, Q_1, \ldots, Q_n$ be two ordered sets of $n + 1$ independent points in an affine n-space A. There is exactly one affine transformation, ϕ, such that $(P_i)\phi = Q_i$, $i = 0, 1, \ldots, n$.

Proof. Let $v_i = \overrightarrow{P_0P_i}$ and $w_i = \overrightarrow{Q_0Q_i}$ for $i = 1, 2, \ldots, n$. There is exactly one linear transformation, f, carrying v_i to w_i, $i = 1, 2, \ldots, n$. Let σ be the affine form of f leaving P_0 fixed, and let $t = \overrightarrow{P_0Q_0}$. Then $\phi = \sigma t$ is the unique affine transformation sending P_i to Q_i for $i = 0, 1, \ldots, n$ (exercise 5.7, problem 1).

Exercise 5.7

1. Complete the proof of theorem 5.28 as follows: (1) Show that σt actually does send P_i to Q_i, $i = 0, 1, \ldots, n$; and (2) show that if ϕ is an affine transformation sending P_i to Q_i, $i = 0, 1, \ldots, n$, then $\phi^i = f$, and hence (theorem 5.26) $\phi = \sigma t$.

2. Use theorem 5.28 to show that there are $f^n(f^n - 1)(f^n - f) \cdots (f^n - f^{n-1})$ affine transformations of an affine n-space over a finite field with f elements.

3. Let ϕ and θ be affine transformations of an affine space A, and let λ^* be a coordinate map for A. If $(x_1, \ldots, x_n)B + (\beta_1, \ldots, \beta_n)$ and $(x_1, \ldots, x_n)C + (\gamma_1, \gamma_2, \ldots, \gamma_n)$ are the analytic forms of ϕ and θ, respectively, then what is the analytic form of $\phi\theta$?

4. Discuss the relation between the analytic forms of an affine transformation ϕ in two affine coordinate systems, λ^* and μ^*. Let s be the translation carrying the λ^* origin to the μ^* origin, and let P be the natural matrix of $\lambda^{-1}\mu$.

5. Assume P and Q are two distinct points in an affine space. Let f be a nonsingular linear transformation of the translations, and let s be the translation $(\overrightarrow{PQ})f + \overrightarrow{QP}$. Show that if ϕ and θ are the affine forms of f leaving P and Q fixed, respectively, then $\phi = \theta s$.

6. Let $\mathcal{A}$ be the group of affine transformations of an affine space A, and let $[\mathcal{A}, P]$ be the subgroup leaving the point P fixed. Show that $[\mathcal{A}, P]$ is isomorphic to $\mathcal{A}/\mathbf{T}$. *Hint*. First show that $[\mathcal{A}, P]$ is isomorphic to $GL_n(\mathbf{F})$, and then find a homomorphism of $\mathcal{A}$ onto $GL_n(\mathbf{F})$ with kernel $\mathbf{T}$.

5.8. *Affine symmetries III: Collineations*

We turn now to the weakest type of symmetry of an affine space. If the affine space is of dimension at most one or if the underlying field has only two elements, then *any* permutation of the points is a symmetry of the type we are about to consider, so we avoid these cases.

Definition. Assume that A is an affine space of dimension at least two over a field with at least three elements. Then a permutation, ϕ, of A is called a *collineation* of A if for all triples, P, Q, and R, of points in A; P, Q, and R are collinear *if and only if* $(P)\phi$, $(Q)\phi$, and $(R)\phi$ are collinear.

Note that while we avoid the field with two elements, we do *not* avoid all fields of characteristic 2. The difficulty with affine spaces over the field with two elements arises not from the fact that $t + t = 0$, but from the fact that each line in such a space contains only two points. In all other cases, lines contain at least three points, and we can prove the following theorem.

Theorem 5.29. Let B be a non-empty subset of an affine space with at least three points on every line. B is an affine subspace if and only if for every pair of points in B the line joining them is entirely contained in B.

Proof. If B is a subspace, say, $B = (P)\mathbf{V}$, and if Q, R are in B, then $\overrightarrow{QR}$ is in $\mathbf{V}$. This implies that the linear subspace, $\mathbf{W}$, generated by $\overrightarrow{QR}$ is contained in $\mathbf{V}$, and hence $Q + R = (Q)\mathbf{W} \subseteq (P)\mathbf{V}$ (theorem 5.3).

Conversely if B is a non-empty subset of A containing $P + Q$ whenever it contains P and Q, then choose a point P in B, and let $\mathbf{V} = \{\overrightarrow{PQ} | Q \in B\}$. Clearly the zero vector, $\overrightarrow{PP}$, is in $\mathbf{V}$, and if a vector $\overrightarrow{PQ}$ is in $\mathbf{V}$, then so is $\alpha\overrightarrow{PQ}$ for any scalar α. Now assume that $v = \overrightarrow{PQ}$ and $w = \overrightarrow{PR}$ are in $\mathbf{V}$. Let $S = (P)v + w$, i.e., $\overrightarrow{PS} = v + w$. If P, Q, and R are collinear, then $v + w = v + \alpha v = (1 + \alpha)v$ is in $\mathbf{V}$. Now assume (Figure 27) that P, Q, and R are not

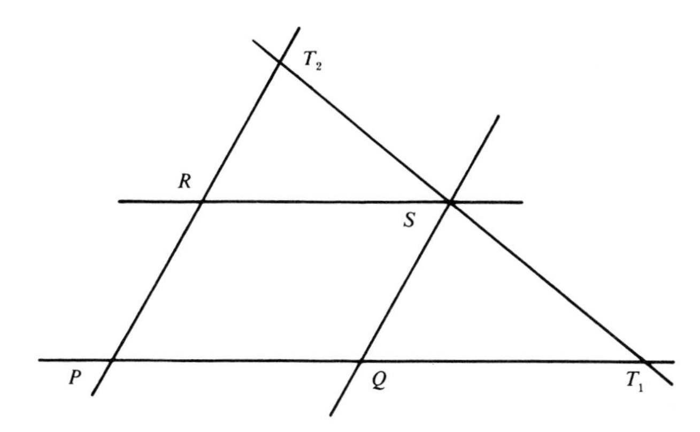

FIG. 27. An affine subspace $B = (P)\mathbf{V}$ defined in terms of lines.

collinear. Since there are at least three points on $P + Q$, we can choose a point T_1 on $P + Q$ and distinct from P or Q. The line $Q + S$ is a line through S parallel to $P + R$; hence (theorem 5.8) the line $S + T_1$ is not parallel to the line $P + R$. The lines $P + R$ and $S + T_1$ are hyperplanes in the plane generated by P, Q, and R. Hence by theorem 5.11 they cannot be skew. Thus, $P + R$ and $S + T_1$ intersect at a point, say, T_2. The points T_1 and T_2 are in B since they are on $P + Q$ and $P + R$, and so the point S on $T_1 + T_2$ must be in B. We have shown that $\mathbf{V}$ is non-empty, closed under addition and multiplication by scalars; hence $\mathbf{V}$ is a linear subspace, and thus, $B = (P)\mathbf{V}$ is an affine subspace.

The theorem above shows that affine subspaces can be defined in terms of lines. It follows immediately that a collineation maps an affine subspace to an affine subspace. We can say more, however, as we see in the following theorem.

Theorem 5.30. Let ϕ be a collineation of the affine space A (with at least three points on each line), and let B be an affine subspace of dimension d. Then $(B)\phi$ is a d-dimensional affine subspace.

Proof. Choose (theorem 5.5) points $P_0, P_1, \ldots, P_d$ which generate B. Then $(B)\phi$ is generated by $(P_0)\phi, (P_1)\phi, \ldots, (P_d)\phi$, and hence $\dim(B)\phi \leq d$ (theorem 5.5). The same argument applied to ϕ^{-1} shows $d = \dim B \leq \dim(B)\phi$. Thus, $\dim(B)\phi = d$.

This property of collineations obviously characterizes them and is the way they should be defined for affine spaces over the field with two elements. An affine transformation is clearly a collineation. In the following we show that a collineation satisfies one of the two requirements for affine transformations.

Theorem 5.31. If ϕ is a collineation of an affine space A, and t is a translation of A, then $\phi^{-1}t\phi = (t)\phi^i$ is also a translation.

Proof. This is trivial if $t = 0$. Nontrivial translations are those dilatations with no fixed points (theorem 5.15), and thus, since $(t)\phi^i$ has no fixed points, it is sufficient to prove that $\phi^{-1}t\phi$ is a dilatation. Given two points, since ϕ is onto, we may call them $(P)\phi$ and $(Q)\phi$. The lines $P + Q$ and $(P)t + (Q)t$ are parallel; hence so are $(P)\phi + (Q)\phi$ and $(P)t\phi + (Q)t\phi$ (exercise 5.8, problem 1). But $\phi^{-1}t\phi$ sends $(P)\phi$ and $(Q)\phi$ to $(P)t\phi$ and $(Q)t\phi$, respectively, and thus, since $(P)\phi$ and $(Q)\phi$ were arbitrary, $\phi^{-1}t\phi$ is a dilatation without fixed points (a translation!).

Corollary 5.32. A permutation, ϕ, of an affine space A (of dimension at least 2 and not over the field with two elements) is an affine transformation if and only if:

(1) It maps lines onto lines.

(2) It preserves ratios of parallel vectors, i.e., for any points P, Q, R, and S, if $\overrightarrow{RS} = \alpha\overrightarrow{PQ}$, then $\overrightarrow{(R)\phi(S)\phi} = \alpha\overrightarrow{(P)\phi(Q)\phi}$.

Proof. The first requirement is another way of saying that ϕ is a collineation. By the theorem above this ensures that ϕ^i leaves $\mathbf{T}$ invariant. The second requirement is equivalent to the second requirement for affine transformations since ϕ^i leaves $\mathbf{T}$ invariant (theorem 5.17).

An affine n-space A with translation group $\mathbf{T}$ over the field $\mathbf{F}$ involves three mathematical systems, A, $\mathbf{T}$, and $\mathbf{F}$. The translations of A permute the points in A but induce the identity transformation on both $\mathbf{T}$ and $\mathbf{F}$. Now we consider those symmetries of A which involve nontrivial symmetries of both $\mathbf{T}$ and $\mathbf{F}$ and relate these symmetries to the collineations. First we must define what we mean by a symmetry of $\mathbf{F}$.

Definition. An *automorphism* of the field **F** is a permutation, λ, of **F** such that for all α, β in F $(\alpha + \beta)^\lambda = \alpha^\lambda + \beta^\lambda$ and $(\alpha \cdot \beta)^\lambda = \alpha^\lambda\beta^\lambda$.

Since the zero element of a field is the only solution of $x + x = x$, and the multiplicative identity is the only nonzero solution of $xx = x$; it follows that zero and the multiplicative identity are always left fixed by an automorphism. The set of elements left fixed by an automorphism of **F** is always a subfield of **F**; hence the only automorphism of $\mathbf{Z}_p$, (p prime) or **Q** are the trivial ones. We show in section 5.11 that **R** has no nontrivial automorphisms either. Complex conjugation in **C** is probably the most familiar automorphism of **C**. It is by no means the only nontrivial automorphism of **C**, but the others [12] are constructed using the axiom of choice, and probably none of them can be described in finite terms. The smallest field with a nontrivial automorphism is the four-element field described in example 1 of section 4.1. The only nontrivial automorphism of this field interchanges b and $1 + b$.

Definition. Let **V** and **W** be vector spaces over the same field **F**, and let λ be an automorphism of **F**. A mapping, g, from **V** into **W** is said to be *semi-linear* (*with respect to* λ) if:

(1) For all $v, w \in \mathbf{V}$, $(v + w)g = (v)g + (w)g$.

(2) For all $v \in \mathbf{V}$, $\alpha \in \mathbf{F}$, $(\alpha v)g = \alpha^\lambda(v)g$.

The two requirements for semi-linearity of g say essentially that g "λ-preserves" all linear combinations, i.e., that

$$(\alpha_1 v_1 + \cdots + \alpha_n v_n)g = \alpha_1^\lambda(v_1)g + \alpha_2^\lambda(v_2)g + \cdots + \alpha_n^\lambda(v_n)g\,.$$

We call a semi-linear map *nonsingular* if and only if it is 1–1 and onto.

Theorem 5.33. Let A be an affine n-space with translation group **T**, and let g be a nonsingular, semi-linear mapping of **T** onto itself. Given a point, P, the mapping γ defined by $(Q)\gamma = R$ if and only if $(\overrightarrow{PQ})g = \overrightarrow{PR}$ is a permutation of A which sends subspaces of A to subspaces of the same dimension and hence is a collineation.

Proof. Since g is a permutation of **T**, γ is a permutation of A by the transitivity axiom. If **V** is a d-dimensional linear subspace of **T**, then $(\mathbf{V})g$ is also (exercise 5.8, problem 2). If $(Q)\gamma = R$ and $(\mathbf{V})g = \mathbf{W}$, then $((Q)\mathbf{V})\gamma = (R)\mathbf{W}$ (exercise 5.8, problem 2); hence g sends affine subspaces to subspaces of the same dimension.

One of the most interesting theorems in geometry is the converse of the theorem above, i.e., if γ is a collineation of an affine space of dimension at least 2 (Interpret "collineation" as preserving subspaces and their dimension if the field only has two elements!), and if γ has a fixed point P, then there is a semi-linear mapping g such that γ is induced by g in the manner above. We prove the projective form of this theorem in Chapter 6 and then derive this as a corollary. For a proof not involving projective spaces, see the appendix by Ebey in [9].

Exercise 5.8

1. Show that collineations preserve parallelism, i.e., assuming ϕ is a collineation and that B and C are parallel subspaces, show that $(B)\phi$ and $(C)\phi$ are parallel.

2. Complete the proof of theorem 5.33.

3. Assuming the converse of theorem 5.33 (proved in section 6.7), discuss the analytic form of collineations. Note carefully how the automorphism of $\mathbf{F}$ enters the picture.

4. Assuming the converse of theorem 5.33, show that there are $2^7 \cdot 3^2 \cdot 5$ collineations of the affine plane over the field with four elements.

5. Assuming the converse of theorem 5.33, prove that if $\mathbf{S}_A$ is the group of all permutations of A and $\mathbf{D}$ is the group of dilatations of A, then the normalizer of $\mathbf{D}$ in $\mathbf{S}_A$ is the group of all collineations.

5.9. *Volume in real affine spaces*

In this section and the next two, we consider concepts and results for affine spaces over the field of real numbers, $\mathbf{R}$. Most of these concepts depend on the fact that the field $\mathbf{R}$ is an ordered field. A few results use the additional fact that $\mathbf{R}$ is a *complete* ordered field.

Definition. Given two points P and Q in a real affine space, the (closed) *line segment*, $\overline{PQ}$, is the set $(P)X$ where $X = \{\alpha\overrightarrow{PQ} | 0 \leq \alpha \leq 1\}$.

Intuitively, we say that the line segment $\overline{PQ}$ is the set of all points, P, Q, and all points between P and Q and on the straight line joining them. An active research area within geometry is the study of convex sets. A subset, B, of an affine space is said to be *convex* if $\overline{PQ} \subseteq B$ for *all* pairs of points, P, Q, in B. We will not take time to present any results about convexity. For more information about convex sets, see [10]. An introductory article about some of the central theorems concerning convexity is [4].

We consider now one of the basic types of convex sets, the parallelepipeds. Usually one considers a parallelepiped to be a three-dimensional object, but we can generalize to other dimensions.

Definition. Let P be a point and $v_1, v_2, \ldots, v_k$ be k independent translations of a real affine space A. The set $(P)X$ with

$$X = \left\{v \middle| v = \sum_{i=1}^{k} \alpha_i v_i, 0 \leq \alpha_i \leq 1\right\}$$

is called the *k-parallelepiped* with *initial vertex* at P and *prime edges* in the direction of $v_1, v_2, \ldots, v_k$. If $k = \dim A$, then we call $(P)X$ a *box*. Single points are sometimes referred to as *0-parallelepipeds*.

Note that a k-parallelepiped is a closed subset of the affine space A. The k-parallelepipeds for $k = 1, 2$, and 3 are line segments, parallelograms, and (ordinary) parallelepipeds. Many authors use the term *parallelotope* in place of parallelepiped. Of course, if $k > 1$, a k-parallelepiped has many edges in addition to its prime edges (exercise 5.9, problem 4).

In euclidean geometry the area of a parallelogram is the length of its base times its height, and the volume of a parallelepiped is the area of its base times its height. Using these concepts, we can compare the size of various parallelograms or parallelepipeds. Interestingly enough, these concepts can be generalized, so that in affine geometry we can compare the size of various boxes in a given affine space. For $k < n$ one cannot compare the size of k-parallelepipeds unless they are in the same (or parallel) k-dimensional subspaces. For example, in a real affine plane we can compare the sizes (areas) of parallelograms but cannot compare the lengths of line segments unless they are on parallel lines. Similarly the volumes of parallelepipeds can be compared in a real affine 3-space, but lengths of line segments (respectively, areas of parallelograms) can be compared only if they are on parallel lines (respectively, planes). The comparison of lengths on nonparallel lines is *not* an affine concept in that it depends on the way in which perpendicularity or distance is defined. To discuss this fully would require a study of bilinear and quadratic forms which we (regretfully) avoid and say simply that there is more than one way to define perpendicularity in a given real affine space, i.e., there *is* more than one euclidean space attached to a given affine space (*cf.* [2], p. 177).

Another way of showing that a given real affine space contains more than one euclidean space is based on the results in section 2.8 concerning the automorphisms of the euclidean group, **E**. Assume that euclidean 3-space is imbedded in the natural way in $A_3(\mathbf{R})$. With this imbedding it is clear that the euclidean group, **E**, and the similarities group, **S**, are both subgroups of the affine group, $\mathfrak{A}$. Since there are affine transformations which are not similarities, neither **E** nor **S** is a normal subgroup of A (exercise 5.9, problem 2). If $\mathbf{E}_1$ and **E** are conjugate subgroups in $\mathfrak{A}$, but $\mathbf{E}_1 \neq \mathbf{E}$, then $\mathbf{E}_1$ determines a second euclidean geometry within $A_3(\mathbf{R})$. This argument is along the lines of the point of view developed by Klein in his "Erlangen Programme."

Now let us develop the method of comparing boxes in a real affine space. To do this we need first to review some properties of determinants. In addition to viewing a determinant as a function defined on n by n matrices, we view it as a function of n variables where the n variables are the vectors (n-tuples) forming the rows of the n by n matrix. With this point of view, we have the following:

(1) $\det(e_1, \ldots, e_n) = 1$ if e_i is the ith unit vector, i.e., the vector with 1 in the ith position and 0 elsewhere.
(2) $\det(v_1, \ldots, v_{i-1}, \alpha v_i, v_{i+1}, \ldots, v_n) = \alpha \det(v_1, \ldots, v_n)$ for all real numbers α and all n-tuples v_j.
(3) If $v_i = u_i + w_i$, then $\det(v_1, \ldots, v_n) = \det(v_1, \ldots, v_{i-1}, u_i, v_{i+1}, \ldots, v_n) + \det(v_1, \ldots, v_{i-1}, w_i, v_{i+1}, \ldots, v_n)$.
(4) If we interchange any two of the v_i, then $\det(v_1, \ldots, v_n)$ changes sign.
(5) If P is an n by n matrix, and if $w_i = v_iP$, $i = 1, \ldots, n$, then $\det(w_1, \ldots, w_n) = \det(v_1, \ldots, v_n)\det(P)$.

To motivate the following definition we suggest that the reader convince himself that the area of any parallelogram in $A_2(\mathbf{R})$ or the volume of any box in $A_3(\mathbf{R})$ is the absolute value of the determinant of its prime edges (exercise 5.9, problem 3).

Definition. If γ^* is a coordinate map for a real affine n-space, then the γ-*volume* of a box, $(P)X$, with initial vertex at P and prime edges $v_1, v_2, \ldots, v_n$ is $|\det(v_1^\gamma, v_2^\gamma, \ldots, v_n^\gamma)|$.

Strictly speaking, the γ-volume of a box is not the volume of the box, but rather the ratio of the volume of that box to the volume of the γ-unit box. In other words, one should think of the box with vertices at the origin and unit points of the γ^* coordinate system as the "unit of volume." The following theorem shows that "volume of boxes" is an affine concept, i.e., that changing coordinates only changes the unit of volume, and not the ratio between volumes of two boxes.

Theorem 5.34. Assume γ^* and δ^* are two coordinate maps of a given real affine n-space onto $A_n(\mathbf{R})$. Then there is a constant α such that the δ-volume of any box is just α times its γ-volume.

Proof. Consider the following pair of commutative diagrams in which A is our given real affine n-space and $\mathbf{T}$ is its translation group.

$$\begin{array}{ccc} A & \xrightarrow{\mathbf{1}_A} & A \\ \gamma^* \downarrow & & \downarrow \delta^* \\ A_n(\mathbf{R}) & \xrightarrow{\gamma^{*-1}\delta^*} & A_n(\mathbf{R}) \end{array} \qquad \begin{array}{ccc} \mathbf{T} & \xrightarrow{\mathbf{1}_T} & \mathbf{T} \\ \gamma \downarrow & & \downarrow \delta \\ \mathbf{V}_n(\mathbf{R}) & \xrightarrow{\gamma^{-1}\delta} & \mathbf{V}_n(\mathbf{R}) \end{array}$$

Let B be the natural matrix of $\gamma^{-1}\delta$. If $(P)X$ is a box in A with edges $t_1, t_2, \ldots, t_n$, then the γ- and δ-volumes of $(P)X$ are $|\det(t_1^\gamma, \ldots, t_n^\gamma)|$ and $|\det(t_1^\delta, \ldots, t_n^\delta)|$, respectively. However, $t_i^\delta = t_i^{\gamma(\gamma^{-1}\delta)} = t_i^{\gamma B}$. Hence by the multiplicative property of determinants [property (5) above], the δ-volume of $(P)X$ is just $|\det B|$ times the γ-volume. Thus, our desired constant is $\alpha = |\det B|$.

Note that the volume of a box is defined in terms of the edges of the box and in no way involves the initial vertex P. In other words, volume is invariant under translation. This is also implied by our notation since we say γ-volume, instead of γ^*-volume, i.e., the origin of the γ^* coordinate map is immaterial when computing volumes of boxes.

What about the volume of k-parallelepipeds if $k < n$? A given k-parallelepiped is a subset of a unique k-dimensional affine subspace and is a box *in this subspace*. Thus, we can consider its volume by viewing this subspace as an affine space in its own right. Again the volume is invariant under translation *whether or not the translation leaves the subspace invariant.* We must be careful not to compare volumes of k-parallelepipeds on nonparallel subspaces, however. For example, consider four points P, Q, R, and S in a real affine 2-space such that $\overrightarrow{PS} = 2\overrightarrow{PQ}$ and P, Q, R are *not* collinear. There is an affine transformation of this plane sending P to P, Q to S, and R to R (theorem 5.28). Thus, it is possible to change lengths on $P + Q$ without changing those on $P + R$.

The generalization of volume to regions which are not boxes can be done as in any of the theories of multiple integration. Depending on the sophistication of the integral used, more or less complicated regions are "measurable," i.e., assigned a volume.

Exercise 5.9

1. Let P, Q, and R be three points in a real affine space. Show that $\overline{QR} = (P)Y$ if $Y = \{\alpha\overrightarrow{PQ} + \beta\overrightarrow{PR} | \alpha \geq 0,\ \beta \geq 0, \text{ and } \alpha + \beta = 1\}$.

2. Assume ϕ is an affine transformation of $A_3(\mathbf{R})$ but is not a similarity transformation. Use corollary 2.29 to prove that $(\mathbf{E})\phi^i \neq \mathbf{E}$ and $(\mathbf{S})\phi^i \neq \mathbf{S}$.

3. Show that the volume of any box in euclidean space is $\pm$ the determinant of its three edges. Since an isometry has determinant ± 1 and does not change volumes, you may assume that the first edge lies on the positive x-axis, the second in the first quadrant of the xy-plane, and the third in the first octant.

4. Define "j-face of a k-parallelepiped" in a way that generalizes the 8 vertices (0-faces), 12 edges (1-faces), and 6 faces (2-faces) of an ordinary parallelepiped. Show that there are $2^{k-j}\binom{k}{j}$, j-faces on a k-parallelepiped. See [5] for a discussion of the special case, $k = 4$.

5.10. Lattices in real affine spaces

A subset of a real affine space is said to be *bounded* if it can be contained in a box. The definition of discrete groups given in Chapter 3 generalizes easily to real affine spaces; namely, a subgroup, **G**, of the group of affine transformations is *discrete* if and only if for any point P and any bounded set B, $(P)\mathbf{G} \cap B$ is finite. A discrete group all of whose elements are translations is called a *lattice group.* If **G** is a lattice group and P is a point, then $(P)\mathbf{G}$ is called a *lattice.* In section 3.3 we promised to give a proof of the following theorem.

Theorem 5.35. Let $t_1, t_2, \ldots, t_n$ be n independent translations, and let $\mathbf{G} = \{a_1t_1 + \cdots + a_nt_n | a_i \text{ are integers}\}$, then **G** is a lattice group. In other words, the subgroup, **G**, of **T** generated by n independent translations is a lattice group.

Proof. There are two main steps to the proof. In the first we count the number of points in the intersection of a **G**-orbit with a box which is positioned carefully. In the second step we show that any bounded subset can be contained in such a carefully positioned box.

Lemma 1. Assume $Q \in (P)\mathbf{G}$ and $w_i = a_it_i$, a_i integers for $i = 1, 2, \ldots, n$. The box, $(Q)Y$, with initial vertex Q and prime edges $w_1, w_2, \ldots, w_n$ has $\prod_{j=1}^{n} (a_j + 1)$ points in common with $(P)\mathbf{G}$.

Proof. Since $Q \in (P)\mathbf{G}$, $(Q)\mathbf{G} = (P)\mathbf{G}$, and thus, $R \in (P)\mathbf{G}$ if and only if $\overrightarrow{QR} \in \mathbf{G}$, i.e., if and only if $\overrightarrow{QR} = \sum_{j=1}^{n} b_jt_j$ with all b_j in **Z**. But R is also in the box $(Q)Y$ if and only if $0 \leq b_j \leq a_j$. By the choice principle there are $\prod_{j=1}^{n} (a_j + 1)$ possible points R in both $(P)\mathbf{G}$ and the box since there are $a_j + 1$ possible integers, b_j, for $j = 1, 2, \ldots, n$.

Lemma 2. Any bounded set can be contained in a box of the type specified in lemma 1.

Proof. Let γ^* be the coordinate map with origin at P and unit vectors $t_1, t_2, \ldots, t_n$. Let B be a bounded set. By definition B can be contained in a box; call the 2^n vertices of this box R_i, $i = 1, 2, \ldots, 2^n$. Let the γ^* coordinates of R_i be $(\alpha_{i_1}, \alpha_{i_2}, \ldots, \alpha_{i_n})$ for $i = 1, 2, \ldots, 2^n$. For each $j = 1, 2, \ldots, n$, examine the 2^n numbers α_{i_j}, and choose integers b_j and c_j such that $b_j \leq \alpha_{i_j} \leq c_j$ for $i = 1, 2, \ldots, 2^n$. Choose Q such that $\overrightarrow{PQ} = \sum_{j=1}^{n} b_jt_j$. The box with initial vertex at Q and prime edges $(c_j - b_j)t_j$ contains all points whose γ^* coordinates have jth component between b_j and c_j, $j = 1, 2, \ldots, n$. Thus, this box at Q contains all of the R_i and hence contains the entire R-box (exercise 5.10, problem 1) and, in particular, contains B. The box at Q is a box of the form specified in lemma 1; hence it is our desired box.

Now let P be any point and B be any bounded set. By lemma 2, B is contained in a box having only a finite number of points in common with $(P)\mathbf{G}$ (lemma 1). Hence $B \cap (P)\mathbf{G}$ is finite, and thus, $\mathbf{G}$ is discrete.

Definition. If $t_1, t_2, \ldots, t_n$ are independent translations, and $\mathbf{G} = \{a_1t_1 + \cdots + a_nt_n | a_i \text{ integers}\}$, then we write $\mathbf{G} = \mathbf{Z}t_1 + \cdots + \mathbf{Z}t_n$ and call $t_1, \ldots, t_n$ an *integral basis* for the lattice group $\mathbf{G}$.

Theorem 5.36. Let $\mathbf{G}$ be a lattice group containing n independent translations $v_1, v_2, \ldots, v_n$. There is an integral basis $t_1, t_2, \ldots, t_n$ for $\mathbf{G}$ which can be chosen such that $t_1, \ldots, t_k$ and $v_1, \ldots, v_k$ generate the same (linear) subspace for $k = 1, 2, \ldots, n$.

Proof. Choose a point P, and let $(P)X$ be the box with initial vertex at P and prime edges $v_1, v_2, \ldots, v_n$. Since $\mathbf{G}$ is discrete, there are only a finite number of points in $(P)\mathbf{G} \cap (P)X$.

We choose the t_j inductively as follows. Assume $t_1, \ldots, t_{k-1}$ have been chosen such that they generate the same linear subspace as do $v_1, v_2, \ldots, v_{k-1}$. Let $(P)\mathbf{V}$ be the affine subspace through P in the direction of $v_1, v_2, \ldots, v_k$, and let γ^* be the coordinate map of $(P)\mathbf{V}$ onto $A_k(\mathbf{R})$ with vertex at P and unit vectors $v_1, v_2, \ldots, v_k$. Consider those boxes in $(P)\mathbf{V}$ (k-parallelepipeds in A) that have initial vertex at P and such that the first $k - 1$ prime edges are $t_1, t_2, \ldots, t_{k-1}$, and the last prime edge, w, satisfies $(P)w \in (P)\mathbf{G} \cap (P)X \cap (P)\mathbf{V}$, i.e., $w \in \mathbf{G} \cap X \cap \mathbf{V}$. There is at least one such box (with $w = v_k$) but only a finite number [since $(P)\mathbf{G} \cap (P)X$ is finite]. Hence we can choose t_k to be a "last prime edge" w such that the corresponding box has minimal (but nonzero!) volume. With this choice $t_1, t_2, \ldots, t_k$ generate the same subspace as $v_1, v_2, \ldots, v_k$, and our choice process can be continued inductively until $k = n$.

Now we must prove that the $t_1, t_2, \ldots, t_n$ chosen above form an integral basis for $\mathbf{G}$. Clearly $\mathbf{Z}t_1 + \mathbf{Z}t_2 + \cdots + \mathbf{Z}t_n \subseteq \mathbf{G}$; hence (exercise 5.10, problem 4) it is enough to show that if $(P)Y$ is the box with initial vertex at P and prime edges $t_1, t_2, \ldots, t_n$, then $(P)\mathbf{G} \cap (P)Y$ is just the set of vertices of $(P)Y$. We prove this inductively in the following lemma.

Lemma. Let $(P)Y_k$ be the k-parallelepiped with initial vertex at P and prime edges $t_1, t_2, \ldots, t_k$. $(P)\mathbf{G} \cap (P)Y_k$ consists only of the vertices of $(P)Y_k$ for $k = 1, 2, \ldots, n$.

Proof. This is clear for $k = 1$ since t_1 was chosen to be the shortest vector in $\mathbf{G}$ of the form αv_1, $0 < \alpha \leq 1$. Now assume the lemma is true for k, and let Q be a point in $(P)\mathbf{G} \cap (P)Y_{k+1}$. Then $\overrightarrow{PQ} \in \mathbf{G}$ and

$$\overrightarrow{PQ} = \alpha_1 t_1 + \cdots + \alpha_k t_k + \alpha_{k+1} t_{k+1},\ 0 \leq \alpha_j \leq 1 .$$

If $\alpha_{k+1} = 0$, then by the induction hypothesis Q is a vertex of $(P)Y_k$ and hence a vertex of $(P)Y_{k+1}$. If $\alpha_{k+1} = 1$, then $\overrightarrow{PQ} - t_{k+1} \in \mathbf{G} \cap Y_k$ and thus by the induction hypothesis carries P to a vertex of $(P)Y_k$. Since t_{k+1} carries vertices of $(P)Y_k$ to vertices of $(P)Y_{k+1}$, this implies that Q is a vertex of $(P)Y_{k+1}$. Finally if $0 < \alpha_{k+1} < 1$, then

$$|\det(t_1, \ldots, t_k, \overrightarrow{PQ})| = |\det(t_1, t_2, \ldots, \alpha_{k+1}t_{k+1})| = |\alpha_{k+1}|\,|\det(t_1, t_2, \ldots, t_{k+1})| < |\det(t_1, t_2, \ldots, t_{k+1})|\,.$$

But this contradicts the choice of t_{k+1}, and so $0 < \alpha_{k+1} < 1$ is impossible. Thus, any point Q in $(P)\mathbf{G} \cap (P)Y_{k+1}$ is a vertex of $(P)Y_{k+1}$, and the induction step is complete, proving the lemma for $k = 1, 2, \ldots, n$. In view of the remarks before the lemma, this also completes the proof of the theorem.

Definition. If $t_1, t_2, \ldots, t_n$ is an integral basis for the lattice group $\mathbf{G}$, then the box with initial vertex at P and prime edges $t_1, \ldots, t_n$ is called a *primitive cell* or *unit cell* of the lattice $(P)\mathbf{G}$.

Theorem 5.37. The volumes of the various primitive cells of a given lattice are all the same.

Proof. Let $t_1, \ldots, t_n$ and $w_1, \ldots, w_n$ be any two integral bases for the same lattice group $\mathbf{G}$, and let γ be the coordinate map of $\mathbf{T}$ with respect to the basis $t_1, \ldots, t_n$. The γ-volume of a unit cell with prime edges $t_1, \ldots, t_n$ is 1 and the γ-volume of a unit cell with prime edges $w_1, \ldots, w_n$ is a positive integer since the γ-coordinates of each of the w_i are integers. Thus, the volume of the t-unit cell is (in *any* coordinate system) less than or equal to the volume of the w-unit cell. Reversing the argument shows that the two-unit cells must have the same volume.

Note that in the preceding definitions and theorems we seem to have implied that the lattice is n-dimensional and that n is the dimension of the underlying affine space. The results are, in fact, true if the lattice is only k-dimensional, i.e., if the lattice group contains only k independent translations, for $k < n$. For if this is the case, instead of using A as our underlying affine space, we use the subspace generated by the lattice points.

If a lattice is one-dimensional, then it has only two primitive cells with given initial vertex P. However, if it has dimension greater than one, then it has infinitely many unit cells with given initial vertex. The relationship between these unit cells is given in the following theorem.

Theorem 5.38. The ordered integral bases for an n-dimensional lattice group $\mathbf{G}$ are in 1–1 correspondence with the group of n by n matrices with integer entries and determinant ± 1 (the extended group of integral, unimodular matrices). Thus, for the lattice $(P)\mathbf{G}$, each unit cell with initial vertex at P corresponds to $n!$ matrices in the group above.

Proof. Let $t_1, \ldots, t_n$ be a fixed integral basis of $\mathbf{G}$, and let γ be the associated coordinate map of $\mathbf{T}$ onto $\mathbf{V}_n(\mathbf{R})$. The correspondences

ordered basis $w_1, w_2, \ldots, w_n \leftrightarrow \delta$, the associated coordinate map,
$\delta \leftrightarrow \gamma^{-1}\delta$,
$\gamma^{-1}\delta \leftrightarrow B$, the natural matrix of $\gamma^{-1}\delta$

are all 1–1, and thus, the composite correspondence, $w_1, \ldots, w_n \leftrightarrow B$, is also 1–1. The ordered basis $w_1, \ldots, w_n$ is an integral basis for $\mathbf{G}$ if and only if the w's can be expressed as integral linear combinations of the t's *and vice versa*, i.e., if and only if both B and B^{-1} have integer entries. But B and B^{-1} both have integer entries if and only if B has integer entries and $\det B = \pm 1$. Thus, the above correspondence between ordered bases and n by n matrices, when restricted to ordered integral bases of $\mathbf{G}$, is a 1–1 correspondence between such bases and the extended group of integral, unimodular matrices.

We call a matrix an *integral* matrix if and only if all of its entries are integers. A matrix is *unimodular* if its determinant is $+1$. The *extended* group of unimodular matrices also includes matrices with determinant -1. In the case of a real affine space, the extended group of unimodular matrices is the natural image within $GL_n(\mathbf{R})$ of those equiaffine, i.e., volume preserving, transformations that leave the origin fixed. The geometric significance of $SL_n(\mathbf{F})$, the group of unimodular matrices, was given in term-paper topic 2 of exercise 5.6.

Corollary 5.39. Let γ^* be a coordinate map with origin P and whose unit vectors form an integral basis for the lattice group $\mathbf{G}$. An affine transformation ϕ leaves $(P)\mathbf{G}$ globally invariant if and only if the γ^*-analytic form of ϕ is

$$(x_1, x_2, \ldots, x_n) \rightarrow (x_1, \ldots, x_n)B + (a_1, \ldots, a_n)$$

with a_i integers and B an integral matrix such that $\det B = \pm 1$.

Proof. ϕ leaves $(P)\mathbf{G}$ globally invariant if and only if it carries P to some point in $(P)\mathbf{G}$ (i.e., $a_i \in \mathbf{Z}$) and ϕ^i carries the integral basis associated with γ to another integral basis (i.e., B is an integral matrix with determinant ± 1).

Exercise 5.10

1. Show that if B and C are parallelepipeds in a real affine space such that every vertex of B is contained in C, then $B \subseteq C$.

2. Inversions in points can be generalized from euclidean space to arbitrary real affine spaces. Do so and prove that any inversion that interchanges two points of a lattice must leave the lattice invariant.

3. Let **G** be a lattice group with integral basis $v_1, v_2, \ldots, v_n$. Show that if Q is in the lattice $(P)\mathbf{G}$, then Q is visible from P (i.e., no point of $(P)\mathbf{G}$ is between P and Q) if and only if $\overrightarrow{PQ} = \Sigma\alpha_i v_i$, and the greatest common divisor of the nonzero α_i is 1.

4. Assume **G** is a *group* of translations containing $t_1, t_2, \ldots, t_n$. Let $(P)Y$ be the box with initial vertex at P and edges along $t_1, \ldots, t_n$. Show that the t_i form an integral basis of **G** if and only if $(P)\mathbf{G} \cap (P)Y =$ the set of vertices of $(P)Y$.

5. (a) Find at least 5 unimodular 2 by 2 matrices with entries all integers.
(b) Prove that the group of unimodular, integral 2 by 2 matrices is infinite and hence that any lattice of dimension at least 2 has infinitely many primitive cells with a given initial vertex.

6. Let $(P)\mathbf{G}$ be an n-dimensional lattice in an n-dimensional real affine space. Show that a box is a primitive cell for this lattice if and only if all of its vertices are lattice points, and its volume is minimal for boxes whose vertices are lattice points.

5.11. Collineations in real affine spaces

We showed earlier that any nonsingular semi-linear map of the group of translations of an affine space induces a collineation (theorem 5.33). The converse of this theorem (see theorem 6.53) is also valid, i.e., given any collineation, ϕ, there is a translation, t, and collineation, θ, induced by a semi-linear map such that $\phi\theta = t$. The collineation ϕ is an affine transformation if and only if the automorphism involved in the semi-linear map inducing θ is the identity. Thus, for affine spaces over fields with no nontrivial automorphisms, the only affine collineations are the affine transformations. Since an automorphism of a field must leave 0 and 1 invariant and preserve all rational algebraic operations, it follows that the fields $\mathbf{Z}_p$ and $\mathbf{Q}$ have no nontrivial automorphisms. A more surprising result is the following theorem.

Theorem 5.40. The only automorphism of the field of real numbers, **R**, is the identity.

Proof. (Outline only!) Any automorphism, λ, of **R** must leave 0 and 1 fixed, and since it preserves sums, products, and quotients, it must, therefore, leave **Q** *pointwise* fixed. A number is positive in **R** if and only if it is the square of a nonzero real number, and $x < y$ if and only if $y - x$ is positive. The "perfect square" property must be preserved by λ; hence $x < y$ if and only if $x^\lambda < y^\lambda$, i.e., λ preserves order. Now if $x \neq x^\lambda$ for some real number x, then we can squeeze a rational number q between x and x^λ. Since $q^\lambda = q$, this would imply that λ reverses the order between x and q. Contradiction! Hence $x^\lambda = x$ for all $x \in \mathbf{R}$, i.e., $\lambda = \mathbf{1}$.

Corollary 5.41. The only collineations of a real affine space are the affine transformations of the space.

The natural way (but not the only way!) to imbed euclidean n-space within $A_n(\mathbf{R})$ is to define the distance between $(a_1, a_2, \ldots, a_n)$ and $(b_1, b_2, \ldots, b_n)$ to be $((b_1 - a_1)^2 + \cdots + (b_n - a_n)^2)^{1/2}$. The standard results of analytic geometry for 3-space generalize to n-space, e.g., the vectors $\langle\alpha_1, \alpha_2, \ldots, \alpha_n\rangle$ and $\langle\beta_1, \beta_2, \ldots, \beta_n\rangle$ are perpendicular if and only if $\alpha_1\beta_1 + \alpha_2\beta_2 + \cdots + \alpha_n\beta_n = 0$. If euclidean n-space is imbedded in affine n-space, then there is an interesting type of affine transformation which is in some ways analogous to the central dilatations.

Definition. Let $(P)\mathbf{H}$ be a hyperplane in euclidean n-space, and let $P + Q$ be the (unique) line through P perpendicular to $(P)\mathbf{H}$. The affine transformation which leaves $(P)\mathbf{H}$ pointwise fixed and sends Q to that point R such that $\overrightarrow{PR} = \mu\overrightarrow{PQ}$ $(\mu > 0)$ is called the *contraction* towards $(P)\mathbf{H}$ with ratio μ.

Note that if $\mu < 1$, then a contraction with ratio μ acts as one would expect a contraction to act, but if $\mu > 1$, then it is more like a stretching. Nevertheless, we call it a contraction in either case. The contraction towards $(P)\mathbf{H}$ with ratio μ exists by theorem 5.28. Moreover, for any point the distance from $(P)\mathbf{H}$ to its image is μ times its distance from $(P)\mathbf{H}$.

Contractions are a type of affine transformation which is easy to visualize. The following result shows that in a certain sense they are the "non-euclidean" factor in affine transformations.

Theorem 5.42. Let ϕ be an affine transformation of a real affine n-space in which a euclidean distance and perpendicularity has been defined. Then ϕ can be factored into a product of an isometry followed by n-contractions towards suitably chosen, mutually perpendicular hyperplanes.

Proof. Choose a point P_0, and let $\mathcal{S}$ be the unit sphere centered at P_0, i.e., $\mathcal{S} = \{P \mid |P_0P| = 1\}$. Among the points in $\mathcal{S}$, choose a point P_1 such that $|(P_0)\phi(P_1)\phi|$ is minimal. Note that the function sending P to $|(P_0)\phi(P)\phi|$ is a continuous function on the compact set $\mathcal{S}$, so it does achieve its minimum on $\mathcal{S}$. Let $(P_1)\mathbf{H}$ be the hyperplane through P_1 perpendicular to $P_0 + P_1$.

Lemma. The image of $(P_1)\mathbf{H}$ under ϕ, $((P_1)\mathbf{H})\phi$, is perpendicular to $(P_0)\phi + (P_1)\phi$.

Proof. If not, let $(Q)\phi$ be the perpendicular projection of $(P_0)\phi$ into $((P_1)\mathbf{H})\phi$ (Figure 28). Then $Q \in (P_1)\mathbf{H}$, and hence the segment $\overline{PQ}$ intersects $\mathcal{S}$ at a point R. But then $|(P_0)\phi(R)\phi| < |(P_0)\phi(Q)\phi| < |(P_0)\phi(P_1)\phi|$ in contradiction to the way P_1 was chosen.

Since the affine transformation ϕ preserves parallelism, the hyperplane $((P_0)\mathbf{H})\phi$ is also perpendicular to $(P_0)\phi + (P_1)\phi$. Essentially the same argument can be repeated to find P_2 within $(P_0)\mathbf{H}$ such that $|(P_0)\phi(P_2)\phi|$ is minimal for points on $\mathcal{S} \cap (P_0)\mathbf{H}$. The hyperplane, $(P_2)\mathbf{K}$, within $(P_0)\mathbf{H}$ perpendicular to

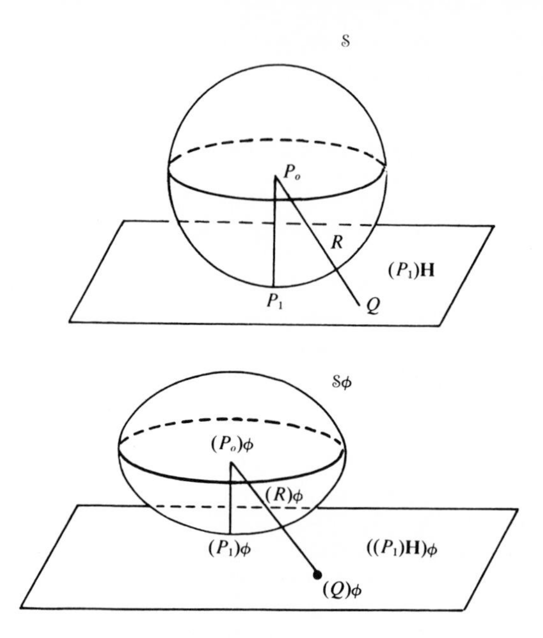

FIG. 28. The perpendicular projection, $(Q)\phi$, of $(P_0)\phi$ into $((P_1)\mathbf{H})\phi$.

$P_0 + P_2$, is such that its ϕ-image is also perpendicular to $(P_0)\phi + (P_2)\phi$, and hence the perpendicularity between $(P_0)\mathbf{K}$ and $P_0 + P_2$ is maintained by ϕ. Continue and find P_3 inside $(P_0)\mathbf{K}$, etc., and we finally obtain a set of $n + 1$ independent points $P_0, P_1, \ldots, P_n$ such that the vectors $\overrightarrow{P_0P_i}$ are all unit vectors ($P_i \in \mathbb{S}$) and are mutually perpendicular. Moreover, the images of these vectors are also mutually perpendicular although they may not be unit vectors. Choose Q_i on the ray starting at $(P_0)\phi$ and containing $(P_i)\phi$ such that $|(P_0)\phi Q_i| = 1$, $i = 1, 2, \ldots, n$. The isometry σ sending P_0 to $(P_0)\phi$ and P_i to Q_i for $i = 1, 2, \ldots, n$ is the desired isometry of the factorization. The ith contraction sends Q_i to $(P_i)\phi$ and leaves the hyperplane generated by $(P_0)\phi$ and the remaining Q_i pointwise fixed. Clearly σ followed by these n-contractions in mutually perpendicular hyperplanes agrees with ϕ on the $n + 1$ independent points $P_0, P_1, \ldots, P_n$. Hence we have the desired factorization of ϕ.

The argument above is essentially that given (for the affine plane only) in [1], p. 49. This reference contains some interesting illustrations of applications of affine transformations, including a simple construction of all triangles with smallest possible area circumscribing a given ellipse.

Exercise 5.11

1. Using any of the results in chapters 4 and 5, show that any matrix $A \in GL_n(\mathbf{R})$ can be factored as follows:

$A = PBC_1C_2 \cdots C_nP^t$, $B^{-1} = B^t$, $P^{-1} = P^t$, and C_j diagonal with all diagonal entries 1, except (perhaps) the jth.

Bibliography and suggestions for further reading

[1] Aleksandrov, A. D., A. N. Kolmogorov, and M. A. Lavrent'ev (eds.), *Mathematics—Its Content, Methods, and Meaning*, Part 2 (Vol. 1 of the *Translations of Math. Monographs*), Am. Math. Soc., Providence, R.I., 1963.

[2] Birkhoff, Garrett, and Saunders MacLane, *A Survey of Modern Algebra*, 3rd ed., Macmillan, New York, 1965.

[3] Coxeter, H. S. M., *Introduction to Geometry*, Wiley, New York, 1961.

[4] Danzer, Ludwig, Branko Grünbaum, and Victor Klee, "Helly's Theorem and its Relatives," *Convexity*, Vol. 7 of *Proc. Symposia in Pure Math.*, Am. Math. Soc., Providence, R.I., 1963, pp. 101–180.

[5] Gardner, Martin, "Is it possible to visualize a four-dimensional figure?," *Sci. Am.* 138–143 (Nov. 1966).

[6] Lang, Serge, *Algebra*, Addison-Wesley, Reading, Mass., 1965.

[7] Mostow, George D., Joseph H. Sampson, and Jean-Pierre Meyer, *Fundamental Structures of Algebra*, McGraw-Hill, New York, 1963.

[8] Nomizu, K., *Fundamentals of Linear Algebra*, McGraw-Hill, New York, 1966.

[9] Snapper, Ernst, *Metric Geometry over Affine Spaces*, Math. Assoc. Am., Buffalo, N.Y., 1964 (notes taken by J. T. Buckley at the first MAA Cooperative Summer Seminar in 1964 at Cornell University).

[10] Valentine, Frederick A., *Convex Sets*, McGraw-Hill, New York, 1964.

[11] Weyl, Hermann, *Symmetry*, Princeton Univ. Press, Princeton, N.J., 1952.

[12] Yale, Paul B., "Automorphisms of the Complex Numbers," *Math. Magazine* **39,** 135–141 (1966).

For background information and help in gaining an intuitive grasp of affine geometry, we recommend section 11 of [1] and chapter 13 of [3]. The elementary applications of affine transformations of the real affine plane given in [1] are very interesting. Notable features of the treatment of the real affine plane and real affine 3-space in [3] are the sections on equiaffine collineations (collineations preserving volume), affine reflections, and areas of lattice polygons.

Treatments, very similar to ours, of affine spaces over arbitrary fields appear in [8] and [9]. Matrices are used extensively in [8].

Chapter 6

Projective Spaces

In previous chapters we focused attention on the symmetries of various geometries. Recall that in the case of euclidean geometry we first studied the group **E** consisting of all symmetries leaving distances invariant, and then we studied **S**, the group of symmetries leaving relative distances invariant. Similarly in the case of affine geometry, we studied the group of dilatations, **D**, then the group of affine transformations, **A**, and finally the group of collineations, **G**. In the case of real affine 3-space, these groups are related as in Figure 29.

In the case of real affine 3-space, note how the amount of transitivity increases as we impose fewer restrictions on the symmetries. $\mathbf{D} \cap \mathbf{E}$ is transitive on points but not on lines or planes. **E** is transitive on points, lines, or planes. **S** is two times transitive on points but not on pairs of intersecting lines or planes. The group **A** comes very close to being two times transitive on planes as well as points. More generally, the group of affine transformations of any affine space is almost two times transitive on hyperplanes, the only restriction being that the two pairs of hyperplanes must be either both parallel or both nonparallel. At this point, we reach an impasse since even collineations, the weakest type of symmetry of interest, must preserve the parallel relationship.

In this chapter we study geometric spaces in which there is complete symmetry between pairs of hyperplanes. Clearly, in order to do this we must change our ideas about points, lines, planes, etc., *but not as much as one might suspect!* We find that these spaces are again very closely associated with vector spaces and that they can be built up from affine spaces simply by adding one new hyperplane of points. Once again our main objective is to study the symmetries of these spaces. A secondary objective is to continue to lead the

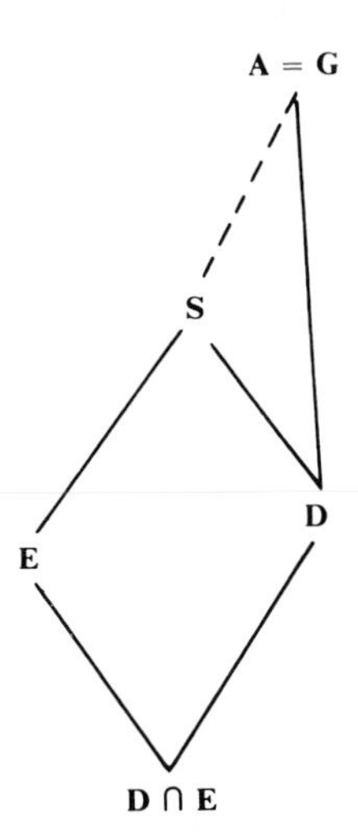

FIG. 29. Subgroups of the group of collineations, **G**, of real affine 3-space.

reader beyond geometries based on the real number system and demonstrate some other possibilities, in particular, the finite projective spaces.

6.1. Extended affine spaces and collapsed vector spaces

In this section we present two common ways of defining projective spaces and show that they lead to essentially the same type of geometry. Both methods are based on the observation that directions of lines in affine geometry have many properties in common with points (exercise 6.1, problem 1). In the first method, we build up an affine space to form a projective space by throwing in the directions of lines as additional points. In the second method, we collapse a vector space to form a projective space by using the lines in that vector space as our projective points. Viewed in a slightly different way, in the second method we use only the directions of lines in an affine space as our projective points. To understand projective geometry one *must* understand both of these approaches to the subject.

Projective spaces as "extended affine spaces." Let A be an n-dimensional affine space with translation group **T**. The points of our new projective space, referred to as *projective points*, are of two types, affine points and ideal points. The *affine points* are, of course, simply the points in A. The *ideal points* are the one-dimensional subspaces of **T**, i.e., the directions of lines. We refer to the set of all projective points as PA; thus, $PA = A \cup \{Q | Q$ is a one-dimensional subspace of $\mathbf{T}\}$.

This is not the complete picture, of course. We must impose a geometrical structure on PA. To do this we define k-dimensional projective subspaces as certain sets of projective points. These projective subspaces again come in two types, one corresponding to affine subspaces, the other consisting only of ideal points.

Definition. Let $(P)\mathbf{V}$ be a k-dimensional (affine) subspace of A. The corresponding *k-dimensional projective subspace* of PA, denoted by $\bar{V} \cup (P)\mathbf{V}$, is $\{Q|Q$ is an ideal point contained in $\mathbf{V}\} \cup (P)\mathbf{V}$. In other words, to form the projective subspace we take the affine $(P)\mathbf{V}$ and throw in all ideal points in the direction of $(P)\mathbf{V}$. The *ideal $(k-1)$-dimensional projective subspace*, $\bar{V}$, determined by $(P)\mathbf{V}$ is simply the set of all ideal points in $\mathbf{V}$, i.e., $\bar{V} = \{Q|Q$ is a one-dimensional subspace of $\mathbf{V}\}$.

There is only one *ideal* $(n-1)$-dimensional projective subspace, namely $\bar{T}$, the set of all ideal points. This projective hyperplane is usually referred to as the *hyperplane at infinity*. This suggestive term is helpful in visualizing the real projective plane or 3-space as being built up from the real affine plane or 3-space by adding points at infinity (ideal points) in such a way that parallel lines always meet at an ideal point. Note that we add one ideal point for each pencil of parallel lines. Some authors let the pencil of lines be the ideal point, but with our presentation of affine spaces it is more convenient to let the ideal point be the corresponding one-dimensional subspace of the translation group.

Example 1. Let A be an affine plane over the field with two elements. The four points and six lines of A are related as in Figure 30a. Since there are three one-dimensional (linear) subspaces of the translation group and one

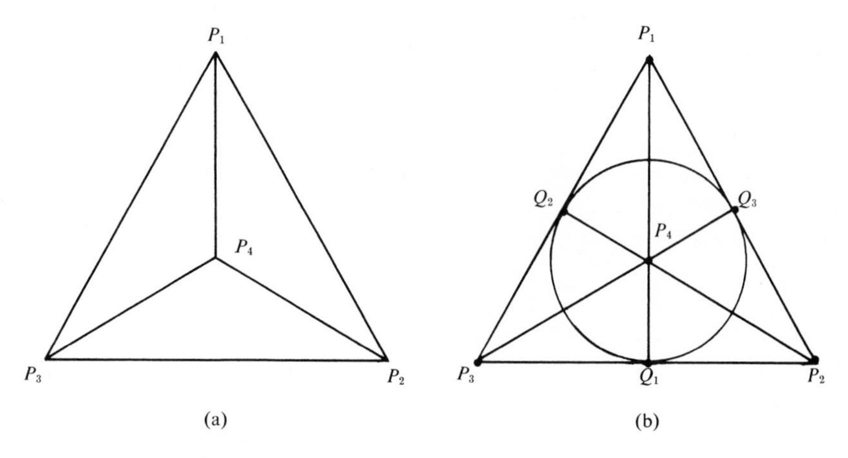

FIG. 30. (a) The affine plane with four points; (b) the projective plane with seven points.

two-dimensional (linear) subspace, we add three ideal points and one ideal line to obtain the projective plane PA. The relationship between the points and lines of PA is shown in Figure 30b.

Example 2. Let A be $A_2(\mathbf{R})$, the standard real affine plane. The translation group has infinitely many one-dimensional subspaces but only one two-dimensional subspace. Each *nontrivial* translation, t, determines a one-dimensional subspace which is the ideal point added to all lines left globally invariant by t. Two translations, t and s, leave the same line globally invariant if and only if they are in the same one-dimensional subspace of $\mathbf{V}_2(\mathbf{R})$, thus once again we add exactly one ideal point to each line and add exactly one ideal line to form the projective space PA.

Projective spaces as "collapsed vector spaces." In extending affine spaces to projective spaces, we introduced ideal points and ideal k-dimensional subspaces. The set of all ideal points formed the hyperplane at infinity. It would be unnatural if the hyperplane at infinity were not itself a projective space. It is; moreover, it is not necessary to mention affine spaces at all in this approach to projective spaces since the ideal points were defined in terms of the translation group. Let us review, being careful to avoid affine concepts and remembering that the dimension drops by one. Let $\mathbf{V}$ be an n-dimensional vector space ($n \geq 1$). The associated ($n - 1$)-dimensional projective space is $\bar{V} = \{Q | Q$ is a one-dimensional subspace of $\mathbf{V}\}$. For each k-dimensional ($k \geq 1$) subspace, $\mathbf{W}$, of $\mathbf{V}$, the corresponding ($k - 1$)-dimensional subspace of $\bar{V}$ is $\bar{W} = \{Q | Q$ is a one-dimensional subspace of $\mathbf{W}\}$.

The connection between the two methods is suggested by considering the two types of projective planes in PA if A is an affine 3-space over the reals. The only ideal plane in PA is $\bar{T}$. Since $\mathbf{T}$ is a three-dimensional vector space, $\bar{T}$ is representative of all real projective planes constructed using the second method. A plane of the form $\bar{V} \cup (P)\mathbf{V}$ is an extended affine plane and is representative of projective planes constructed using the first method. Choose two points, P and Q, in A and a plane, $(P)\mathbf{V}$, that does not contain Q. The points in $\bar{T}$ are clearly in 1–1 correspondence with the lines of A through Q. The lines of A through Q are also in 1–1 correspondence with the points in $\bar{V} \cup (P)\mathbf{V}$, namely, if m is a line through Q, then m corresponds to $m \cap (P)\mathbf{V}$ if $(P)\mathbf{V}$ and m are not parallel, and if $m \mathbin{/\!/} (P)\mathbf{V}$, then m corresponds to its direction. The composition of these two 1–1 correspondences is the desired 1–1 correspondence between $\bar{T}$ and $\bar{V} \cup (P)\mathbf{V}$ via the set of lines of A through Q.

The following theorem generalizes the construction above to arbitrary projective spaces, and thus, we prove that PA and $\bar{V}$ are essentially the same if they have the same dimension over a common field of scalars. We prove another, more precise version (theorem 6.20) in section 6.4 after we have introduced homogeneous coordinates.

Theorem 6.1. Let A be an n-dimensional affine space and $\mathbf{V}$ an $(n+1)$-dimensional vector space over the same field $\mathbf{F}$. Then there is a 1–1 correspondence between PA and $\bar{V}$ in which k-dimensional subspaces of PA correspond to k-dimensional subspaces of $\bar{V}$.

Proof. Convert $\mathbf{V}$ into an $(n+1)$-dimensional affine space over $\mathbf{F}$ as in the example given in section 5.1. Since points in the affine space V are also vectors in the vector space $\mathbf{V}$, we use the letters u, v, and w to denote points in V instead of the traditional P, Q, R. Note also that in V, $(u)v = u + v$.

Choose an affine hyperplane, $(u)\mathbf{H}$, in V such that $(u)\mathbf{H}$ does not contain the origin, i.e., $0 \notin (u)\mathbf{H}$ or equivalently $u \notin \mathbf{H}$. The affine spaces A and $(u)\mathbf{H}$ have the same dimension. Hence their translation groups, $\mathbf{T}$ and $\mathbf{H}$, respectively, are both n-dimensional vector spaces over $\mathbf{F}$. Let σ be a nonsingular linear transformation from $\mathbf{T}$ onto $\mathbf{H}$. Using σ, we define a mapping, ϕ, from $PA = \bar{T} \cup A$ onto $\bar{H} \cup (u)\mathbf{H}$ as follows. Choose a point P to serve as the "origin" in A. The ϕ-image of any one-dimensional subspace of $\mathbf{T}$ is just its σ-image. For each $Q \in A$, let $(Q)\phi = u + v$ if and only if $(\overrightarrow{PQ})\sigma = v$.

Since σ is a nonsingular linear transformation, it is not difficult to check that ϕ is a 1–1 correspondence between PA and $\bar{H} \cup (u)\mathbf{H}$. It is also clear that if $\mathbf{W}$ is a (linear) subspace of $\mathbf{T}$ and if $(\mathbf{W})\sigma = \mathbf{U}$, then $(\bar{W})\phi = \bar{U}$ and $\bar{W}$ and $(\bar{W})\phi$ have the same (projective) dimension.

Now we define a mapping, θ, from $\bar{H} \cup (u)\mathbf{H}$ to $\bar{V}$. On $\bar{H}$, θ reduces to the identity, and if $v \in (u)\mathbf{H}$, then $(v)\theta$ is the one-dimensional subspace of $\mathbf{V}$ generated by v. The mapping θ is a 1–1 correspondence between $\bar{H} \cup (u)\mathbf{H}$ and $\bar{V}$ (exercise 6.1, problem 4).

We now have a 1–1 correspondence, $\phi\theta$, between PA and $\bar{V}$. To complete the proof we must show that X is a k-dimensional projective subspace of PA if and only if $(X)\phi\theta$ is a k-dimensional projective subspace of $\bar{V}$.

Assume X is a k-dimensional projective subspace of PA. There are two cases to consider, X ideal and X an extended affine subspace.

Case 1. $X = \bar{W}$, $\mathbf{W}$ a $(k+1)$-dimensional (linear) subspace of $\mathbf{T}$. Let $(\mathbf{W})\sigma = \mathbf{U}$, then $(\bar{W})\phi\theta = (\bar{U})\theta$ since $(\bar{W})\phi = \bar{U}$. But θ is the identity on $\bar{H}$, and since $\bar{U} \subseteq \bar{H}$ (σ maps $\mathbf{T}$ onto $\mathbf{H}$ and hence ϕ maps $\bar{T}$ onto $\bar{H}$), $(\bar{W})\phi\theta = (\bar{U})\theta = \bar{U}$. Both $\bar{W}$ and $\bar{U}$ have projective dimension k; hence $(\bar{W})\phi\theta = \bar{U}$ is a k-dimensional subspace of $\bar{V}$.

Case 2. $X = \bar{W} \cup (Q)\mathbf{W}$, $\mathbf{W}$ a k-dimensional (linear) subspace of $\mathbf{T}$. Let $\mathbf{W}_1 = (\mathbf{W})\sigma$ be the corresponding k-dimensional (linear) subspace of $\mathbf{H}$, and let $(Q)\phi = u + v$, i.e., $(\overrightarrow{PQ})\sigma = v$. Let $\mathbf{W}_2$ be the subspace generated by $(Q)\phi = u + v$ and $\mathbf{W}_1$. We claim that $\bar{W}_2$ is the $\phi\theta$-image of $\bar{W} \cup (Q)\mathbf{W}$ and that the (projective) dimension of $\overline{W}_2$ is k. We support this claim in the following three steps.

1. What is the dimension of $\overline{W}_2$? Since $v = (\overrightarrow{PQ})\sigma \in (\mathbf{T})\sigma = \mathbf{H}$ but $u \notin \mathbf{H}$, it follows that $u + v = (Q)\phi \notin \mathbf{H}$. But $\mathbf{W}_1 \subseteq \mathbf{H}$; hence $u + v \notin \mathbf{W}_1$. Thus, the (linear) dimension of $\mathbf{W}_2$ is $k + 1$, and the projective dimension of $\overline{W}_2$ is k.

2. Is $(X)\phi\theta \subseteq \overline{W}_2$? Let $R \in X$. If R is an ideal point, then $(R)\phi\theta \in \overline{W}_1 \subseteq \overline{W}_2$. If R is an affine point, then $\overrightarrow{QR} \in \mathbf{W}$ so $(\overrightarrow{QR})\sigma = w_1 \in \mathbf{W}_1$. But $(\overrightarrow{PR})\sigma = (\overrightarrow{PQ})\sigma + (\overrightarrow{QR})\sigma = v + w_1$, so $(R)\phi = u + v + w_1 \in \mathbf{W}_2$, and thus, $(R)\phi\theta \subseteq \overline{W}_2$.

3. Is $\overline{W}_2 \subseteq (X)\phi\theta$? Let $S \in \overline{W}_2$, then S is the one-dimensional subspace generated by a vector in $\mathbf{W}_2$, i.e., by some vector of the form $\alpha(u + v) + w_1$, $\alpha \in \mathbf{F}$, $w_1 \in \mathbf{W}_1$. If $\alpha = 0$, then $S \in \overline{W}_1 = (\overline{W})\phi\theta \subseteq (X)\phi\theta$. If $\alpha \neq 0$, then $(u + v) + (1/\alpha)w_1$ also generates S. Choose the point R in A such that $(\overrightarrow{QR})\sigma = (1/\alpha)w_1$, then since $(\overrightarrow{QR})\sigma \in \mathbf{W}_1$, $\overrightarrow{QR} \in \mathbf{W}$, and hence $R \in (Q)\mathbf{W}$. But $(R)\phi\theta = S$, so $S \in (X)\phi\theta$. Thus, any point S in $\overline{W}_2$ is the $\phi\theta$-image of either an ideal or affine point in X. This completes the proof that $(X)\phi\theta$ is a k-dimensional subspace whenever X is. The proof that $(\overline{W})\theta^{-1}\phi^{-1}$ is a k-dimensional subspace of PA whenever $\overline{W}$ is a k-dimensional subspace of $\bar{V}$ is left to the reader.

Although the two approaches to projective spaces given above are equivalent, it is clear that the "collapsed vector space" approach is less cumbersome since one does not have to continually discuss two cases—one for ideal and the other for affine points or subspaces. For this reason we *henceforth consider any projective space to be a space of the form* $\bar{V}$.

Exercise 6.1

1. (a) Prove the affine form of Desargues theorem, i.e., assume that A_1, B_1, C_1 and A_2, B_2, C_2 are triangles (with corresponding vertices distinct and corresponding sides nonparallel) such that the lines $A_1 + A_2$, $B_1 + B_2$, and $C_1 + C_2$ meet at a point P, and show that the points $X = (B_1 + C_1) \cap (B_2 + C_2)$, $Y = (A_1 + C_1) \cap (A_2 + C_2)$, and $Z = (A_1 + B_1) \cap (A_2 + B_2)$ are collinear (Figure 31). Note that you are to prove this in any affine space over any field. The proof is fairly simple if the two triangles are in distinct planes in the affine space. If they are in the same plane (or if the space itself is only two-dimensional), then choose coordinates judiciously, and complete the proof as in ordinary analytic geometry.
(b) Desargues theorem is valid if any of the points are ideal points. Prove the special cases illustrated in Figures 32a, b, and c.
(c) Illustrate the special case of Desargues theorem in which the points A_1, B_1, and Z are ideal points.

2. Let $\mathbf{F}$ be a finite field with f elements. An affine n-dimensional space, A, over $\mathbf{F}$ contains f^n points. Show (by induction on n; review the proof of theorem 6.1 to

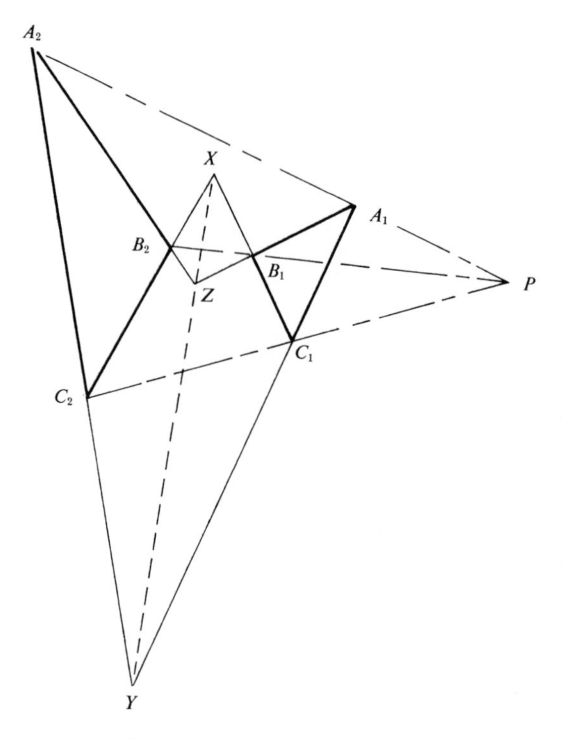

FIG. 31. Desargues theorem.

see how to carry out the induction step) that there are $1 + f + f^2 + \cdots + f^{n-1}$ directions for lines in A, and hence PA contains $\sum_{j=0}^{n} f^j$ points.

3. Let **V** be an $(n + 1)$-dimensional vector space over a finite field, **F**, with f elements. Show that the number of elements in $\bar{V}$ is $(f^{n+1} - 1)/(f - 1)$ and that this result agrees with the number of elements in the n-dimensional projective space PA as determined in problem 2.

4. Prove that the mapping θ, as defined in the proof of theorem 6.1, is a one-to-one correspondence between $\bar{H} \cup (u)\mathbf{H}$ and $\bar{V}$.

6.2. *Projective subspaces*

Let $\bar{V}$ be a projective n-dimensional space over the field **F**. Any k-dimensional subspace of $\bar{V}$ is, by definition, of the form $\overline{W}$, with **W** a $(k + 1)$-dimensional vector space, so the analog of theorem 5.1 is trivial for projective spaces. The next concept discussed for affine subspaces, which is also relevant to projective subspaces, is that of a subspace generated by a certain set of

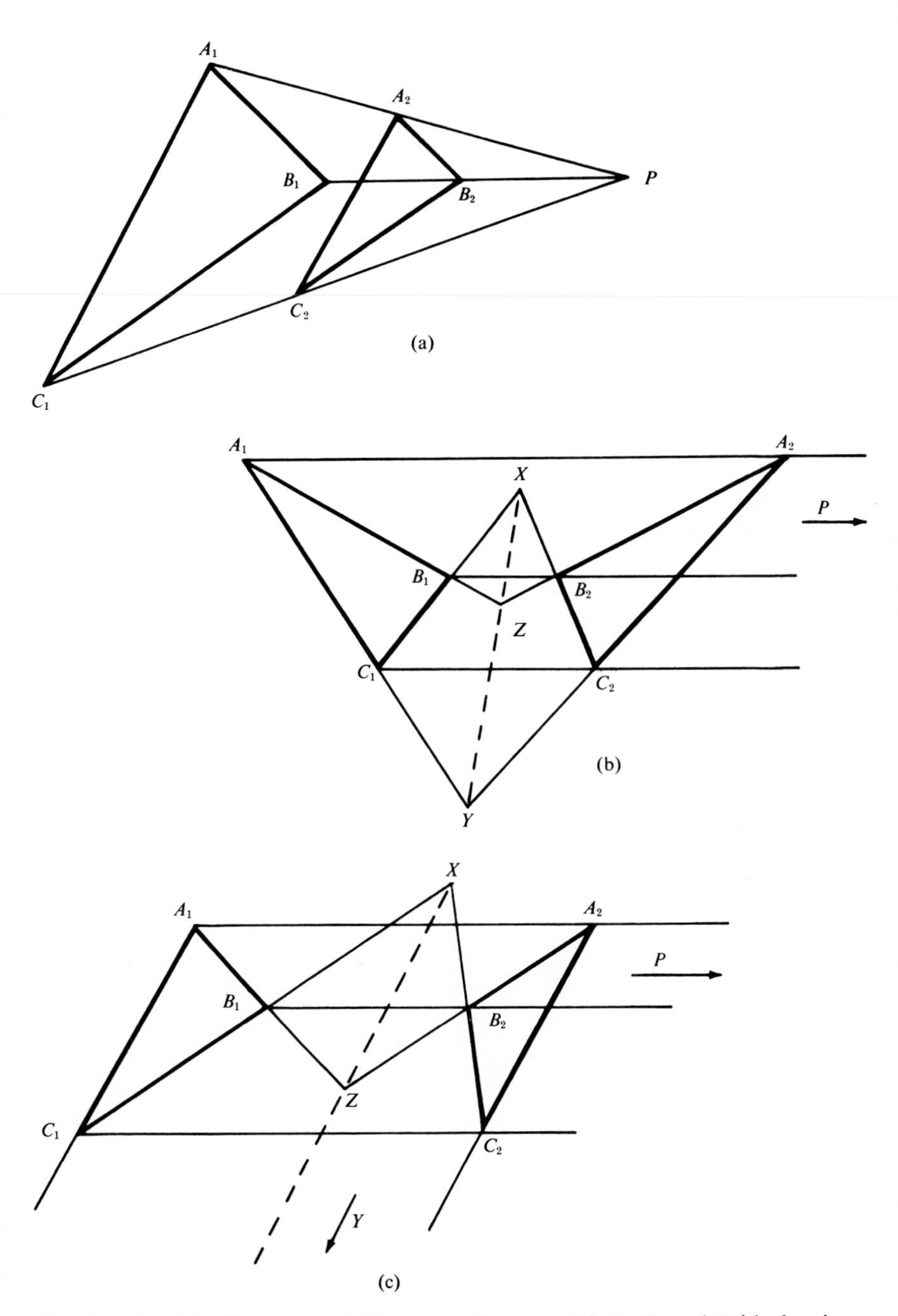

FIG. 32. Special affine cases of Desargues theorem. (a) X, Y, and Z ideal points; (b) P an ideal point; (c) P and Y ideal points.

points. To discuss this for projective spaces we must first introduce some notation.

Definition. For each nonzero vector $v \in \mathbf{V}$, we let $[v]$ denote the one-dimensional subspace of $\mathbf{V}$ generated by v. Thus, $[v]$ is a point in $\bar{V}$, and we say that the vector v *represents* $[v]$. Assume that v and w are nonzero vectors in $\mathbf{V}$ such that $[v] \neq [w]$, then $[v] + [w]$ denotes the line $\bar{W}$, where $\mathbf{W}$ is the (linear) subspace of $\mathbf{V}$ generated by v and w. We say that $[v] + [w]$ is the *line joining* $[v]$ and $[w]$. Since αv and βw generate the same subspace as v and w generate (if $\alpha, \beta \neq 0$) the line $[v] + [w]$ is independent of the representatives chosen from $[v]$ and $[w]$.

Clearly $[v] + [w]$ is the unique line containing both $[v]$ and $[w]$; moreover, it is the smallest subspace of $\bar{V}$ containing both $[v]$ and $[w]$. Note that $[v] = [w]$ if and only if $v = \alpha\, w$ has a nonzero solution α in the field $\mathbf{F}$.

Theorem 6.2. Let X be a non-empty subset of the projective space $\bar{V}$. X is a subspace if and only if for all pairs of distinct points, $[v]$ and $[w]$, in X, $[v] + [w] \subseteq X$.

Proof. If X is a subspace, then it clearly satisfies the condition. To prove the converse let $Y = \{v \in \mathbf{V} | v = 0 \text{ or } [v] \in X\}$. Since $\bar{Y} = X$, we need only show that Y is a (linear) subspace of $\mathbf{V}$ and that $Y \neq \{0\}$. Since $X \neq \varnothing$, we must have at least one nonzero vector in Y. Y is closed with respect to multiplication by scalars. Now assume $v \in Y$ and $w \in Y$. This implies that $[v]$ and $[w]$ are in X. If $[v] = [w]$, then $v = \alpha w$ and $v + w = (\alpha + 1)w \in Y$. If $[v] \neq [w]$, then $[v + w] \in [v] + [w] \subseteq X$, so $v + w \in Y$. This shows that Y is closed under addition and completes the proof.

The theorem above is the analog of theorem 5.29. Note that here we do *not* have to avoid the case in which $\mathbf{F} = \mathbf{Z}_2$. The difficulty in the affine situation arose since lines in an affine space over $\mathbf{Z}_2$ contain only two points. In a projective space, however, any line has at least three points.

The line joining two points is one of the simplest cases of a subspace generated by a set of points. With the following theorem we can generalize to much more complicated sets.

Theorem 6.3. Let $\mathfrak{F}$ be a family of projective subspaces of $\bar{V}$, then $\bigcap_{\bar{W} \in \mathfrak{F}} \bar{W}$ is also a projective subspace unless it is empty. Moreover,

$$\bigcap_{\bar{W} \in \mathfrak{F}} \bar{W} = \overline{\bigcap_{\bar{W} \in \mathfrak{F}} \mathbf{W}}.$$

Proof. See exercise 6.2, problem 1.

Definition. If X is a non-empty subset of $\bar{V}$, then the intersection of all projective subspaces of $\bar{V}$ containing X is called the *subspace generated by X.* In the special case that $X = \bar{U} \cup \bar{W}$, we call the subspace generated by X, $\bar{U} + \bar{W}$, and call it the *join* of $\bar{U}$ and $\bar{W}$. Note that if $[u]$ and $[v]$ are distinct points, then the join of $[u]$ and $[v]$ is $[u] + [v]$, the line joining $[u]$ and $[v]$.

Clearly the join of two subspaces $\bar{U}$ and $\bar{W}$ is just $\overline{U + W}$, and, by theorem 6.3, if $\bar{U} \cap \bar{V} \neq \varnothing$, then $\bar{U} \cap \bar{V} = \overline{U \cap V}$. This suggests that we ought to have a projective analog to the dimension count theorem (4.4) for vector spaces.

Theorem 6.4. If we adopt the convention that the projective dimension of the empty set is -1, then for any two subspaces $\bar{U}$ and $\bar{W}$ of $\bar{V}$ all three of the following hold:

(1) $\dim \bar{U} + \dim \bar{W} = \dim(\bar{U} + \bar{W}) + \dim(\bar{U} \cap \bar{W})$.
(2) $-1 \leq \dim(\bar{U} \cap \bar{W}) \leq \min\{\dim \bar{U}, \dim \bar{W}\}$.
(3) $\max\{\dim \bar{U}, \dim \bar{W}\} \leq \dim(\bar{U} + \bar{W}) \leq \dim \bar{V}$.

Proof. $\bar{U} \cap \bar{W} = \varnothing$ if and only if $U \cap W = \{0\}$, $\bar{U} \cap \bar{W} = \overline{U \cap W}$, and $\bar{U} + \bar{W} = \overline{U + W}$. Thus, part (1) follows from theorem 4.4 simply by subtracting 1 from each of the four terms of the equation in 4.4. Parts (2) and (3) follow from the corresponding inequalities for vector subspaces in the same manner.

Corollary 6.5. If $\bar{U}$ is a hyperplane in $\bar{V}$, and $\bar{W}$ is a subspace of $\bar{V}$ not contained in $\bar{U}$, then $\dim(\bar{U} \cap \bar{W}) = (\dim \bar{W}) - 1$.

This is usually referred to by saying that intersecting with a hyperplane chops down the dimension by one.

In affine spaces we found that disjoint subspaces fell into two categories, parallel subspaces and skew subspaces. By introducing ideal points in projective spaces we have eliminated parallel subspaces, hence in projective spaces skew subspaces are simply subspaces that are disjoint as sets.

Theorem 6.6. An n-dimensional projective space, $\bar{V}$, contains skew subspaces of dimensions j and k if and only if $j + k + 1 \leq n$.

Proof. The necessity of $j + k + 1 \leq n$ is an immediate consequence of theorem 6.4. If $j + k + 1 \leq n$, then choose a basis of $\mathbf{V}$, and let $\mathbf{U}$ be the (linear) subspace generated by the first $j + 1$ basis vectors and $\mathbf{W}$ the subspace generated by the next $k + 1$ basis vectors. Since $\dim \mathbf{V} = n + 1 \geq (j + 1) + (k + 1)$, we do not have a collision. The projective subspaces $\bar{U}$ and $\bar{W}$ are the desired skew subspaces.

Exercise 6.2

1. Use theorem 6.2 to show that the intersection (if non-empty) of a family of projective subspaces is a projective subspace. Give another proof based on the proof of the "moreover" clause of theorem 6.3.

2. Review theorem 5.5 and its proof. State, and prove, an analogous theorem for projective subspaces.

3. Prove or disprove that skew subspaces in an affine space A extend to skew subspaces of PA.

4. Prove that two lines of a projective space which are distinct but coplanar intersect in exactly one point.

5. Prove that if a line in a projective space is not contained in a hyperplane, then it intersects that hyperplane in exactly one point. Thus, each line in PA is either entirely "at infinity" or contains exactly one point at infinity.

6.3. Projective planes

Let $\bar{V}$ be a two-dimensional projective space, i.e., a *projective plane over a field.* The following three statements are true in $\bar{V}$:

(1) Two points are contained in one and only one line.

(2) Two lines intersect in exactly one point.

(3) There are four points such that no three are in the same line.

These statements are almost universally recognized as the bare minimum of requirements for a mathematical system to be called a projective plane. Therefore, we adopt the standard convention and *distinguish carefully between a projective plane and a projective plane over a field.* The precise definition of a projective plane is as follows.

Definition. Any set (we refer to its members as *points*) with certain distinguished subsets (called *lines*) is called a *projective plane*, provided these points and lines satisfy axioms (1), (2), and (3) above.

It would appear that the three axioms are not very stringent and that one could invent very peculiar projective planes. In a certain sense, this is true as one may verify by consulting the references cited at the end of this section. Yet there are many interesting results about projective planes, and it is surprising how little one needs to assume in addition to axioms (1), (2), and (3) in order to ensure that a projective plane is a projective plane over a field. We derive a few of the basic results, but omit the problem of the geometric construction of the field of scalars and the discussion of the assumptions needed to guarantee its existence. We recommend [15], Chapters 3 and 9, or [8], Chapter 20, for this topic. Perhaps the most elegant way to introduce coordinates appears in Chapter 2 of [1]. Artin considers an affine instead of a projective plane, but this is not a significant change since it is easy to show that affine planes and projective planes with one line singled out as the line at infinity are equivalent.

Theorem 6.7. In any projective plane there are at least seven points and at least seven lines satisfying the incidences shown in Figure 33, with the possible exception of $Q_3 \in m$.

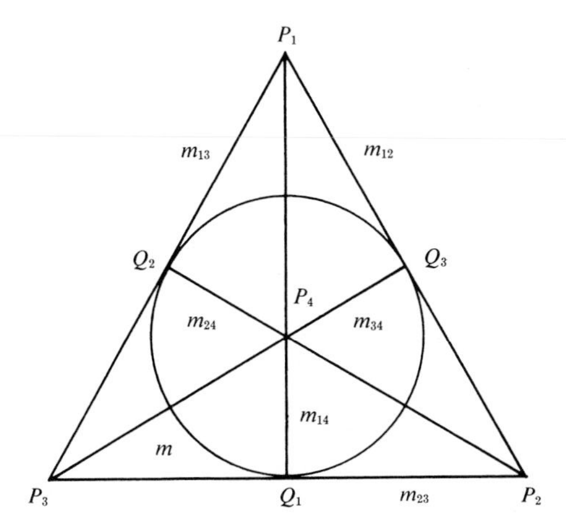

FIG. 33. Seven points and seven lines in a projective plane.

Proof. By axiom (3) there are four points, P_1, P_2, P_3, and P_4, such that no three are collinear. By axiom (1) if $\{a, b\} \subseteq \{1, 2, 3, 4\}$ and $a < b$, then there is a unique line containing both P_a and P_b. Call this line m_{ab}. If a, b, and c are distinct, then $P_a \notin m_{bc}$ since P_a, P_b, and P_c are noncollinear. Thus, there are six *distinct* lines of the form m_{ab}. If a, b, c, and 4 are all distinct and $a < b$, then by axiom (2) m_{ab} and m_{c4} share exactly one point. This cannot be one of the P_i; hence there is a new point $Q_c \in m_{ab} \cap m_{c4}$. It is easy to show that this point Q_c is not in any of the other lines m_{ij}, e.g., $Q_1 \notin m_{13}$ since axiom (1) implies that not both m_{13} and m_{14} can contain P_1 and Q_1. From this it follows that Q_1, Q_2, and Q_3 are distinct, and we have our seven distinct points. By axiom (1) there is a line, m, containing both Q_1 and Q_2, and since m cannot be any of the m_{ij}, we have our seventh line. The only incidence shown in Figure 33 which need not occur is $Q_3 \in m$. This is, indeed, an ambiguous situation because it does occur (Compare Figures 30 and 33!) in the seven-point projective plane (the projective plane over $\mathbf{Z}_2$), but it does not occur in the real projective plane nor in any projective plane over a field of characteristic not 2.

Corollary 6.8. In any projective plane there are at least four lines, no three of which contain a point in common.

Proof. Lines m_{12}, m_{34}, m_{24}, and m_{13} form such a collection of four lines.

Corollary 6.9. Any line in a projective plane contains at least three points.

Proof. If m is a line such that $P_a \notin m$, then m_{ab}, m_{ac}, and m_{ad} (a, b, c, and d being a reordering of 1, 2, 3, and 4) intersect m in three distinct points. No line m contains more than two of the P_a. Hence each line has at least three points.

Definition. A *pencil* of lines is a set of lines in a projective plane consisting of *all* lines through a point called the *center* of the pencil. If P is a point, then the pencil centered at P is denoted by P^*.

Theorem 6.10. If $P \notin m$, then the correspondence $Q \leftrightarrow P + Q$ is a 1–1 correspondence between m and P^*.

Proof. If $Q \in m$, then $Q \neq P$; hence there is a unique line, $P + Q$, containing both P and Q. Conversely, if $n \in P^*$, then there is a unique point $Q \in n \cap m$ and then $n = P + Q$. Thus, the correspondence $Q \leftrightarrow P + Q$ is a function from m onto P^*. It is 1–1 since $Q_1 \neq Q_2$ clearly implies $P + Q_1 \neq P + Q_2$.

Corollary 6.11. There is a 1–1 correspondence between any two lines of a projective plane.

Proof. Given lines m_1 and m_2, then we can choose a point P such that $P \notin m_1 \cup m_2$ (exercise 6.3, problem 1). The composite of the correspondences between m_1 and P^*, and between P^* and m_2 is a 1–1 correspondence between m_1 and m_2. This correspondence, in which $Q \in m_1$ corresponds to $(Q + P) \cap m_2$, is usually referred to as a *perspectivity* from m_1 to m_2 with *center* P.

If the reader compares axiom (1) with axiom (2), and axiom (3) with corollary 6.8, he will notice a striking symmetry in the use of point and line. This is exploited in the following theorem.

Theorem 6.12. Let π be a projective plane. π^*, the set of all lines in π, is also a projective plane if the distinguished subsets of π^* used as lines in π^* are the pencils of lines in π.

Proof. We note that $m \in P^*$ if and only if $P \in m$. Hence axioms (1), (2), and (3) for π^* translated into statements about π are simply axiom (2), axiom (1), and corollary 6.8. Since these statements are valid in π, their translations are valid in π^*, and thus, π^* is a projective plane.

The fact that π^* is a projective plane implies that any theorem valid for *all* projective planes must be valid in π^*. When we translate this statement about π^* into a statement about π, it is still valid, and we have the following result about theorems concerning projective planes.

Meta-theorem 6.13 (the principle of duality). If we start with a theorem valid for *all* projective planes and interchange the roles of points and lines and reverse inclusions, then the resulting theorem, called the *dual* of the original theorem, is valid for all projective planes. If the original statement was valid in π, but not necessarily in other projective planes, then the dual statement will be valid in π^* but not necessarily in other projective planes (in particular, not necessarily in π).

Since corollaries 6.9 and 6.11 are valid in any projective plane, so are their duals, and we find that any point in a projective plane is in at least three lines, and there is a 1–1 correspondence between any two pencils of lines.

The number of points on a line (which is also the number of lines in a pencil) in a given finite projective plane is an important constant for that plane. In the special case in which the plane is of the form $\bar{V}$, $\mathbf{V}$ a three-dimensional vector space over a field $\mathbf{F}$ with $\#F = f$, the number of points on a line in $\bar{V}$ is $f + 1$. Since these projective planes are the most important ones, it is customary to define the order of a finite projective plane as follows.

Definition. If there are $n + 1$ points on each line in a finite projective plane, then n is said to be the *order* of that plane.

Since there are finite fields with n-elements for any n of the form p^k, p prime, it follows that there are projective planes of any prime power order. One of the major unsolved problems concerning projective planes is whether or not there is a projective plane whose order is not a prime power. The first two integers (greater than 1) which are not prime powers are 6 and 10. It is known that there are no projective planes whose orders are in a certain infinite sequence beginning with 6, but the question for 10 is still open as of early 1967.

Two projective planes of the same order need not be isomorphic. They are if the order is 2, 3, 4, 5, 7, or 8, but at order 9 we meet the smallest projective planes which are not isomorphic to a projective plane over a field. For a discussion of these planes, see [9] and [17], and for information about their symmetries see [11]. For a complete exposition of all results through 1955 on projective planes, see [13].

The question of axiomatizing projective spaces of dimension greater than 2 is covered adequately in [15], chapters 5 and 7 and in [2], chapter 7.

Exercise 6.3

1. Prove carefully that if m_1 and m_2 are two lines in a projective plane, then there is at least one point P such that $P \notin m_1 \cup m_2$.

2. Assume X is a set with certain distinguished subsets, called lines, such that axioms (1) and (2) are satisfied. Show that axiom (3) is satisfied if and only if each line contains at least 3 points and not all points are in one line.

3. Prove that a projective plane of order n contains $n^2 + n + 1$ points and $n^2 + n + 1$ lines.

4. Let $\overline{V}$ be a projective space of dimension k. Show that $\overline{V}$ is a projective plane if and only if $k = 2$. We assume, of course, that "lines" in $\overline{V}$ are to be the same in both interpretations.

5. Show that a projective plane, π, of order 3 cannot contain a subplane of order 2, i.e., show that there is no subset, X, of π such that if we define lines in X to be the non-empty intersections of X with lines in π, then X is a projective plane of order 2. *Hint*. Consider Figure 33 (within X), and show that some line has too many points.

6. Let S_2 be the unit sphere in euclidean 3-space, i.e., $S_2 = \{P|\text{distance between origin and } P \text{ is } 1\}$. Let π be S_2 with opposite points identified, i.e., $\pi = \{P'|P' = \{P, -P\}, P \in S_2\}$. For each great circle m on S_2, let $m' = \{P'|P' = \{P, -P\}, P \in m\}$. Show that π with the various m' as its lines is a projective plane and is isomorphic to $\overline{\mathbf{V}}$, $\mathbf{V}$ a three-dimensional vector space over the reals. This model of the real projective plane is the favorite of topologists and is called the spherical model.

7. Show that a finite projective plane of order n can be partitioned into two sets such that neither set contains an entire line if and only if $n > 2$. This appears as problem A–5345 in the *Am. Math. Monthly* **72,** 1136 (1965). You can probably find the solution in a more recent issue of the *Monthly*.

6.4. Homogeneous coordinates

In previous chapters we introduced the analytic geometry of vector spaces [$\mathbf{V}_n(\mathbf{F})$ and linear coordinate maps], and of affine spaces [$A_n(\mathbf{F})$ and affine coordinate maps]. In this section we resume the study of projective spaces over fields and develop the analytic geometry of projective spaces. The main tasks are to introduce a set of "natural" projective spaces (one for each dimension over a given field), discuss the coordinate maps onto these spaces, explain how to change coordinates effectively, and finally, discuss subspaces from the coordinate viewpoint.

Definition. $P_n(\mathbf{F}) = \overline{\mathbf{V}_{n+1}(\mathbf{F})}$ is the *standard projective n-space over the field* $\mathbf{F}$. We use our previous notation for points in $P_n(\mathbf{F})$, i.e., if $v \in \mathbf{V}_{n+1}(\mathbf{F})$, $v \neq 0$, then $[v]$ denotes the one-dimensional subspace of $\mathbf{V}_{n+1}(\mathbf{F})$ generated by v; however, we modify the notation so that if $v = \langle \alpha_0, \alpha_1, \ldots, \alpha_n \rangle$, then $[v] = [\alpha_0, \alpha_1, \ldots, \alpha_n]$. In other words, we use the square bracket notation but drop the extra diamond brackets whenever convenient. The $(n + 1)$-tuple $\alpha_0, \alpha_1, \ldots, \alpha_n$ is called a set of *homogeneous coordinates* of $[\alpha_0, \alpha_1, \ldots, \alpha_n]$, and the vector $\langle \alpha_0, \ldots, \alpha_n \rangle$ is said to *represent* $[\alpha_0, \ldots, \alpha_n]$.

The word "homogeneous" refers to the fact that the homogeneous coordinates of a projective point may be multiplied by any nonzero scalar, e.g., $[1, 1, 1] = [-1/3, -1/3, -1/3]$ in $P_2(\mathbf{Q})$. Since homogeneous coordinates of a point are not unique, the reader must exercise caution. For example, in the following definition, it is not obvious that the mapping $[\phi]$ is unambiguously defined.

Definition. Let $\bar{V}$ be an n-dimensional projective space over the field $\mathbf{F}$, and let ϕ be a (linear) coordinate map of $\mathbf{V}$ onto $\mathbf{V}_{n+1}(\mathbf{F})$. The (projective) *coordinate map* of $\bar{V}$ onto $P_n(\mathbf{F})$ *induced by* ϕ is the mapping $[\phi]$ defined by $[v]^{[\phi]} = [v^\phi]$ for each $[v] \in \bar{V}$.

When we ask if $[\phi]$ is unambiguously defined, we need to know if $[v]^{[\phi]} = [w]^{[\phi]}$ whenever $[v] = [w]$. Since ϕ is a nonsingular linear transformation, there is no problem in checking this nor is there any great difficulty in verifying that $[\phi]$ is a 1–1 map of $\bar{V}$ onto $P_n(\mathbf{F})$ which sends k-dimensional subspaces of $\bar{V}$ onto k-dimensional subspaces of $P_n(\mathbf{F})$. We leave these details to the reader.

Theorem 6.14. Assume $[\phi]$ and $[\theta]$ are both coordinate maps of $\bar{V}$ onto $P_n(\mathbf{F})$. Then $[\phi] = [\theta]$ if and only if there is a nonzero scalar $\alpha \in \mathbf{F}$ such that $\phi = \alpha\theta$.

Proof. If $\alpha \neq 0$ and $\phi = \alpha\theta$, then for any nonzero vector $v \in \mathbf{V}$, $v^\phi = \alpha v^\theta$, so $[v^\phi] = [v^\theta]$, i.e., $[v]^{[\phi]} = [v]^{[\theta]}$. Since this is true for any $[v] \in \bar{V}$, this means that $[\phi] = [\theta]$.

To prove the converse, note that $[v^\phi] = [v^\theta]$ implies that $v^\phi = \alpha v^\theta$ for some nonzero $\alpha \in \mathbf{F}$. Thus, if $[\phi] = [\theta]$, then for each nonzero $v \in \mathbf{V}$, there is a corresponding α such that $v^\phi = \alpha v^\theta$. To complete the proof we must show that α is the same for all v. Given two nonzero vectors v and w, there are two cases to consider.

Case 1. v and w are dependent, i.e., $v = \beta w$, $\beta \neq 0$. Then if $v^\phi = \alpha v^\theta$, we have

$$\beta(w^\phi) = (\beta w)^\phi = v^\phi = \alpha v^\theta = \alpha(\beta w)^\theta = (\alpha\beta)w^\theta .$$

Since β is nonzero, we can cancel it from the first and last members of this string, and then $w^\phi = \alpha w^\theta$ as desired.

Case 2. v and w are independent. Consider the α's for v, w, and $v + w$, say, $v^\phi = \alpha v^\theta$, $w^\phi = \beta w^\theta$, and $(v + w)^\phi = \gamma(v + w)^\theta$. Since ϕ and θ are both linear,

$$\gamma v^\theta + \gamma w^\theta = \gamma(v + w)^\theta = (v + w)^\phi = \alpha v^\theta + \beta w^\theta ,$$

i.e., $(\gamma - \alpha)v^\theta + (\gamma - \beta)w^\theta = 0$. Since θ is nonsingular, and v and w are independent, this implies $\gamma = \alpha$ and $\gamma = \beta$.

This theorem shows that projective coordinate maps are determined as soon as we know the corresponding linear coordinate map up to a scalar multiple. It also settles the question of how much freedom we have in choosing the linear map inducing a given coordinate system. A more difficult question is: How much freedom do we have in preassigning homogeneous coordinates to points in $\bar{V}$? Since there is no point with homogeneous coordinates $0, 0, \ldots, 0$, it is not appropriate to say that we must choose an origin. The simplest points in $P_n(\mathbf{F})$ are $[1, 0, \ldots, 0]$, $[0, 1, 0, \ldots, 0]$, $\ldots$, $[0, 0, \ldots, 0, 1]$. We should be able to preassign these as homogeneous coordinates of $n + 1$ points in $\bar{V}$—avoiding collisions, of course. This is possible and almost determines the coordinate system. First we explain how to "avoid collisions."

Theorem 6.15. Let $[w_0], \ldots, [w_k]$ be $k + 1$ points in the projective space $\bar{V}$. These $k + 1$ points are in a $(k - 1)$-dimensional subspace of $\bar{V}$ if and only if the vectors $w_0, \ldots, w_k$ are linearly dependent in $\bar{V}$.

Proof. If $[w_0], \ldots, [w_k]$ are all in $\bar{W}$, and the projective dimension of $\bar{W}$ is $k - 1$, then $w_0, \ldots, w_k$ are all in the k-dimensional linear subspace $\mathbf{W}$ and hence must be linearly dependent in $\mathbf{V}$. Conversely, if $w_0, \ldots, w_k$ are linearly dependent, then there is at least one k-dimensional subspace of $\mathbf{V}$ containing all $k + 1$ vectors and hence at least one $(k - 1)$-dimensional subspace of $\bar{V}$ containing the corresponding $k + 1$ points.

Definition. A set of $k + 1$ points in $\bar{V}$ is said to be *independent* if and only if these points are not contained in a $(k - 1)$-dimensional subspace of $\bar{V}$.

Note the cases for small values of k; two points are independent if and only if they are distinct, three points are independent if and only if they are not collinear, and four points are independent if and only if they are not coplanar. At the other extreme, $n + 1$ points in an n-dimensional projective space are independent if and only if they are not contained in a hyperplane.

Theorem 6.16. Let $\bar{V}$ be an n-dimensional projective space, and let $P_0, P_1, \ldots, P_{n+1}$ be $n + 2$ points such that no $n + 1$ of them are contained in a hyperplane. There is a unique coordinate map $[\phi]$ of $\bar{V}$ onto $P_n(\mathbf{F})$ such that $P_0^{[\phi]} = [1, 0, \ldots, 0]$, $P_1^{[\phi]} = [0, 1, 0, \ldots, 0]$, $\ldots$, $P_n^{[\phi]} = [0, 0, \ldots, 1]$, and $P_{n+1}^{[\phi]} = [1, 1, \ldots, 1]$.

Proof. Choose a vector $v_i \in P_i$ for $i = 0, 1, \ldots, n + 1$. Since the $n + 1$ points $P_0, P_1, \ldots, P_n$ do not lie in a hyperplane, they are independent. Hence the vectors $v_0, v_1, \ldots, v_n$ are independent and therefore form a basis of $\mathbf{V}$. Then $v_{n+1} = \Sigma\alpha_i v_i$, $\alpha_i \in \mathbf{F}$. None of the α_i are zero since if one were, then deleting the corresponding P_i would leave us with $n + 1$ of the P_i in a hyperplane. Let $w_i = \alpha_i v_i$, $i = 0, 1, \ldots, n$, and then the w_i still form a basis of $\mathbf{V}$ and $v_{n+1} = w_0 + w_1 + \cdots + w_n$. Let ϕ be the linear coordinate map of $\mathbf{V}$ onto $\mathbf{V}_n(\mathbf{F})$ associated with this new basis. The coordinate map $[\phi]$ is clearly a coordinate map of $\bar{V}$ onto $P_n(\mathbf{F})$ with the desired action on the P_i.

Now assume $[\theta]$ is also a coordinate map with the desired action on

$P_0, P_1, \ldots, P_{n+1}$. We must show that $[\phi] = [\theta]$, i.e., that $\theta = \alpha\phi$ for some nonzero α in $\mathbf{F}$. Consider the following:

$$[w_0]^{[\theta]} = [w_0]^{[\phi]} = [1, 0, \ldots, 0]; \text{ hence } w_0^\theta = \langle \alpha_0, 0, 0, \ldots, 0\rangle$$
$$[w_1]^{[\theta]} = [w_1]^{[\phi]} = [0, 1, \ldots, 0]; \text{ hence } w_1^\theta = \langle 0, \alpha_1, 0, 0, \ldots, 0\rangle$$
$$\vdots$$
$$[w_n]^{[\theta]} = [0, 0, \ldots, 1]; \text{ hence } w_n^\theta = \langle 0, 0, \ldots, \alpha_n\rangle$$
$$[w_{n+1}]^{[\theta]} = [1, 1, \ldots, 1]; \text{ hence } w_{n+1}^\theta = \langle \alpha, \alpha, \ldots, \alpha\rangle .$$

Since θ is linear and $w_{n+1} = \Sigma w_i$, we must have $\langle \alpha, \ldots, \alpha\rangle = \langle \alpha_0, \alpha_1, \ldots, \alpha_n\rangle$; hence $\alpha_i = \alpha$, and thus, θ and $\alpha\phi$ agree on a basis of $\mathbf{V}$. Thus, $\theta = \alpha\phi$ and $[\theta] = [\phi]$.

Definition. If $P_0, P_1, \ldots, P_{n+1}$, and $[\phi]$ are as described in the theorem above, then P_{n+1} is called the *unit point* for $[\phi]$, and $P_0, P_1, \ldots, P_n$ are called the *vertices* for $[\phi]$. In a projective plane (respectively 3-space), the vertices form the *reference triangle* (respectively *tetrahedron*) for $[\phi]$.

Note that the proof of theorem 6.16 is constructive, i.e., it is a recipe that shows how to compute the homogeneous coordinates of a point if one is given the vertices and unit point of the coordinate system. Namely, first express a vector representing the unit point as a linear combination of vectors representing the vertices, and adjust the representatives of the vertices so that the new representatives added together yield the chosen representative of the unit point. Then the homogeneous coordinates of any point are simply the coefficients used to express a representative of the point in terms of the new representatives of the vertices. The most common situation in which one must carry this out is the case in which $\bar{V} = P_n(\mathbf{F})$, and there is need to consider a coordinate map distinct from the natural one, $[\mathbf{1}]$. This is a special case of "changing coordinates." Commutative diagrams again help to keep track of which matrices to use and where to use them.

Theorem 6.17. Let $[\phi]$ and $[\theta]$ be coordinate maps for an n-dimensional projective space, $\bar{V}$, over $\mathbf{F}$. If A is the natural matrix of $\phi^{-1}\theta$, i.e., the rows of A are the θ-coordinates of the ϕ-basis vectors, then the diagram (on a triangular prism) of Figure 34 is commutative, and hence we can convert homogeneous coordinates in the $[\phi]$ system to $[\theta]$-coordinates simply by multiplying by A.

Before starting the proof, we call the reader's attention to the fact that $[\phi]^{-1}[\theta] = [A]$, not A. Thus, any nonzero scalar multiple of A can be used to change $[\phi]$-coordinates to $[\theta]$-coordinates. This is especially useful when converting back from $[\theta]$ to $[\phi]$-coordinates, in which case we use A^{-1}. The often

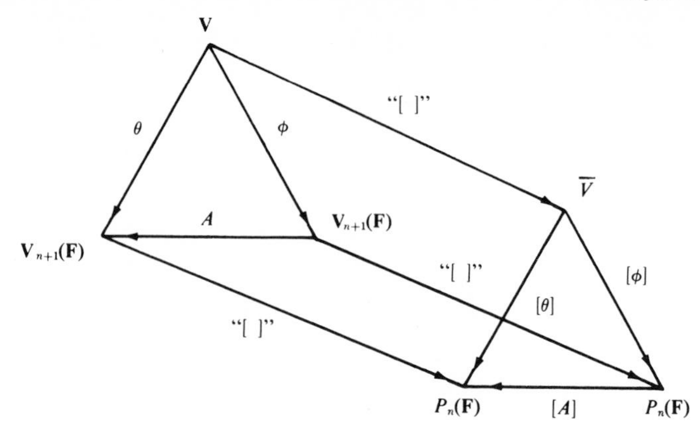

FIG. 34. Changing coordinates in a projective space.

tiresome denominators in A^{-1} can be multiplied out! In particular, if the dimension is small and A^{-1} is computed using determinants, then one may use the (classical) adjoint of A in place of A^{-1}.

Proof. The three rectangular diagrams on the sides of the prism are commutative by definition of $[\phi]$, $[\theta]$, and A. The upper triangle was shown to be commutative in section 4.5. Finally, to show that the lower triangle is commutative, let $[v] \in P_n(\mathbf{F})$, then

$$[v]^{[A]} = [v^A] = [v^{\phi^{-1}\theta}] = [v^{\phi^{-1}}]^{[\theta]} = [v]^{[\phi^{-1}][\theta]} .$$

Clearly $[\phi^{-1}] = [\phi]^{-1}$, so $[A]$ and $[\phi]^{-1}[\theta]$ agree everywhere on $P_n(\mathbf{F})$.

Example. Let $\bar{V} = P_3(\mathbf{R})$. Consider the natural coordinate map $[\mathbf{1}]$ and the coordinate map $[\theta]$ with vertices $[1, 0, 0, 0]$, $[1, 1, 0, 0]$, $[1, 0, 1, 0]$, $[1, 0, 0, 1]$, and unit point $[1, 1, 1, 1]$. In order to draw a figure illustrating this situation we treat the hyperplane with equation $x_0 = 0$ as the plane at infinity. If we then adjust the homogeneous coordinates of all points off that plane so that their first (x_0) coordinates are 1, then we claim (see theorem 6.20) that the remaining three coordinates are affine coordinates in the affine space obtained by deleting the plane at infinity from $P_3(\mathbf{R})$. We can then sketch these points using the traditional cartesian coordinates, noting that the projective point $[0, a, b, c]$ is the ideal point corresponding to affine lines with direction numbers a, b, and c. Figure 35 is drawn using these conventions.

To compute $[\theta]$-coordinates we use theorems 6.16 and 6.17. The representatives $\langle 1, 0, 0, 0\rangle$, $\langle 1, 1, 0, 0\rangle$, $\langle 1, 0, 1, 0\rangle$, $\langle 1, 0, 0, 1\rangle$ of the $[\theta]$-vertices do not add to $\langle 1, 1, 1, 1\rangle$ so they must be adjusted to find a suitable θ-basis of $\mathbf{V}_4(\mathbf{R})$. Since $\langle 1, 1, 1, 1\rangle = -2\langle 1, 0, 0, 0\rangle + \langle 1, 1, 0, 0\rangle + \langle 1, 0, 1, 0\rangle + \langle 1, 0, 0, 1\rangle$, we need only replace $\langle 1, 0, 0, 0\rangle$ by $\langle -2, 0, 0, 0\rangle$ as the representative of $[1, 0, 0, 0]$. The natural matrix for $\theta^{-1}\mathbf{1}$ is much easier to find than the

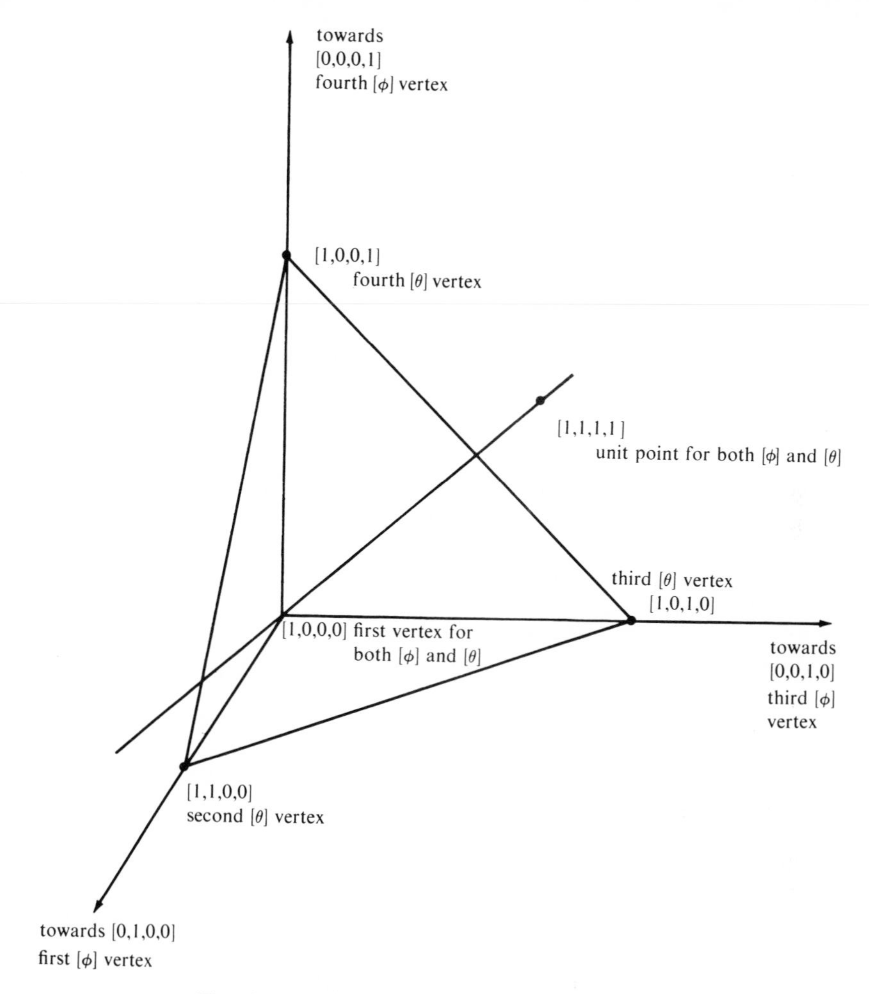

FIG. 35. Vertices and unit point for $[\phi]$ and $[\theta]$.

matrix for $\mathbf{1}^{-1}\theta$ since we already know the **1**-coordinates of the θ-basis vectors. Thus, by theorem 6.17 the following diagrams are commutative if

$$A = \begin{bmatrix} -2 & 0 & 0 & 0 \\ 1 & 1 & 0 & 0 \\ 1 & 0 & 1 & 0 \\ 1 & 0 & 0 & 1 \end{bmatrix}.$$

$$\begin{array}{ccc} & P_3(\mathbf{R}) & \\ {\scriptstyle [\mathbf{1}]}\swarrow & & \searrow{\scriptstyle [\theta]} \\ P_3(R) & \xleftarrow{\;A\;} & P_3(R) \end{array} \qquad \begin{array}{ccc} & P_3(R) & \\ {\scriptstyle [\mathbf{1}]}\swarrow & & \searrow{\scriptstyle [\theta]} \\ P_3(R) & \xrightarrow{\;A^{-1}\;} & P_3(R) \end{array}$$

We find that

$$A^{-1} = \begin{bmatrix} -\frac{1}{2} & 0 & 0 & 0 \\ \frac{1}{2} & 1 & 0 & 0 \\ \frac{1}{2} & 0 & 1 & 0 \\ \frac{1}{2} & 0 & 0 & 1 \end{bmatrix},$$

and thus, we can convert natural coordinates to θ-coordinates in $P_3(\mathbf{R})$ by using A^{-1} or αA^{-1}. For example,

$$[3, 1, 1, 1]^{[\theta]} = [3, 1, 1, 1]A^{-1} = [0, 1, 1, 1]$$

$$[0, 1, 1, 1]^{[\theta]} = [0, 1, 1, 1]\, \tfrac{2}{3}\, A^{-1} = [1, \tfrac{2}{3}, \tfrac{2}{3}, \tfrac{2}{3}]\,,$$

and, in general, on the line joining these two points,

$$[a, b, b, b]^{[\theta]} = [a, b, b, b]2A^{-1} = [3b - a, 2b, 2b, 2b]\,.$$

It is interesting to note how the ratios b/a and $2b/(3b - a)$ vary as one moves along the line. This is illustrated in Figure 36.

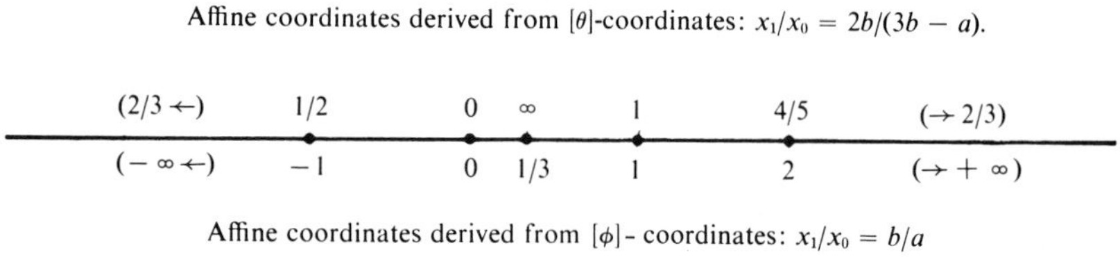

Affine coordinates derived from $[\theta]$-coordinates: $x_1/x_0 = 2b/(3b - a)$.

$(2/3 \leftarrow)$ $1/2$ 0 ∞ 1 $4/5$ $(\rightarrow 2/3)$

$(-\infty \leftarrow)$ -1 0 $1/3$ 1 2 $(\rightarrow +\infty)$

Affine coordinates derived from $[\phi]$-coordinates: $x_1/x_0 = b/a$

FIG. 36. Two affine coordinate systems on the line joining [3, 1, 1, 1] to [0, 1, 1, 1] in $P_3(\mathbf{R})$.

In analytic geometry one of the central topics is the graphing of equations. Since we use homogeneous coordinates in projective geometry, we consider only a certain type of equation in the analytic geometry of projective spaces.

Definition. Let $f(x_0, x_1, \ldots, x_n) = 0$ be a polynomial equation in the $n + 1$ unknowns $x_0, \ldots, x_n$ with coefficients in the field $\mathbf{F}$. This equation is said to be *homogeneous of degree* k if $f(tx_0, \ldots, tx_n) = t^k f(x_0, \ldots, x_n)$. Homogeneous equations of degree 1 are *homogeneous linear equations.*

A homogeneous equation has the crucial property that if any one representative of a projective point is a solution of the equation, then all representatives of that point are solutions. Thus, we can state the following definition without fear of confusion.

Definition. Let $f_i(x_0, x_1, \ldots, x_n) = 0$, $i = 1, 2, \ldots, k$ be a system of homogeneous equations, perhaps of varying degrees. The *solution set* or *graph* of this system in $P_n(\mathbf{F})$ is the set of all points $[\alpha_0, \ldots, \alpha_n]$ in $P_n(\mathbf{F})$ such that $f_i(\alpha_0, \ldots, \alpha_n) = 0$, $i = 1, 2, \ldots, k$.

When the system of equations is a linear system, then we have a standard way of abbreviating the notation; namely, let $x = (x_0, \ldots, x_n)$, and let A be the $(n + 1)$ by k matrix whose jth column consists of the coefficients of the jth equation. Then we can write down a single matrix equation, $xA = 0$, to abbreviate the system of k scalar equations. In this case the solution set in $\mathbf{V}_{n+1}(\mathbf{F})$ is just the kernel of A and is hence a subspace. Thus, the solution set in $P_n(\mathbf{F})$ is a subspace of $P_n(\mathbf{F})$. Using the standard result that rank A + nullity $A = n + 1$, where nullity A is the (linear) dimension of the kernel of A, we obtain the following result for projective subspaces. Note that in passing from linear results to projective results we reduce all dimensions by one, but we do *not* reduce ranks.

Theorem 6.18. A subset of $P_n(\mathbf{F})$ is a k-dimensional subspace if and only if it is the graph in $P_n(\mathbf{F})$ of a system of homogeneous linear equations of rank $n - k$.

Corollary 6.19. Hyperplanes in $P_n(\mathbf{F})$ are graphs of single homogeneous linear equations.

For example, lines in a projective plane are graphs of homogeneous linear equations in three unknowns. The familiar computational rule for finding cross products may be used to find the equation of the line joining two points or the point on two lines (exercise 6.4, problem 1).

The intersection of the graphs of two systems of equations is the graph of the combined system, so that when we add a new linear equation to a system the dimension of the graph drops by one unless the new equation is redundant. This is simply a restatement of corollary 6.5.

To complete this section we exploit homogeneous coordinates to derive another, more precise version of theorem 6.1 in which we proved that projective spaces are essentially affine spaces with a hyperplane of points "at infinity" adjoined.

Theorem 6.20. Let A and $\bar{V}$ be, respectively, an affine and a projective n-dimensional space over the field $\mathbf{F}$. For any hyperplane $\bar{H}$ in $\bar{V}$, we can imbed A in $\bar{V}$ such that $\bar{H}$ plays the role of the hyperplane at infinity. More precisely, there is a map k: $A \rightarrow \bar{V}$ and a linear transformation j:$\mathbf{T} \rightarrow \mathbf{H}$ ($\mathbf{T}$ the translation group of A) such that:

(1) j is one-to-one and maps $\mathbf{T}$ onto $\mathbf{H}$.
(2) k is one-to-one with range $\bar{V} - \bar{H}$.
(3) If P and Q are distinct affine points, then k maps the affine line $P + Q$ into the projective line joining $(P)k$ and $(Q)k$. Moreover, the only projective point in $(P)k + (Q)k$, but not in $(P + Q)k$, is the point at infinity, and it is represented by the vector $(\overrightarrow{PQ})j$.

Proof. Choose a linear coordinate map $\phi:\mathbf{V} \rightarrow \mathbf{V}_{n+1}(\mathbf{F})$ such that $\mathbf{H}$ is sent by ϕ to the graph of $x_0 = 0$. The associated projective coordinate map $[\phi]:\bar{V} \rightarrow P_n(\mathbf{F})$ sends $\bar{H}$ to

$$(\bar{H})^{[\phi]} = \{[x_0, x_1, \ldots, x_n] \in P_n(\mathbf{F}) | x_0 = 0\} .$$

Let $\theta:\mathbf{T} \rightarrow \mathbf{V}_n(\mathbf{F})$ be any linear coordinate map, and let $\theta^*:A \rightarrow A_n(\mathbf{F})$ be an associated affine coordinate map. We define $J:\mathbf{V}_n(\mathbf{F}) \rightarrow \mathbf{V}_{n+1}(\mathbf{F})$ and $K:A_n(\mathbf{F}) \rightarrow P_n(\mathbf{F})$ as follows:

$$\langle x_1, \ldots, x_n \rangle J = \langle 0, x_1, \ldots, x_n \rangle$$
$$(\alpha_1, \ldots, \alpha_n)K = [1, \alpha_1, \ldots, \alpha_n] .$$

J and K are both one-to-one, range $J = \mathbf{H}^{\phi}$, and range $K = P_n(\mathbf{F}) - \bar{H}$. Thus, J and K satisfy (in coordinate form) conditions (1) and (2). We leave to the

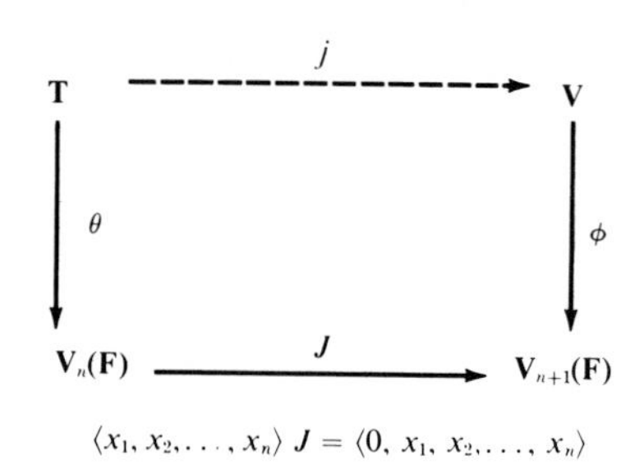

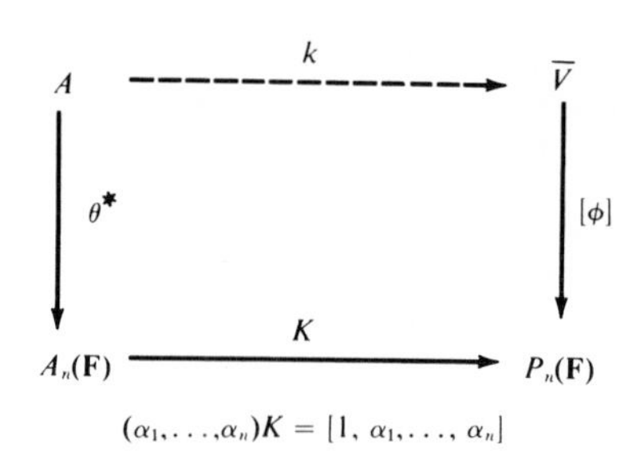

FIG. 37. Lifting J and K back to $\mathbf{T}$ and A to obtain desired mapping.

reader the elementary computations needed to verify that J and K satisfy condition (3) relative to $A_n(\mathbf{F})$, $P_n(\mathbf{F})$, and $\bar{H}^{[\phi]}$. Now we simply lift J and K back to $\mathbf{T}$ and A (see Figure 37) to obtain our desired mapping, $j = \theta J \theta^{-1}$ and $k = \theta^* K[\phi]^{-1}$.

Exercise 6.4

1. Let $[a, b, c]$ and $[a', b', c']$ be distinct points in $P_2(\mathbf{F})$. Show that if

$$d = \det \begin{bmatrix} b & c \\ b' & c' \end{bmatrix}, \quad e = \det \begin{bmatrix} c & a \\ c' & a' \end{bmatrix}, \quad f = \det \begin{bmatrix} a & b \\ a' & b' \end{bmatrix},$$

then an equation for the line joining the two points is $dx + ey + fz = 0$. Also show that d, e, f is the point of intersection of the two lines whose equations are $ax + by + cz = 0$ and $a'x + b'y + c'z = 0$.

2. Find coordinates for all seven points and equations for all seven lines in the 7-point projective plane shown in Figure 30b if the reference triangle is P_1, P_2, P_3. Note that there is only one point which can be the unit point (see problem 8).

3. Consider the linear functions f and g: $\mathbf{V}_3(\mathbf{Q}) \to \mathbf{Q}$ defined by $(\langle x, y, z\rangle)f = 2x + 3y - z$ and $(\langle x, y, z\rangle)g = 5x - 7y + 11z$. Let $\bar{H}_1$ and $\bar{H}_2$ be the lines in $P_2(\mathbf{Q})$ consisting of all points whose homogeneous coordinates are in the kernel of f or g, respectively. Show that a line passes through $\bar{H}_1 \cap \bar{H}_2$ if and only if its equation can be put in the form $(2\alpha + 5\beta)x + (3\alpha - 7\beta)y + (-\alpha + 11\beta)z$ with not both α and β zero.

4. Generalize the result above, i.e., show that if $\mathbf{V}$ is an $(n + 1)$-dimensional vector space over $\mathbf{F}$, and if f, g belong to $\mathbf{V}^t$ (the space of all linear functions from $\mathbf{V}$ to $\mathbf{F}$) with $\mathbf{H}_1 = \text{kernel } f$, $\mathbf{H}_2 = \text{kernel } g$, then the hyperplane $\bar{H}$ in $\bar{V}$ contains $\bar{H}_1 \cap \bar{H}_2$ if and only if H is the kernel of $\alpha f + \beta g$ for some scalars α and β not both zero. We return to this in section 6.8.

5. Prove the "dual" of the result above, i.e., show that if $[v_1]$, $[v_2]$ are points in the projective space $\bar{V}$, then $[v]$ is in the line $[v_1] + [v_2]$ if and only if $v = \alpha v_1 + \beta v_2$ for some scalars α and β not both zero.

6. Prove Desargues theorem in an arbitrary projective plane over a field. For the statement of Desargues theorem, see exercise 6.1, problem 1. The computations needed are much simpler if one exploits the results of the two previous exercises.

7. Prove the theorem of Pappus in an arbitrary projective plane over a field. The statement of the theorem is: If A, B, C and A', B', C' are two triples of distinct points on two lines, as in Figure 38, then the points $(A + B') \cap (A' + B)$, $(A + C') \cap (A' + C)$, and $(B + C') \cap (B' + C)$ are collinear. This theorem is basic in the study of the foundations of projective geometry because if it is valid in a projective plane, then that plane is isomorphic to a projective plane over a field.

8. Show that the vertices uniquely determine the coordinate map if and only if $\#\mathbf{F} = 2$.

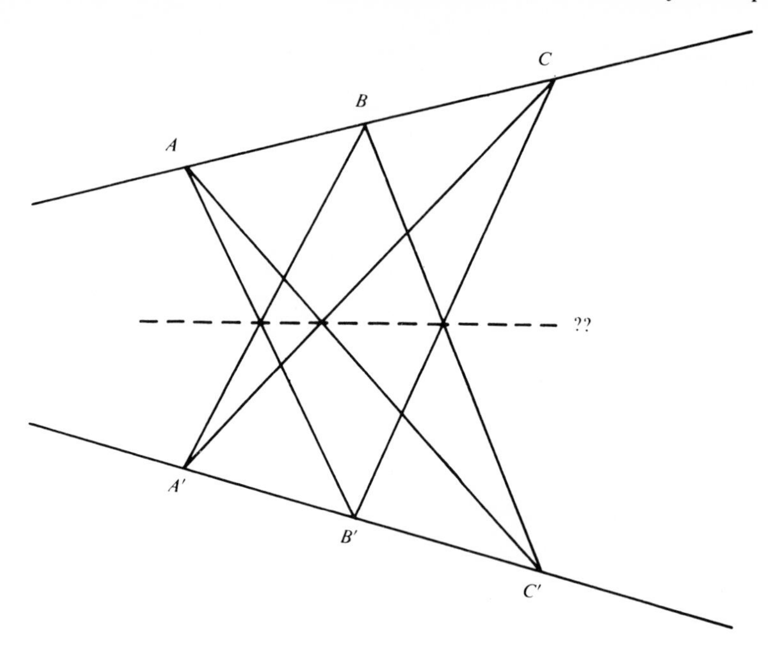

FIG. 38. The theorem of Pappus.

9. Let **F** be a finite field and $f = \#\mathbf{F}$. How many different coordinate maps are there for a given n-dimensional projective space over **F**?

6.5. Projective symmetries I: Perspectivities

In this section we consider those symmetries of projective spaces which are the analogs of dilatations of affine spaces. Viewing projective space, for the moment, as an extended affine space, the dilatations of the affine space leave all ideal points fixed since they preserve directions. The set of ideal points is a hyperplane. Thus, the dilatations correspond to projective symmetries leaving a hyperplane pointwise fixed. Formalizing this observation, and treating all hyperplanes alike, we arrive at the following definition.

Definition. Let $\bar{V}$ be a projective space of dimension at least two. A *perspectivity* of $\bar{V}$ is a permutation, f, of $\bar{V}$ such that:

(1) f leaves exactly one hyperplane, called the *axis*, pointwise fixed. *Caution!* We do not require that *all* fixed points lie in one hyperplane!

(2) If P and Q are distinct points not on the axis, then $(P + Q)$ and $((P)f + (Q)f)$ intersect the axis at the same point.

We avoid one-dimensional projective spaces since *any* permutation of such a space with exactly one fixed point would satisfy this definition. Often perspectivities are referred to as central or perspective collineations. In the case of a three-dimensional space, the axis of the perspectivity is often called the mirror of f. We are using the term axis in analogy with its customary use for perspectivities in projective planes.

Note that since a perspectivity f is assumed to be one-to-one, property (2) makes sense, i.e., if $P \neq Q$, then $(P)f \neq (Q)f$, so the line $(P)f + (Q)f$ is defined whenever $P + Q$ is. Note also that property (2) is the straightforward translation of the property $P + Q \mathbin{/\!/} (P)f + (Q)f$ for dilatations.

In addition to the analogy between dilatations and perspectivities, there is another reason for considering perspectivities as the "basic" symmetries of a projective space. This reason is historical and is related to the layman's concept of perspective. A rough translation into mathematical terms is as follows. Assume our projective space $\bar{V}$ is embedded as a hyperplane in a larger projective space (imagine a two-dimensional picture to be copied by an artist). Choose another hyperplane, $\bar{V}'$ (the new canvas), and a point P not on either $\bar{V}$ nor $\bar{V}'$ (the artist's eye, or center of perspective). We project $\bar{V}$ onto $\bar{V}'$ through P, i.e., we define a mapping $p{:}\bar{V} \to \bar{V}'$ by $(Q)p = (Q + P) \cap \bar{V}'$ for each $Q \in \bar{V}$. Choose another point P' (the viewer's point of perspective), and project $\bar{V}'$ back onto $\bar{V}$ via P', i.e., $p'{:}\bar{V}' \to \bar{V}$ is defined by $(Q')p' = (Q' + P') \cap \bar{V}$. If $P = P'$, then clearly $pp' = \mathbf{1}$ (two people with the same perspective see the same picture), but if $P \neq P'$, then pp' is a perspectivity of $\bar{V}$ with axis $\bar{V} \cap \bar{V}'$.

Before proving that every projective space has many perspectivities, we first prove a sequence of theorems showing that perspectivities are, in fact, all very much like dilatations. In the proofs of these theorems, we use many times the following facts about lines. Both of these facts follow easily from theorem 6.4 and were stated as problems in exercise 6.2.

(1) If two lines are distinct, they intersect in at most one point. Distinct lines are co-planar if and only if they intersect in exactly one point.

(2) If a line is not contained in a hyperplane, then it intersects that hyperplane in exactly one point.

Theorem 6.21. A permutation, f, of $\bar{V}$ with exactly one hyperplane of fixed points is a perspectivity if and only if $(L)f$ is a line whenever L is a line.

Proof. Let f be a perspectivity and L be a line. If L is contained in the axis, then $(L)f = L$ is a line. If not, let P be the point of intersection of L and the axis, and choose a second point Q on L. We claim that $(L)f$ is the line $(Q)f + P$. Given $R \in L, Q + R = L$ meets the axis at P. Hence $(Q)f + (R)f$ must, too, i.e., $(R)f \in (Q)f + P$. Conversely, given $X \in (Q)f + P$, we can choose S such that $(S)f = X$. But then $(S)f + (Q)f$ meets the axis at P; hence $S + Q$ must also meet it at P, i.e., $S \in Q + P = L$. Thus, $(L)f$ is a line.

Now assume that f is a permutation with exactly one hyperplane, $\bar{H}$, of fixed points, and that $(L)f$ is a line whenever L is. Given distinct points P and Q not on $\bar{H}$, the line $P+Q$ meets $\bar{H}$ in exactly one point, call it X. The point X is left fixed by f, but also $(X)f \in ((P)f + (Q)f)$. Since f is one-to-one, neither $(P)f$ nor $(Q)f$ is in $\bar{H}$, and thus, $X = (X)f$ is the unique point in $\bar{H}$ on both $P+Q$ and $(P)f + (Q)f$.

Definition. A permutation, f, of $\bar{V}$ with the property that $(L)f$ is a line whenever L is a line is called a *collineation* of $\bar{V}$.

Note that a collineation not only carries collinear points to collinear points, but also the images of the points on a line "fill up" a complete line. This means that f^{-1} is a collineation whenever f is. Since fg is also a collineation whenever f and g are, the collineations of $\bar{V}$ form a group. We study this group later in section 6.7. It is easy to show (Do so!) that no nontrivial collineation can leave a hyperplane $\bar{H}$ pointwise fixed and also have more than one fixed point off $\bar{H}$. Thus the theorem above implies the following result.

Corollary 6.22. The subgroup of the collineation group consisting of those collineations leaving the hyperplane $\bar{H}$ pointwise fixed consists simply of **1** and the perspectivities with axis $\bar{H}$.

Proof. If a collineation leaves more than the one hyperplane $\bar{H}$ fixed then it certainly has at least two fixed points off $\bar{H}$ and is therefore the identity.

Corollary 6.23. If f is a perspectivity and $P \neq (P)f$ then the line $P + (P)f$ is left globally fixed by f.

Proof. If $P \neq (P)f$ then (since f is one to one) neither P nor $(P)f$ are on the axis. Hence $P + (P)f$ meets the axis at a point Q. Then $(P+Q) = (P + (P)f) = ((P)f + Q)$, hence $(P + (P)f)f = (P+Q)f = ((P)f + Q) = P + (P)f$.

Theorem 6.24. A perspectivity is uniquely determined by its axis and its action on two points off the axis.

Proof. This follows immediately from the remark just before corollary 6.22, but we present another proof to show how one uses $\bar{H}, P, Q, (P)f, (Q)f$, and R to locate $(R)f$. If $R \in \bar{H}$ then $(R)f = R$, so we assume $R \notin \bar{H}$.

Case 1. $R \notin P+Q$. The lines $R+P$ and $R+Q$ are distinct; hence the points X and Y, $X = (R+P) \cap \bar{H}$ and $Y = (R+Q) \cap \bar{H}$ are distinct. Since neither $(P)f$ nor $(Q)f$ is in $\bar{H}$ and $X \neq Y$, it follows that the two lines $(P)f + X$ and $(Q)f + Y$ are distinct and so intersect in at most one point. By condition (2) on perspectivities, $(R)f$ must be on both lines; hence there is at most one possibility for $(R)f$.

Case 2. $R \in P+Q$. Since $\dim \bar{V} \geq 2$, we can choose $S \notin P+Q$, $S \notin \bar{H}$. There is at most one possibility for $(S)f$ (case 1) and so (again applying case 1 to R with P and Q replaced by P and S) at most one possibility for $(R)f$.

Note that we have been very cautious not to claim that a perspectivity with axis $\bar{H}$ exists for all possible choices of P, Q, $(P)f$, $(Q)f$ off $\bar{H}$ such that $P + Q$ and $(P)f + (Q)f$ meet in the hyperplane $\bar{H}$. This is not true, as one can see by examining the 7-point projective plane (Figure 30b or 33) and noting that no perspectivity of that plane could have axis $P_2 + P_3$, leave P_4 fixed, and send Q_2 to Q_3.

Recall that in the case of dilatations there is always a point left "linewise" fixed by the dilatation; this point was the center in the case of a central dilatation and was the direction (an ideal point!) in the case of a translation. For this reason we expect that a perspectivity always leaves one point "linewise" fixed. In addition to this motivation, we have the added background of interpreting a perspectivity as the composite of two projections, $p:\bar{V} \to \bar{V}'$ via P and $p':\bar{V}' \to \bar{V}$ via P'. Clearly the point $(P + P') \cap \bar{V}$ is left "linewise" fixed. We now carefully define "linewise fixed" and prove that each perspectivity does indeed leave a (unique!) point linewise fixed.

Definition. Let f be a permutation of the projective space $\bar{V}$. A point P is *linewise fixed* by f if and only if every line through P is left globally fixed by f. It is said to be *hyper-fixed* if every hyperplane through P is left globally fixed.

Note that "f leaves $\bar{H}$ pointwise fixed" and "f leaves P hyper-fixed" are very similar concepts. In section 6.8 we study duality and find that it is natural to expect hyperplanes and points to be analogous.

Proposition 6.25. Let f be a permutation of a projective space $\bar{V}$ of dimension at least 2. A point P is left hyper-fixed by f if and only if it is left linewise fixed.

Proof. Any line through P is the intersection of $n - 1$ hyperplanes through P. Since any permutation preserves intersections, f must leave each line through P globally fixed if it leaves P hyper-fixed. On the other hand, f need not preserve joins so it is not enough to simply say a hyperplane through P is the join of $n - 1$ lines to prove "linewise implies hyper-fixed." We can circumvent this difficulty as follows. Any hyperplane through P is the *union* of lines through P. Thus, a permutation f leaves P hyper-fixed whenever it leaves P linewise fixed, since a permutation preserves unions (finite or infinite!).

Corollary 6.26. If f is a permutation of $\bar{V}$ leaving two points, P and Q, hyper-fixed, then f leaves fixed every point not on $P + Q$. If, in addition, $\dim \bar{V} \geq 2$ and f is a collineation, then f is the identity.

Proof. If $R \notin P + Q$, then R is the unique point of intersection of $R + P$ and $R + Q$. Both of these lines are left fixed by f; hence R is also left fixed. If f is a collineation and f leaves P, Q, and every point off $P + Q$ fixed, then f leaves *every* line in $\bar{V}$ fixed; hence (if $\dim \bar{V} \geq 2$) it leaves every point fixed.

Theorem 6.27. Let f be a perspectivity with axis $\bar{H}$. There is exactly one point left hyper-fixed by f. Moreover, this point is on every line (except perhaps those in $\bar{H}$) left fixed by f.

Proof. By theorem 6.24 there is at most one point off $\bar{H}$ left fixed by f, and by corollary 6.26 there is at most one point left hyper-fixed.

Case 1. P is the unique fixed point off $\bar{H}$. Every line, L, through P contains two fixed points, P and $L \cap \bar{H}$. Hence P is left linewise (and hence hyper-)fixed by f. Moreover, P is on every line of the form $Q + (Q)f$ since $P + Q$ is a line through P, so $(Q)f \in P + Q$.

Case 2. There are no fixed points off $\bar{H}$. If Q and R are distinct points off $\bar{H}$ then Q, R, $(Q)f$, and $(R)f$ are coplanar since $Q + R$ and $(Q)f + (R)f$ meet in $\bar{H}$. Thus $Q + (Q)f$ and $R + (R)f$ coincide or intersect in a single point. Now fix Q and let $P = (Q + (Q)f) \cap \bar{H}$. By corollary 6.23, any line of the form $X + (X)f$ is left globally invariant by f. Thus, if $R \notin Q + (Q)f$ and $R \neq (R)f$, then $(R + (R)f)$ intersects $(Q + (Q)f)$ in an invariant point. Since there are no invariant points off $\bar{H}$ this implies that $P \in X + (X)f$ for every $X \notin \bar{H}$. In other words, $(X)f \in X + P$ whenever $X \notin \bar{H}$. Clearly P is left linewise fixed.

In each case we have shown that the unique point left linewise fixed is on every line of the form $Q + (Q)f$. Since every line contains at least three points, and there is at most one fixed point off $\bar{H}$, every invariant line not in $\bar{H}$ is of this form. Thus, the proof is complete.

Definition. Let f be a perspectivity. The unique point left hyper-fixed by f is called the *center* of f. The perspectivity is called an *elation* or *homology* according as the center is or is not on the axis.

Corollary 6.28. A perspectivity is uniquely determined by its axis, center, and action on one other point, Q. The center, Q, and the image of Q must be collinear.

Proof. This is an obvious consequence of theorem 6.24 in the case of homologies. The proof for elations is left to the reader. Simply mimic the proof of theorem 6.24 using the fact (derived above) that the center is on every line of the form $X + (X)f$.

This result limits the possible perspectivities. There are several ways to prove the converse, i.e., to show that *all* possible perspectivities exist. The first method formalizes our observations concerning the analogies between dilatations and perspectivities. Namely, we know that all possible dilatations of an n-dimensional affine space A exist. Thus, we know that if the axis is to be the hyperplane of ideal points in PA, then all possible perspectivities of the projective space PA (with this axis) exist. If $\bar{V}$ is a projective space, and $\bar{H}$ is *any* hyperplane in $\bar{V}$, then (by theorem 6.20) $\bar{V}$ is essentially the same as the projective space PA with $\bar{H}$ playing the role of the hyperplane of ideal points. Thus, all possible perspectivities of $\bar{V}$ with axis $\bar{H}$ exist. $\bar{H}$ is an arbitrary hyperplane in $\bar{V}$ so all possible perspectivities in $\bar{V}$ exist.

Another, more direct method is based on Desargues theorem (exercise 6.5, problem 1), for this theorem is precisely the condition needed to avoid collisions and to be sure the result *is* a perspectivity when one uses the proof of theorem 6.24 or corollary 6.28 as a guide for defining a mapping with the desired axis, center, and action on one point. This approach is the only one available in the axiomatic study of projective spaces. For a short discussion of the interesting ramifications for imbedding projective planes in higher dimensional projective spaces, see chapter 20 of [8]. For complete details of this method, see chapter 7 of [2].

Still a third method involves coordinates. In order to review homogeneous coordinates and to provide simple examples for the methods to be used in the next section we carry out this method in detail.

Theorem 6.29. Let $\bar{V}$ be a projective space of dimension n, $n \geq 2$. Let $\bar{H}$ be a hyperplane in $\bar{V}$, and let P, Q, and R be distinct collinear points with neither Q nor R in $\bar{H}$. There is a unique perspectivity, f, with axis $\bar{H}$ and center P such that $(Q)f = R$.

Proof. By corollary 6.28 there is at most one such perspectivity.

The outline of the proof that there is at least one is as follows. We expect homologies and elations to behave like central dilatations and translations if we treat $\bar{H}$ as the hyperplane at infinity. Therefore we restrict ourselves to coordinate maps, $[\phi]$, such that $\bar{H}$ is the graph of $x_0 = 0$. Then each point off $\bar{H}$ has a *unique* set of $[\phi]$-coordinates with $x_0 = 1$. The other n $[\phi]$-coordinates can be viewed as its affine coordinates. If $P \notin \bar{H}$ we choose $[\phi]$ so that the affine coordinates of P are $(0, 0, \ldots, 0)$, i.e., we choose our "origin" at P. A central dilatation with center at the origin sends $(x_1, x_2, \ldots, x_n)$ to $(\beta x_1, \beta x_2, \ldots, \beta x_n)$; thus, in this case, we try to find a perspectivity, f, such that $f^{[\phi]}$ behaves this way on the last n $[\phi]$-coordinates. Similarly, if $P \in \bar{H}$ we choose our origin at Q and look for a perspectivity whose $[\phi]$-coordinate form sends (if $x_0 = 1$) the last n coordinates, $(x_1, x_2, \ldots, x_n)$, to $(x_1 + b_1, x_2 + b_2, \ldots, x_n + b_n)$, where $(b_1, \ldots, b_n)$ are the affine coordinates of R. With this outline available we are ready to present the details of the construction.

Case 1. $P \notin \bar{H}$. Choose a vector $v_0 \in \mathbf{V}$ such that $P = [v_0]$, and let $v_1, v_2, \ldots, v_n$ be a basis of $\mathbf{H}$. Since $P \notin \bar{H}$, the vectors $v_0, v_1, \ldots, v_n$ form a basis of $\mathbf{V}$. Establish a homogeneous coordinate system $[\phi]: \bar{V} \to P_n(\mathbf{F})$ with vertices $[v_0], [v_1], \ldots, [v_n]$. In this coordinate system, $\bar{H}$ is the set of all points with coordinates of the form $[0, x_1, x_2, \ldots, x_n]$, i.e., $\bar{H}$ is the graph of the equation $x_0 = 0$. For any point not on $\bar{H}$, we can choose homogeneous coordinates such that $x_0 = 1$, in particular, $P^{[\phi]} = [1, 0, 0, \ldots, 0]$ and $Q^{[\phi]} = [1, a_1, a_2, \ldots, a_n]$ with not all of the $a_i = 0$. Since $R \in P + Q$,

$$R^{[\phi]} = [\alpha\langle 1, 0, \ldots, 0\rangle + \beta\langle 1, a_1, \ldots, a_n\rangle] = [\alpha + \beta, \beta a_1, \ldots, \beta a_n]$$

with α, β, and $\alpha + \beta$ not zero since $R \neq Q$, $R \neq P$, and $R \notin \bar{H}$. By proposition 4.9 there is a unique linear transformation $g:\mathbf{V} \to \mathbf{V}$ such that $(v_0)g = (\alpha + \beta)v_0$ and $(v_i)g = \beta v_i$, $i = 1, 2, \ldots, n$. Define $f:\bar{V} \to \bar{V}$ by $([v])f = [(v)g]$. Since g is linear (in particular, homogeneous!), f is well defined, i.e., $[w] = [v]$ implies $w = \beta v$, and hence $([v])f = ([w])f$ since $(w)g = \beta((v)g)$.

$X^{[\phi]} = [x_0, x_1, \ldots, x_n]$ implies

$$(X)f^{[\phi]} = [(\alpha + \beta)x_0, \beta x_1, \ldots, \beta x_n]\,,$$

i.e., the $[\phi]$-coordinate form of f, $f^{[\phi]} = [\phi]^{-1}f[\phi]$, sends $[x_0, x_1, \ldots, x_n]$ to $[(\alpha + \beta)x_0, \beta x_1, \ldots, \beta x_n]$. From this it follows that:

(1) $X \in \bar{H}$ implies $(X)f = X$ since $[0, \beta x_1, \ldots, \beta x_n] = [0, x_1, \ldots, x_n]$.

(2) If $X \notin \bar{H}$, then $(X)f = X$ only if $X = P$ since $\alpha \neq 0$, and, therefore, if $x_0 \neq 0$, then $[(\alpha + \beta)x_0, \beta x_1, \ldots, \beta x_n] = [x_0, x_1, \ldots, x_n]$ only if all the $x_i = 0$, $i = 1, 2, \ldots, n$.

(3) If X is as in 2, and Y is another point off $\bar{H}$, then $(X + Y) \cap \bar{H} = ((X)f + (Y)f) \cap \bar{H}$ since we can choose homogeneous coordinates for Y, $[y_0, y_1, \ldots, y_n]$, such that $y_0 = x_0$, and then $(X + Y) \cap \bar{H}$ has coordinates $[0, x_1 - y_1, \ldots, x_n - y_n]$, whereas $((X)f + (Y)f) \cap \bar{H}$ has coordinates $[0, \beta(x_1 - y_1), \ldots, \beta(x_n - y_n)]$.

(4) If $X = (X)f$, then $P \in (X + (X)f)$ since $[1, 0, \ldots, 0] = [\beta\langle x_0, \ldots, x_n\rangle - \langle(\alpha + \beta)x_0, \beta x_1, \ldots, \beta x_n\rangle]$.

From (1) and (2) it follows that $\bar{H}$ is the unique hyperplane of fixed points, and hence, by (3), f is a perspectivity with axis $\bar{H}$. The center of f is P by (4). Clearly f sends Q to R, so we have found the desired f for this case.

Case 2. $P \in \bar{H}$. In this case we choose a vector $v_0 \in \mathbf{V}$ such that $Q = [v_0]$, and let $v_1, v_2, \ldots, v_n$ be a basis of $\mathbf{H}$. Since $Q \notin \bar{H}$, $v_0, v_1, \ldots, v_n$ form a basis for $\mathbf{V}$. Again we establish a homogeneous coordinate system $[\phi]:\bar{V} \to P_n(\mathbf{F})$ with vertices $[v_0], [v_1], \ldots, [v_n]$. Note that, as in case 1, we do not bother to fix the unit point, thus (unless $\#\mathbf{F} = 2$), there is more than one possibility for $[\phi]$. Since $\bar{H}$ is the graph of $x_0 = 0$, and $R \notin \bar{H}$,

$$R^{[\phi]} = [1, b_1, b_2, \ldots, b_n]\,.$$

Not all of the $b_i = 0$, because $R \neq Q$. Since $P = (Q + R) \cap \bar{H}$, $P^{[\phi]} = [0, b_1, b_2, \ldots, b_n]$.

Let $h:\mathbf{V} \to \mathbf{V}$ be the linear transformation such that

$$(v_0)h = v_0 + b_1v_1 + \cdots + b_nv_n$$

and $(v_i)h = v_i$, $i = 1, 2, \ldots, n$. As in case 1, define $f:\bar{V} \to \bar{V}$ such that $([v])f = [(v)h]$. Since $[x_0, x_1, \ldots, x_n]$ are the $[\phi]$-coordinates of $[x_0v_0 + x_1v_1 + \cdots + x_nv_n]$, it follows that the coordinate form of f, $f^{[\phi]}$, sends $[x_0, x_1, \ldots, x_n]$ to $[x_0, b_1x_0 + x_1, \ldots, b_nx_0 + x_n]$. From this it follows that:

(1) $(X)f = X$ if and only if $X \in \bar{H}$ since $\langle x_0, \ldots, x_n\rangle$ and $\langle x_0, b_1x_0 + x_1, \ldots, b_nx_0 + x_n\rangle$ are dependent if and only if $x_0 = 0$ (at least one $b_i \neq 0$).

(2) If X and Y are distinct points, neither on $\bar{H}$, then $(X + Y) \cap \bar{H} = ((X)f + (Y)f) \cap \bar{H}$ since we can choose coordinates for Y, $[y_0, y_1, \ldots, y_n]$ with $y_0 = x_0$, and then

$$\begin{aligned}\langle 0, x_1 - y_1, \ldots, x_n - y_n\rangle &= \langle x_0, x_1, \ldots, x_n\rangle - \langle y_0, y_1, \ldots, y_n\rangle \\ &= \langle x_0, b_1x_0 + x_1, \ldots, b_nx_0 + x_n\rangle - \langle y_0, b_1y_0 + y_1, \ldots, b_ny_0 + y_n\rangle .\end{aligned}$$

(3) If $X \notin \bar{H}$, then $P \in (X + (X)f)$ since

$$\begin{aligned}[0, b_1, \ldots, b_n] = [0, b_1x_0, \ldots, b_nx_0] = [-&\langle x_0, x_1, \ldots, x_n\rangle \\ &+ \langle x_0, b_1x_0 + x_1, \ldots, b_nx_0 + x_n\rangle] .\end{aligned}$$

Thus, f is a perspectivity with axis $\bar{H}$ (by 1 and 2) and with center P (by 3). Since $R^{[\phi]} = (Q)f^{[\phi]}$, f sends Q to R and therefore is the desired perspectivity for this case.

With the assurance that many perspectivities exist, we can prove our previous assertion that all hyperplanes are essentially the same; in fact, we can now show that we can project any one hyperplane onto another by a perspectivity of $\bar{V}$.

Corollary 6.30. If $\bar{H}_1$ and $\bar{H}_2$ are hyperplanes in $\bar{V}$, then there is a perspectivity of $\bar{V}$ sending $\bar{H}_1$ to $\bar{H}_2$.

Proof. If $\bar{H}_1 = \bar{H}_2$, this is trivial. If $\bar{H}_1 \neq \bar{H}_2$, then $\dim(\bar{H}_1 \cap \bar{H}_2) = n - 2$, so we can choose P_1 and P_2 with $P_1 \in \bar{H}_1$, $P_1 \notin \bar{H}_2$ and $P_2 \in \bar{H}_2$, $P_2 \notin \bar{H}_1$. The line $P_1 + P_2$ has at least three points; hence there is a point R such that $R \in P_1 + P_2$, but $R \notin \bar{H}_1$ or $\bar{H}_2$. Let $\bar{H}$ be the join of $(\bar{H}_1 \cap \bar{H}_2)$ and R. $\bar{H}$ is clearly a hyperplane with $\bar{H} \cap \bar{H}_1 = \bar{H} \cap \bar{H}_2 = \bar{H}_1 \cap \bar{H}_2$. Let f be the perspectivity with axis $\bar{H}$, center R, sending P_1 to P_2. Clearly f sends $\bar{H}_1$ to $\bar{H}_2$, so it is the desired perspectivity.

The following theorem lends added weight to our intuition that homologies, i.e., perspectivities with centers off their axes, are like central dilatations, and elations, i.e., those with centers on their axes, are the projective analogs of translations. It is the projective analog to the fact that in affine spaces dilatations and translations form groups **D** and **T**, with **T** an abelian, normal subgroup of **D**.

Theorem 6.31. Let $\bar{H}$ be a hyperplane in the projective space $\bar{V}$, and let $\mathbf{D}_H = \{f \mid f = \mathbf{1}$ or f is a perspectivity with axis $\bar{H}\}$, and $\mathbf{T}_H = \{f \mid f = \mathbf{1}$ or

f is an elation with axis $\bar{H}\}$. $\mathbf{D}_H$ and $\mathbf{T}_H$ are groups; moreover, $\mathbf{T}_H$ is an abelian, normal subgroup of $\mathbf{D}_H$.

Proof. Rather than appealing (via *PA*) to the affine facts, we give a direct proof using the properties of perspectivities established in this section.

By corollary 6.22, $\mathbf{D}_H$ is a group.

Now assume f and g are both elations with axis $\bar{H}$. If fg^{-1} is not an elation, then, since it is in $\mathbf{D}_H$, it must leave a point P not on $\bar{H}$ hyper-fixed. Then $(P)f = (P)g$, so

$$\text{the center of } f = (P + (P)f) \cap \bar{H} = (P + (P)g) \cap \bar{H} = \text{the center of } g\,.$$

The common center of f and g is clearly left hyper-fixed by fg^{-1}, so fg^{-1} leaves two points hyper-fixed, i.e., $fg^{-1} = \mathbf{1}$ whenever fg^{-1} is not an elation. Thus, $\mathbf{T}_H$ is a subgroup of $\mathbf{D}_H$.

If $f \in \mathbf{T}_H$ and $g \in \mathbf{D}_H$, then $g^{-1}fg$ is either $\mathbf{1}$ (if $f = \mathbf{1}$) or else has all of its fixed points in $(\bar{H})g = \bar{H}$ (if f is an elation). Thus, $\mathbf{T}_H$ is normal in $\mathbf{D}_H$.

Why is $\mathbf{T}_H$ abelian? Let f and g be elations in $\mathbf{T}_H$ with distinct centers. Then (exercise 6.5, problem 2) $fgf^{-1}g^{-1}$ leaves both centers hyper-fixed; hence $fgf^{-1}g^{-1} = \mathbf{1}$, i.e., f and g commute. If f and g have the same center, then by theorem 6.29 we can choose an elation h in $\mathbf{T}_H$ such that f and h have distinct centers. But then f and gh also have distinct centers (exercise 6.5, problem 4), and hence f commutes with both h and gh. From this it follows easily (Verify!) that f must commute with g.

If perspectivities have distinct axes, their product may or may not be a perspectivity. In exercise 6.5, problem 6 there is one situation to be studied in which the product is a perspectivity. The following example shows that the product of two elations with distinct axes is not always a perspectivity.

Example. Let $\bar{V}$ be the seven-point projective plane (Figure 30). There are no homologies of this plane (exercise 6.5, problem 8). Let f be the elation with axis $Q_1 + Q_3$, center Q_2, sending P_2 to P_4, and let g be the elation with axis $P_1 + P_2$, center Q_3, sending P_3 to P_4. Using the fact that f and g are collineations, it is easy to determine the action of f and g on the remaining points. Thus, in cycle notation, $f = (P_1, P_3)(P_2, P_4)$, $g = (Q_1, Q_2)(P_3, P_4)$, and $fg = (P_1, P_4, P_2, P_3)(Q_1, Q_2)$. Since Q_3 is the only fixed point of fg, it cannot be a perspectivity.

In the next section we consider the group of collineations generated by the perspectivities.

Exercise 6.5

Caution! Assume $\dim \bar{V} \geq 2$ in each of these problems.

1. (Use Desargues theorem to prove perspectivities exist.) Assume $\bar{H}$ is a hyperplane in $\bar{V}$ and that P, Q, and R are distinct collinear points with neither Q nor R in

$\bar{H}$. The natural way to try to build a perspectivity, f, of $\bar{V}$ with axis $\bar{H}$, center P, and $(Q)f = R$ is as follows:

(1) For $S \in \bar{H} \cup \{P\}$, let $(S)f = S$.

(2) For $S \notin (P + Q) \cup \bar{H}$, let $(S)f = (P + S) \cap (R + (\bar{H} \cap (Q + S)))$.

(3) For $S \in (P + Q)$, $S \notin \bar{H} \cup \{P\}$, choose $Q_1 \notin (P + Q) \cup \bar{H}$ and determine $R_1 = (Q_1)f$ as in (2), and then determine $(S)f$ as in (2) with Q and R replaced by Q_1 and R_1.

Use Desargues theorem to show first that (3) is unambiguous, i.e., $(S)f$, for $S \in P + Q$ is independent of the choice of Q_1, and second to show that f is, in fact, a perspectivity.

2. Let f and g be elations with the same axis but distinct centers. Prove that $fgf^{-1}g^{-1}$ leaves both centers hyper-fixed.

3. Prove the dual of problem 2, i.e., prove that if f and g are elations with the same center but distinct axes, then $fgf^{-1}g^{-1}$ leaves both axes pointwise fixed.

4. Let f, g, and h be elations with axis $\bar{H}$ such that f and g have the same center, but f and h have distinct centers. Show that f and gh are elations with axis $\bar{H}$ but with distinct centers.

5. In the classical study of projective planes, a perspectivity between two lines was defined as follows: Given two lines, L_1 and L_2, and a point, $P \notin L_1 \cup L_2$, in a projective plane, $f{:}L_1 \to L_2$ is a perspectivity with center P if and only if P, Q, and $(Q)f$ are collinear for each $Q \in L_1$. The standard notation for such a perspectivity is $Q_1Q_2 \cdots Q_n \overset{P}{\overline{\wedge}} Q_1'Q_2' \cdots Q_n'$, meaning, of course, that $(Q_i)f = Q_i'$, $Q_i \in L_1$, $Q_i' \in L_2$.

(1) Prove that if the projective plane is a projective plane over a field, then a classical perspectivity is simply the restriction to a line of a perspectivity in our sense.

(2) Discuss the converse.

6. Let f and g be the following elations of the seven-point projective plane—using the notation of Figure 33:

$$f\colon \text{axis } m, \text{ center } Q_2, (P_2)f = P_4$$
$$g\colon \text{axis } m_{13}, \text{ center } Q_2, (P_4)f = P_2\,.$$

Show that $fg = gf$ is an elation with center Q_2.

7. Let P, P_1, Q, Q_1, and R be five points in a projective plane (over a field) such that no three are collinear. Show that there is a unique perspectivity, f, such that $(P)f = P_1$, $(Q)f = Q_1$, and $R \in$ axis f.

8. Show that there are no homologies of a projective space over the field with two elements. Compare, but do not use, problem 9, exercise 6.4.

9. Show that every elation of a projective space over a field of characteristic 2 is an involution.

10. Assume that there exist homologies of $\bar{V}$, i.e., assume $\bar{V}$ is a projective space over a field with at least three elements. Show that $\mathbf{D}_H$ is not abelian for any hyperplane $\bar{H}$ in $\bar{V}$.

11. Let $\bar{V}$ be a projective space of dimension n over a finite field with f elements. Show that for any hyperplane $\bar{H}$ in $\bar{V}$ there are $f^n(f-2)$ homologies with axis $\bar{H}$ and $f^n - 1$ elations with axis $\bar{H}$.

12. Let f and g be perspectivities of $\bar{V}$. Show that $fg = gf$ implies that either the center of each is on the axis of the other, or else both are homologies with the same center and axis.

6.6. Projective symmetries II: Projective transformations

Since we have defined a projective space, $\bar{V}$, in terms of a vector space, $\mathbf{V}$, we should expect the symmetries of $\mathbf{V}$ to induce symmetries of $\bar{V}$. In this section we study such symmetries of $\bar{V}$ and show that they can be represented as products of perspectivities.

Definition. A permutation, f, of $\bar{V}$ is called a *projective transformation* or *projectivity* if and only if there is a linear transformation $g:\mathbf{V} \to \mathbf{V}$ such that $[v]f = [(v)g]$ for all $[v] \in \bar{V}$.

Proposition 6.32. Every perspectivity is a projective transformation, and every projectivity is a collineation.

Proof. Let f be a perspectivity with center P, axis $\bar{H}$, sending Q to R, $Q \notin \{P\} \cup \bar{H}$. In proving theorem 6.29, we showed that there is a linear transformation $g:\mathbf{V} \to \mathbf{V}$ such that g induces a perspectivity of $\bar{V}$ with center P, axis $\bar{H}$, sending Q to R. By corollary 6.28, $[v]f = [(v)g]$ for every $[v] \in \bar{V}$; hence f is a projectivity.

If f is a projectivity, say $[v]f = [(v)g]$ for all $[v] \in \bar{V}$, with g a linear transformation of $\mathbf{V}$, then g must be nonsingular, for otherwise $(v)g$ would be zero for some nonzero vector v, and then $[v]f$ would not be defined. Thus, g maps any subspace of (linear) dimension 2 in $\mathbf{V}$ onto a subspace of dimension 2, i.e., f maps any line in $\bar{V}$ onto a line in $\bar{V}$.

Since we defined a projectivity to be a special *permutation* of $\bar{V}$, only nonsingular linear transformations of $\mathbf{V}$ can induce projectivities of $\bar{V}$. We now investigate the manner in which this occurs.

Theorem 6.33. Let $\mathbf{G}$ be the group of nonsingular linear transformations of a vector space $\mathbf{V}$. For each $g \in \mathbf{G}$, there is a permutation, $\bar{g}$, of $\bar{V}$ such that $[v]\bar{g} = [(v)g]$ for all $[v] \in \bar{V}$. The mapping "–" from $\mathbf{G}$ into the group of permutations of $\bar{V}$ is a homomorphism, and thus, the projectivities of $\bar{V}$ form a group. The kernel of "–" is the group of all nonzero scalar transformations of $\mathbf{V}$, i.e., if $g, h \in \mathbf{G}$, then $\bar{g} = \bar{h}$ if and only if there is a nonzero scalar λ such that $g = \lambda h$.

Proof. If g is a nonsingular linear transformation, then $(v)g \neq 0$ whenever $v \neq 0$, and thus, $[(v)g]$ is defined whenever $[v]$ is. Moreover, $(\alpha v)g =$

$\alpha(v)g$, and thus, $[v] = [w]$ implies $[(v)g] = [(w)g]$. Thus, the transformation, $\bar{g}:\bar{V} \to \bar{V}$, defined by $[v]\bar{g} = [(v)g]$ is well defined. Since $(v)g$ and $(w)g$ are linearly independent whenever (v) and (w) are, $\bar{g}$ is one-to-one. Moreover, given $w \neq 0$, there is a $v \neq 0$ such that $(v)g = w$, i.e., $[v]g = [w]$; thus, $\bar{g}$ maps $\bar{V}$ onto $\bar{V}$. Since $\bar{g}$ is one-to-one and onto whenever $g \in \mathbf{G}$, "–" maps $\mathbf{G}$ into the group of permutations of $\bar{V}$. It is clearly a homomorphism, and since the projectivities of $\bar{V}$ are the images of the elements of $\mathbf{G}$ under this homomorphism, the projectivities must form a group.

If f is a nonzero scalar transformation of $\mathbf{V}$, i.e., if $f = \lambda\mathbf{1}$, $\lambda \neq 0$, then $(v)f = \lambda v$, and hence $[v]\bar{f} = [v]$ for all $[v] \in \bar{V}$. Thus, the nonzero scalar transformations are in the kernel of "–". Conversely, if $\bar{g}$ is the identity on $\bar{V}$, then for each nonzero vector $v \in \mathbf{V}$ there is a corresponding nonzero scalar λ_v such that $(v)g = \lambda_v v$. However, this implies (problem 2 in either exercise 4.5 or 4.6) that $g = \lambda\mathbf{1}$.

Definition. If g is a nonsingular linear transformation of $\mathbf{V}$, then $\bar{g}$ (as defined above) is called the *projectivity induced by g*. The group of all projectivities of $\bar{V}$ is called the *projective group* on $\bar{V}$.

Recall that if $\mathbf{V}$ is an $(n+1)$-dimensional vector space over $\mathbf{F}$, then each coordinate mapping ϕ from $\mathbf{V}$ onto $\mathbf{V}_{n+1}(\mathbf{F})$ determines an isomorphism between $\mathbf{G}$, the group of nonsingular linear transformations of $\mathbf{V}$, and $GL_{n+1}(\mathbf{F})$, the group of nonsingular $(n+1)$ by $(n+1)$ matrices over $\mathbf{F}$. In each of these isomorphisms, the scalar transformations correspond to the scalar matrices. Let $\mathbf{F}^*$ be the group of nonzero $(n+1)$ by $(n+1)$ scalar matrices. The homomorphism "–" from $\mathbf{G}$ to the projective group is represented in coordinate form by a homomorphism of $GL_{n+1}(\mathbf{F})$ with kernel $\mathbf{F}^*$, i.e., we can represent any projectivity in terms of homogeneous coordinates as follows: $[y_0, y_1, \ldots, y_n]$ are the $[\phi]$-coordinates of $(P)\bar{g}$ if $P^{[\phi]} = [x_0, \ldots, x_n]$, $g^{\phi'} = A$, and

$$\langle y_0, y_1, \ldots, y_n\rangle = \langle x_0, x_1, \ldots, x_n\rangle A.$$

Note that the matrices A and λA, $\lambda \neq 0$, represent the same projectivity in the same coordinate system, $[\phi]$. The factor group $GL_{n+1}(\mathbf{F})/\mathbf{F}^*$ is customarily denoted by $PGL_{n+1}(\mathbf{F})$; i.e., $PGL_{n+1}(\mathbf{F})$ is the group obtained from $GL_{n+1}(\mathbf{F})$ by "factoring out" or "ignoring" elements in $\mathbf{F}^*$, that is, by identifying nonsingular matrices which are scalar multiples of each other. In theorem 4.29 we proved that $\mathbf{F}^*$ is the center of $GL_{n+1}(\mathbf{F})$. Thus, $PGL_{n+1}(\mathbf{F})$ is obtained from $GL_{n+1}(\mathbf{F})$ by factoring out its center.

Caution! Remember that $PGL_n(\mathbf{F})$ is isomorphic to the group of projectivities of an $(n-1)$-dimensional projective space, not an n-dimensional one.

Note that the fixed points of a projectivity $\bar{f}$ correspond not only to the fixed points of f but also to the characteristic vectors of f. Thus, if $[v]$ and $[w]$ are distinct fixed points of $\bar{f}$, then (exercise 6.6, problem 1) the line $[v] + [w]$ is left pointwise fixed if and only if v and w correspond to the same characteristic value of f. We generalize this observation in the following theorem.

Theorem 6.34. Let $\bar{g}$ be a projectivity of $\bar{V}$, and let $\bar{W}$ be a k-dimensional subspace of $\bar{V}$. The following three conditions on $\bar{g}$ and $\bar{W}$ are equivalent:

(1) There is a nonzero scalar, λ, such that $(w)g = \lambda w$ for all $w \in \mathbf{W}$.
(2) $\bar{g}$ leaves $\bar{W}$ pointwise fixed.
(3) $\bar{g}$ leaves $k + 2$ independent points in $\bar{W}$ fixed.

Remark. Note that if $k = 1$, i.e., if $\bar{W}$ is a line, then if a projectivity leaves three distinct points on the line fixed, it leaves the line pointwise fixed. This is one version of the fundamental theorem of projective geometry. We meet another version just after proving this theorem and meet still a third version in the next section. All three versions are descendants of the classical version which asserts that if a collineation of the real projective plane leaves three points on a line fixed then it leaves the line pointwise fixed.

Proof. Clearly condition (1) implies (2) and (2) implies (3). Now assume that $[v_0]$, $[v_1]$, . . . , $[v_{k+1}]$ are independent points in $\bar{W}$ left fixed by $\bar{g}$, i.e., there are nonzero scalars, λ_i, such that $(v_i)g = \lambda_i v_i$, $i = 0,1, \ldots, n + 1$. Since these points are independent, $v_0, \ldots, v_k$ form a basis of $\mathbf{W}$, and

$$v_{k+1} = \alpha_0 v_0 + \alpha_1 v_1 + \cdots + \alpha_k v_k$$

with all $\alpha_i \neq 0$. (Review the proof of theorem 6.16 if this is not clear to you.) On the one hand,

$$(v_{k+1})g = \lambda_{k+1} v_{k+1} = \lambda_{k+1}\alpha_0 v_0 + \lambda_{k+1}\alpha_1 v_1 + \cdots + \lambda_{k+1}\alpha_k v_k,$$

but on the other hand,

$$(v_{k+1})g = \Sigma\alpha_i(v_i)g = \alpha_0\lambda_0 v_0 + \alpha_1\lambda_1 v_1 + \cdots + \alpha_k\lambda_k v_k.$$

Since $v_0, v_1, \ldots, v_k$ are linearly independent, there can only be one way to express $(v_{k+1})g$ as a linear combination of $v_0, v_1, \ldots, v_k$, and thus, $\lambda_{k+1}\alpha_j = \lambda_j\alpha_j$, $j = 0,1, \ldots, k$. Since the α_j are all nonzero, this means that all the λ_j's equal λ_{k+1}, and hence $(w)g = \lambda_{k+1}(w)$ for all $w \in \mathbf{W}$. This shows that condition (3) implies (1), and the proof is complete.

Theorem 6.35 (fundamental theorem of projective geometry). If P_0, $P_1, \ldots, P_{n+1}$ and $Q_0, Q_1, \ldots, Q_{n+1}$ are two sets of independent points in $\bar{V}$, then there is exactly one projectivity, $\bar{g}$, such that $(P_i)\bar{g} = Q_i$, $i = 0,1, \ldots, n + 1$.

Proof. Uniqueness follows directly from the theorem above since the projectivities form a group. To establish existence we appeal to theorem 6.16 and establish two homogeneous coordinate systems, $[\phi]$ and $[\theta]$, in $\bar{V}$ such that the unit points of $[\phi]$ and $[\theta]$ are P_{n+1} and Q_{n+1}, respectively, and the vertices of $[\phi]$ and $[\theta]$ are the P_i and Q_i $(i = 0,1, \ldots, n)$, respectively. The nonsingular linear transformation $\phi\theta^{-1}$ mapping $\mathbf{V}$ to $\mathbf{V}$ via $\mathbf{V}_{n+1}(\mathbf{F})$ induces the desired projectivity.

The classical use of the term perspectivity is that a perspectivity is the restriction to a hyperplane of a perspective collineation as we have defined it. In other words, a mapping, $p{:}\bar{H}_1 \rightarrow \bar{H}_2$, from one hyperplane to another is a classical perspectivity if there is a point $C \notin \bar{H}_1 \cup \bar{H}_2$ such that for each $P \in \bar{H}_1$, $(P)p = Q$ if and only if $Q = (P + C) \cap \bar{H}_2$. A projectivity in classical terms is a finite sequence (product!) of perspectivities. This is also true in our sense of perspectivities and projectivities as we now show. Note that since $(P)\bar{g} = P$ and $(Q)\bar{g} = Q$ does not imply that the projectivity $\bar{g}$ leaves $P + Q$ pointwise fixed, we cannot assume that the fixed points of $\bar{g}$ form a subspace. For this reason we focus on the fixed points of g rather than $\bar{g}$ in the following proof.

Theorem 6.36. The projective group of $\bar{V}$ is generated by the perspectivities of $\bar{V}$. More precisely, if dim $\bar{V} = n$ and $\bar{g}$ is a projectivity of $\bar{V}$, then we need at most $n + 1$ perspectivities in order to represent $\bar{g}$ as a product of perspectivities.

Proof. Let g be a nonsingular linear transformation of $\mathbf{V}$. We attempt to find $n + 1$ (or fewer) linear transformations, $f_0, f_1, \ldots, f_n$, such that $gf_0f_1 \ldots f_n = \mathbf{1}$, and such that $\bar{f}_i$ is a perspectivity. If we can do this, then bv theorem 6.33 and corollary 6.22, $\bar{g} = \bar{f}_n^{-1}\bar{f}_{n-1}^{-1} \cdots \bar{f}_0^{-1}$ is the desired representation of $\bar{g}$ as a product of at most $n + 1$ perspectivities. Clearly we can find the f_i if we can always increase by at least one the (linear) dimension of the space of fixed points (in $\mathbf{V}$!) as we pass from g to gf_0 or from $gf_0 \cdots f_j$ to $gf_0 \cdots f_jf_{j+1}$. Therefore, the following lemma completes the proof.

Lemma. Let h: $\mathbf{V} \rightarrow \mathbf{V}$ be a nonsingular linear transformation with $\mathbf{W} = \{w | (w)h = w\}$, i.e., $\mathbf{W}$ is the space of fixed vectors of h. If $\mathbf{W} \neq \mathbf{V}$, then there is a linear transformation $f{:}\mathbf{V} \rightarrow \mathbf{V}$ such that $\bar{f}$ is a perspectivity, and the dimension of the space of fixed vectors of hf is greater than the dimension of $\mathbf{W}$.

Proof. Let $w_0, w_1, \ldots, w_k$ be a basis of $\mathbf{W}$. Since $\mathbf{W} \neq \mathbf{V}$, we can choose $v \notin \mathbf{W}$, and let $v' = (v)h$. Since h is one-to-one, $v' \notin \mathbf{W}$, and we need only consider the following two cases.

Case 1. $w_0, w_1, \ldots, w_k, v$, and v' are linearly dependent, but neither v nor v' is in $\mathbf{W}$. In this case v is a linear combination of the independent vectors $w_0, w_1, \ldots, w_k, v'$. We complete this set of $k + 2$ vectors to a basis of $\mathbf{V}$, say, with the vectors $x_1, x_2, \ldots, x_j$ $(n = k + j + 1)$. By proposition 4.9 there is a linear transformation, f, such that f leaves each of the w_i and x_i fixed and sends v' to v. Since the (projective) hyperplane spanned by the $[w_i]$ and $[x_i]$ is left pointwise fixed by $\bar{f}$, $\bar{f}$ is a perspectivity by theorem 6.21. The fixed points of hf include, in addition to **W**, the vector v. Hence the space of fixed points of hf properly includes **W** and therefore has greater dimension. This case is therefore closed.

Case 2. $w_0, w_1, \ldots, w_k, v$, and v' are linearly independent. Complete this set of vectors to a basis of **V**, say with $y_1, \ldots, y_j$ ($n = k + j + 2$). If we replace one of v or v' by $v + v'$, we still have a basis of **V**. Hence $w_0, \ldots, w_k, v + v', y_1, \ldots, y_j$ span a (linear) hyperplane **H** containing neither v nor v'. There is a linear transformation f:**V** → **V** such that f leaves each of the w_i and y_i fixed and interchanges v and v' (proposition 4.9). But f also leaves $v + v'$ fixed, so it leaves **H** pointwise fixed. Thus, $\bar{f}$ is a perspectivity. Once again the fixed vectors of hf include v as well as all vectors in **W**, so this case is closed, and the proof of the lemma (and hence of the theorem) is complete.

To complete this section and illustrate some of the interesting differences between $GL_n(\mathbf{F})$ and $PGL_n(\mathbf{F})$ we study involutions of projective spaces. Clearly any linear involution of **V**, except $-\mathbf{1}$, induces a projective involution of $\bar{V}$; however, since $\bar{f}^2 = \mathbf{1}$ (on $\bar{V}$) need not imply $f^2 = \mathbf{1}$ (on **V**), there may be other involutions of $\bar{V}$.

Definition. Let $\bar{f}$ be a *projective involution* of $\bar{V}$, i.e., a projectivity that is also an involution. We call $\bar{f}$ an *involution of the first or second kind* according as there is or is not a linear involution, g, of **V** such that $\bar{f} = \bar{g}$.

Note that we do not insist that f be an involution of **V** in order for $\bar{f}$ to be an involution of the first kind. We avoided this since $g^2 = \mathbf{1}$ and $\bar{g} = \bar{f}$ implies only that $f = \mu g$, and hence $f^2 = \mu^2\mathbf{1}$, not **1**. The following theorem clarifies the status of involutions of the second kind.

Theorem 6.37. The projective group of $\bar{V}$ contains involutions of the second kind if and only if there is a scalar that is not a perfect square, and the dimension of $\bar{V}$ is odd, i.e., the linear dimension of **V** is even. An involution of the second kind has no fixed points in $\bar{V}$.

Proof. If $\bar{f}$ is a projective involution, then $f^2 = \lambda\mathbf{1}$. If λ is a perfect square in **F**, say, $\lambda = \mu^2$, then $g = (1/\mu)f$ is an involution in **V** such that $\bar{g} = \bar{f}$. Thus, if $\bar{f}$ is of the second kind, $f^2 = \lambda\mathbf{1}$, and λ is a nonsquare in **F**.

Let v be a nonzero vector in **V**, then $(v)f^2 = \lambda v$. If $[v]$ is a fixed point of $\bar{f}$, then $(v)f = \mu v$ for some scalar μ. This would imply that $\lambda v = (v)f^2 = \mu^2 v$, and hence that $\lambda = \mu^2$ (since $v \neq 0$). Thus, an involution of the second kind has no fixed points.

If $f^2 = \lambda\mathbf{1}$, λ being a nonsquare, choose a basis of **V** as follows. Let the first basis vector be any nonzero vector, v_1, and let the second be $(v_1)f$. Since

$[v_1]\bar{f} \neq [v_1]$, we are assured that v_1 and $(v_1)f$ are independent. Continue inductively, i.e., if $v_1, (v_1)f, \ldots, v_j, (v_j)f$ form a basis of the subspace $\mathbf{W}_j \neq \mathbf{V}$, then choose $v_{j+1} \notin \mathbf{W}_j$. Since $[v_{j+1}]\bar{f} \neq [v_{j+1}]$, v_{j+1} and $(v_{j+1})f$ are independent. Since the two sets of $2j$ vectors $\{v_1, (v_1)f, \ldots, v_j, (v_j)f\}$ and $\{(v_1)f, \lambda v_1, \ldots, (v_j)f, \lambda v_j\}$ span the same subspace, $(\mathbf{W}_j)f = \mathbf{W}_j$. Thus neither v_{j+1} nor $(v_{j+1})f$ is in $\mathbf{W}_j$, since v_{j+1} is not and $\lambda v_{j+1} = (v_{j+1})f^2$. If at least one of α or β is not zero, and if $w \doteq \alpha v_{j+1} + \beta(v_{j+1})f$, then at least one of v_{j+1} and $(v_{j+1})f$ is a linear combination of w and $(w)f$. But neither v_{j+1} nor $(v_{j+1})f$ is in $\mathbf{W}_j$, hence $w \notin \mathbf{W}_j$. Thus $v_1, (v_1)f, \ldots, v_j, (v_j)f, v_{j+1}, (v_{j+1})f$ are still independent. This process must terminate with some $\mathbf{W}_k = \mathbf{V}$. Thus, $n + 1 = \dim \mathbf{V} = \dim \mathbf{W}_k = 2k$ is even, and the only hope for involutions of the second kind is in projective spaces of odd dimension.

Now assume that $\bar{V}$ is a projective space of odd dimension, n, and that λ is a nonsquare scalar. Let $x_i, y_i, i = 1, 2, \ldots, (n + 1)/2$ be a basis of $\mathbf{V}$, and let f be the linear transformation of $\mathbf{V}$ such that $(x_i)f = y_i, (y_i)f = \lambda x_i$ for all i. Clearly $f^2 = \lambda\mathbf{1}$, and $\bar{f}$ is an involution of the second kind.

We outline a different proof for those familiar with minimal and characteristic polynomials. One shows that $\bar{f}$ is an involution of the second kind if and only if $f^2 = \lambda\mathbf{1}$ with λ a nonsquare. The minimal polynomial of f is then $x^2 - \lambda$, which is irreducible over $\mathbf{F}$. The characteristic polynomial must be $(x^2 - \lambda)^k$ and must also be of degree $n + 1$. Hence $n + 1 = 2k$ is even. Since f has no characteristic roots, it has no characteristic vectors, and thus, f has no fixed points.

In order to determine the nature of all involutions of the first kind we digress to discuss the projective analog of "midpoints." One of the simplest constructions of the midpoint of a segment $\overline{Q_1Q_2}$ is to locate it as the intersection of the diagonals of a parallelogram. In terms of ideal points, this construction can be phrased as follows. Let P_1 be the ideal point on $Q_1 + Q_2$, and choose two other ideal points R_1 and R_2 such that P_1, R_1, and R_2 are on the same ideal line in the hyperplane at infinity. The midpoint of $\overline{Q_1Q_2}$ is then $P_2 = (Q_1 + Q_2) \cap (S_1 + S_2)$ where $S_1 = (Q_1 + R_1) \cap (Q_2 + R_2)$ and $S_2 = (Q_1 + R_2) \cap (Q_2 + R_1)$, Figure 39a. The projective analog of this construction is as follows.

Definition. Let P_1, Q_1, and Q_2 be distinct collinear points in a projective space $\bar{V}$ of dimension at least 2. There are at least two lines through P_1 and at least three points on each line, so choose R_1, R_2 not on $Q_1 + Q_2$ such that P_1, R_1, R_2 are distinct and collinear. The point $P_2 = (Q_1 + Q_2) \cap (S_1 + S_2)$ (Figure 39b) with $S_1 = (Q_1 + R_1) \cap (Q_2 + R_2)$ and $S_2 = (Q_1 + R_2) \cap (Q_2 + R_1)$ is called the *harmonic conjugate* of P_1 with respect to Q_1 and Q_2.

From our experience with midpoints, we should expect two things. First, the harmonic conjugate should be independent of the choice of R_1 and R_2, and,

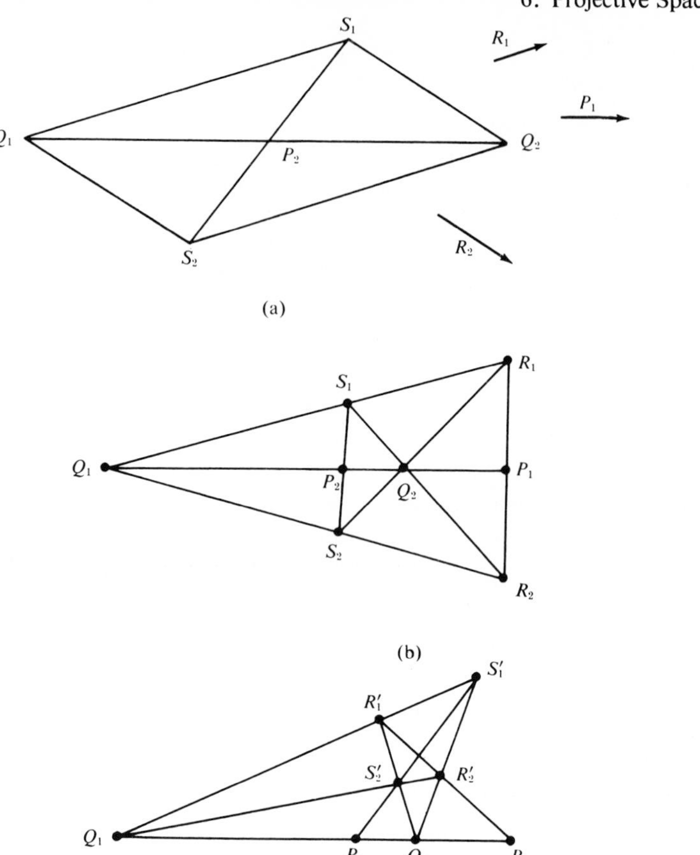

FIG. 39. Midpoints and harmonic conjugates. (a) Affine construction of the midpoint of $\overline{Q_1Q_2}$; (b) two projective constructions of the harmonic conjugate.

second, we probably will have trouble if the characteristic of **F** is 2. The following two results verify that our intuition is correct.

Theorem 6.38. The harmonic conjugate of P_1 with respect to Q_1 and Q_2 is independent of the choice of R_1 and R_2.

Proof. Let R_1' and R_2' be another pair of points not on the line $Q_1 + Q_2$ such that P_1, R_1', R_2' are distinct and collinear. The two sets of points Q_1, Q_2, R_1, R_2 and Q_1, Q_2, R_1', R_2' are both independent. Since, if necessary, these sets can each be enlarged to a set of $n + 2$ independent points in $\bar{V}$, there is a projectivity, $\bar{g}$, leaving Q_1 and Q_2 fixed and such that $(R_1)\bar{g} = R_1'$, $(R_2)\bar{g} = R_2'$. Since $\bar{g}$ is a collineation, it is clear that $(P_1)\bar{g} = P_1$, and $\bar{g}$ carries the harmonic conjugate using R_1, R_2 to the harmonic conjugate using R_1', R_2'. But $\bar{g}$ leaves

Q_1, Q_2, and P_1 fixed and hence leaves $Q_1 + Q_2$ pointwise fixed. Thus, the harmonic conjugate is independent of the choice of R_1 and R_2.

Caution! If a projective plane is not a projective plane over a field, then the above theorem need not be true. The connections between the uniqueness of the harmonic conjugate and Desargues theorem have been studied in detail; see [13], pp. 129–130.

Theorem 6.39. Let P_1, Q_1, and Q_2 be distinct collinear points, i.e., assume $Q_1 = [w_1]$, $Q_2 = [w_2]$ and $P_1 = [\alpha w_1 + \beta w_2]$, α and β nonzero. Then P_2 is the harmonic conjugate of P_1 with respect to Q_1 and Q_2 if and only if $P_2 = [\alpha w_1 - \beta w_2]$. In particular, $P_2 = P_1$ if and only if the characteristic of $\mathbf{F}$ is 2.

Proof. Let $R_1 = [x]$. All the points under consideration are of the form $[y]$, with y a linear combination of w_1, w_2, and x. Since R_2, R_1, and P_1 are collinear and distinct, $R_2 = [\alpha w_1 + \beta w_2 + \gamma x]$, $\gamma \neq 0$. $S_1 = [\alpha w_1 + \gamma x]$ since $\alpha w_1 + \gamma x$ is simultaneously a linear combination of w_1 and x and also of w_2 and $\alpha w_1 + \beta w_2 + \gamma x$. Similarly (interchange 1's and 2's) $S_2 = [\beta w_2 + \gamma x]$. Finally $P_2 = [\alpha w_1 - \beta w_2]$ since this is a linear combination of $\alpha w_1 + \gamma x$ and $\beta w_2 + \gamma x$ in which the coefficient of x is zero.

Definition. If $\bar{V}$ is only a one-dimensional space, then we define the *harmonic conjugate* of P_1 with respect to Q_1 and Q_2 in accordance with the result above.

As a final preparation for classifying involutions of the first kind, we define complementary (projective) subspaces and derive two results concerning them.

Definition. Two subspaces, $\bar{W}_1$ and $\bar{W}_2$, are *complementary* in $\bar{V}$ if and only if $\mathbf{W}_1$ and $\mathbf{W}_2$ are complementary in $\mathbf{V}$, i.e., $\mathbf{W}_1 \cap \mathbf{W}_2 = \{0\}$ and $\mathbf{W}_1 + \mathbf{W}_2 = \mathbf{V}$.

Proposition 6.40. Subspaces $\bar{W}_1$ and $\bar{W}_2$ are complementary in $\bar{V}$ if and only if they are skew (i.e., disjoint) subspaces such that $\dim \bar{W}_1 + \dim \bar{W}_2 = \dim \bar{V} - 1$.

Remark. Thus, two subspaces of a projective plane are complementary if and only if one is a line and the other is a point not on the line. Similarly complementary subspaces of a three-dimensional projective space are skew lines or else a plane and a point not on that plane. Two distinct points are the only complementary subspaces of a projective line.

Proof. $\bar{W}_1$ and $\bar{W}_2$ are skew if and only if $\mathbf{W}_1 \cap \mathbf{W}_2 = \{0\}$. But in this case, $\mathbf{W}_1 + \mathbf{W}_2 = \mathbf{V}$ if and only if $\dim \mathbf{V} = \dim \mathbf{W}_1 + \dim \mathbf{W}_2$, i.e., if and only if $\dim \bar{V} + 1 = (\dim \bar{W}_1 + 1) + (\dim \bar{W}_2 + 1)$. The proof follows easily from these facts.

Proposition 6.41. Let $\bar{W}_1$ and $\bar{W}_2$ be complementary subspaces in $\bar{V}$, and assume $P \notin \bar{W}_1 \cup \bar{W}_2$. There is a unique line, L, through P meeting both $\bar{W}_1$ and $\bar{W}_2$. Moreover, $L \cap \bar{W}_1$ is a point as is $L \cap \bar{W}_2$.

Proof. Let $P = [v]$. There is a unique representation of v in the form $v = w_1 + w_2$, $w_1 \in \mathbf{W}_1$, $w_2 \in \mathbf{W}_2$ (exercise 4.2, problem 1). The line $[w_1] + [w_2]$ is clearly the desired line L and is unique. If it met either $\bar{W}_1$ or $\bar{W}_2$ in more than one point, then P would be in one of the two subspaces, but we assumed $P \notin \bar{W}_1 \cup \bar{W}_2$.

Theorem 6.42. Let $\bar{f}$ be a projectivity of a projective space, $\bar{V}$, over a field of characteristic not two. The following are then equivalent conditions on $\bar{f}$:

(1) $\bar{f}$ is an involution with at least one fixed point.

(2) $\bar{f}$ is an involution of the first kind.

(3) There are complementary subspaces, $\bar{W}_1$ and $\bar{W}_2$, both left pointwise fixed by $\bar{f}$ such that for $P \notin \bar{W}_1 \cup \bar{W}_2$, $(P)\bar{f}$ is the harmonic conjugate of P with respect to $L \cap \bar{W}_1$, $L \cap \bar{W}_2$, where L is the unique line through P meeting both $\bar{W}_1$ and $\bar{W}_2$.

Moreover, for any pair of complementary subspaces in $\bar{V}$, the permutation defined as in (3) is, in fact, an involution of the first kind.

Proof. If $\bar{f}$ is a projective involution with a fixed point, then it must be of the first kind by theorem 6.37. Thus, (1) implies (2). If $\bar{f}$ is a projective involution of the first kind, then $\bar{f} = \bar{g}$ with $g^2 = \mathbf{1}$, but $g \neq \pm\mathbf{1}$. But then (exercise 4.3, problem 4) $\mathbf{W}_1 = \{v|(v)g = v\}$ and $\mathbf{W}_2 = \{v|(v)g = -v\}$ are complementary subspaces of $\mathbf{V}$, neither of which is $\{0\}$. Thus, $\bar{W}_1$ and $\bar{W}_2$ are complementary subspaces of $\bar{V}$, both of which are left pointwise fixed by $\bar{g} = \bar{f}$. If $P \notin \bar{W}_1 \cup \bar{W}_2$, then $P = [w_1 + w_2]$, $w_1 \in \mathbf{W}_1$, $w_2 \in \mathbf{W}_2$, and $(P)\bar{f} = [(w_1 + w_2)g] = [w_1 - w_2]$, so $(P)\bar{f}$ is the harmonic conjugate of P with respect to $[w_1]$ and $[w_2]$. Thus, (2) implies (3).

Now assume $\bar{W}_1$ and $\bar{W}_2$ are complementary in $\bar{V}$, and define a permutation $\phi:\bar{V} \to \bar{V}$ as in (3). The linear transformation $f:\mathbf{V} \to \mathbf{V}$ defined by $(w_1 + w_2)f = w_1 - w_2$ for all $w_1 \in \mathbf{W}_1$, $w_2 \in \mathbf{W}_2$ is an involution of $\mathbf{V}$ and is not $\pm\mathbf{1}$ since neither $\mathbf{W}_1$ nor $\mathbf{W}_2$ is $\mathbf{V}$. Clearly $\phi = \bar{f}$ so ϕ is an involution of the first kind with more than one fixed point. This proves the "moreover" clause and also proves that (3) implies (1).

Corollary 6.43. If $\bar{V}$ is a projective plane over a field of characteristic not two then any projective involution is a *harmonic homology*, i.e., is a homology, $\bar{f}$, with axis L and center C such that $P \notin L \cup \{C\}$ implies $(P)\bar{f}$ is the harmonic conjugate of P with respect to C and $(P + C) \cap L$.

Proof. If $\bar{f}$ is a projective involution, then it must be of the first kind since $\dim \bar{V} = 2$ is even. The only complementary subspaces of $\bar{V}$ consist of a line L and a point C not on L. The rest is obvious.

In classical projective geometry, a projectivity of a line is called *elliptic*, *parabolic*, or *hyperbolic* according as the number of fixed points is 0, 1, or 2. Using this terminology, we state the following result.

Corollary 6.44. A projective involution of a space over a field of characteristic not two is either elliptic on all lines left globally but not pointwise fixed, or else hyperbolic on all lines left globally but not pointwise fixed.

Proof. Any line left globally but not pointwise fixed by $\bar{f}$ is of the form $P + (P)f$ for some P. By theorem 6.42, $\bar{f}$ is hyperbolic on this line if and only if it has at least one fixed point somewhere in $\bar{V}$. The rest is clear.

For a direct proof of the results above (for projective planes over a field), see [5], pp. 55–56. For the theory of involutions of projective planes in general (not necessarily of the form $\bar{V}$), see [3].

Exercise 6.6

1. Let v and w be characteristic vectors belonging to *distinct* characteristic values of a linear transformation f. Show that the only characteristic vectors of f in the plane spanned by v and w are the nonzero multiples of v and the nonzero multiples of w.

2. Let A and $\bar{V}$ be, respectively, n-dimensional affine and projective spaces over the field **F**. Let $\bar{H}$ be a hyperplane in $\bar{V}$. Show that the group of all projectivities of $\bar{V}$ leaving $\bar{H}$ globally fixed is isomorphic to the group of affine transformations of A.

3. Find a matrix inducing the projectivity of $P_3(\mathbf{Q})$ carrying [1, 1, 1, 0], [1, 1, 0, 1], [1, 0, 1, 1], [0, 1, 1, 1], and [1, 1, 1, 1] to the points [1, 0, 0, 0], [0, 1, 0, 0], [0, 0, 1, 0], [0, 0, 0, 1], and [1, 1, 1, 1], respectively.

4. What is the order of $PGL_n(\mathbf{F})$ if $\#\mathbf{F} = f$? *Caution!* This is not the same as the order of the projective group of $P_n(\mathbf{F})$. What is the relation between the two orders?

5. Review the proof of theorem 6.36, and then show how to represent any projectivity of any projective plane over a field as a product of at most three perspectives.

6. Show that any projectivity of the real projective plane has at least one fixed point. Exhibit a projectivity of the rational projective plane with no fixed point. Show that every projectivity of $P_2(\mathbf{F})$ has a fixed point if and only if every cubic polynomial over **F** has a root in **F**.

7. Let $\bar{W}_1$, $\bar{W}_2$ and $\bar{W}_3$, $\bar{W}_4$ be two pairs of disjoint subspaces in the projective space $\bar{V}$. Show that there is a projectivity, $\bar{f}$, such that $(\bar{W}_1)\bar{f} = \bar{W}_3$ and $(\bar{W}_2)\bar{f} = \bar{W}_4$ if and only if $\dim \bar{W}_1 = \dim \bar{W}_3$ and $\dim \bar{W}_2 = \dim \bar{W}_4$.

8. Let $\bar{V}$ be a projective space over a field of characteristic not two. Let P_2 be the harmonic conjugate of P_1 with respect to Q_1 and Q_2. How many permutations are there of (P_1, P_2, Q_1, Q_2) such that the fourth point is the harmonic conjugate of the third with respect to the first two? Why?

9. Let $\bar{f}$ and $\bar{g}$ be two involutions of the first kind. When do $\bar{f}$ and $\bar{g}$ commute?

10. Assume $\bar{V}$ is a projective space over a field of characteristic not two and that $\bar{f}$ is an involution of the first kind whose pair of complementary spaces are of dimension k_1 and k_2. We call $\bar{f}$ a j-involution if and only if $j = \min(k_1, k_2)$. Prove that two involutions of the first kind are conjugate in the projective group of $\bar{V}$ if and only if they are both j-involutions (for the same j).

11. Using problem 3 of exercise 4.3, investigate involutions of the first kind in a projective space over a field of characteristic two. Show that there are $(n + 1)/2$ conjugacy classes of such involutions if n is the (projective) dimension of the space.

12. Let $\bar{f}$, $\bar{g}$, and $\bar{h}$ be three harmonic homologies (of a projective plane over a field of characteristic not two) whose centers and axes form the three vertices and sides of a traingle. Describe fgh.

13. How many harmonic homologies are there of a projective plane over a field with f elements? *Hint*. Consider f odd or even separately!

6.7. *Projective symmetries III: Collineations*

Recall that in section 6.5 we defined a collineation of a projective space $\bar{V}$ to be a permutation of the points in $\bar{V}$ which maps lines onto lines. Clearly any permutation of $\bar{V}$ is a collineation if $\dim \bar{V} = 1$, so *we assume* $\dim \bar{V} \geq 2$, i.e., $\dim \mathbf{V} \geq 3$, *in this section*. Our basic objective is to prove that all collineations arise from combining symmetries of the underlying vector space and field of scalars. In other words, we show that if $\bar{V}$ is a projective space over a field $\mathbf{F}$ such that its geometrical structure is nontrivial ($\dim \bar{V} \geq 2$), then all of the geometric symmetries of $\bar{V}$ induce (and are induced by) algebraic symmetries of $\mathbf{V}$ and $\mathbf{F}$.

Many authors adopt the point of view that a projective space is the set of *all* subspaces of a vector space rather than simply a set of points, i.e., the set of all one-dimensional subspaces of a vector space. This viewpoint is closely associated with the work of the nineteenth-century mathematician Grassmann; hence the set of all subspaces of a vector space is often called a *Grassmann space* or a *Grassmann manifold.* In these terms it would be more reasonable to define a collineation to be a permutation of the subspaces of $\bar{V}$ that preserves inclusions. The following theorem shows that it is immaterial which viewpoint we adopt.

Theorem 6.45. If θ is any permutation of the subspaces of $\bar{V}$ that preserves inclusions, then $\dim(\bar{W})\theta = \dim \bar{W}$ for any subspace $\bar{W}$, and θ restricted to the points of $\bar{V}$ is a collineation. Conversely, if ϕ is a collineation of $\bar{V}$, then there is a unique way to extend ϕ to an inclusion preserving permutation of the subspaces of $\bar{V}$.

Proof. Let θ be an inclusion preserving permutation of the subspaces of $\bar{V}$, and let $\bar{W}$ be a k-dimensional subspace. Choose a strictly increasing chain of subspaces $\bar{W}_0 \subseteq \bar{W}_1 \subseteq \cdots \subseteq \bar{W}_n = \bar{V}$ that includes $\bar{W}$. In such a chain the dimensions of the subspaces must also be strictly increasing from 0 to n; thus, $\dim \bar{W}_j = j$ and $\bar{W} = \bar{W}_k$. Since θ is a *permutation* that preserves inclusions, $(\bar{W}_0)\theta \subseteq \cdots \subseteq (\bar{W}_n)\theta$ is also a strictly increasing chain of subspaces, and thus, $\dim(\bar{W})\theta = \dim(\bar{W}_k)\theta = k$. Hence θ permutes the subspaces of dimension k.

If L is a line, then (since θ permutes lines) so is $(L)\theta$, but to show that θ (restricted to points) is a collineation we must show that $(L)\theta = \{(P)\theta | P \in L\}$. Clearly $\{(P)\theta | P \in L\} \subseteq (L)\theta$ since θ preserves inclusions. If $Q \in (L)\theta$, then (since θ permutes zero-dimensional subspaces) we can choose P such that $(P)\theta = Q$. We must show that $P \in L$. Choose $P' \in L$ with $P' \neq P$, and let $L' = P + P'$. Then

$$(L)\theta = (P)\theta + (P')\theta \subseteq (L')\theta ,$$

which implies $(L)\theta = (L')\theta$ since they are both lines. Thus, $L = L'$, so $P \in L$ and the proof that θ (restricted to points) is a collineation is complete.

Assume θ_1 and θ_2 are inclusion-preserving permutations of subspaces with the same action on points. Since θ_1 and θ_2 (restricted to points) are both collineations, it follows from theorem 6.2 that

$$\{(P)\theta_1 | P \in \bar{W}\} = \{(P)\theta_2 | P \in \bar{W}\}$$

is a subspace whenever $\bar{W}$ is. The same argument as above shows that $\{(P)\theta_1 | P \in \bar{W}\}$ has the same dimension as $\bar{W}$. Since $\{(P)\theta_1 | P \in \bar{W}\}$ is contained in $(\bar{W})\theta_1$ and $(\bar{W})\theta_2$, and all three spaces have the same dimension, $(\bar{W})\theta_1 = \{(P)\theta_1 | P \in \bar{W}\} = (W)\theta_2$. Thus, $\theta_1 = \theta_2$ whenever they have the same action on points.

From the paragraph above, it is clear that the only possible way to extend a collineation, ϕ, to an inclusion-preserving permutation of the subspaces is to let $(\bar{W})\phi = \{(P)\phi | P \in \bar{W}\}$. This is (by theorem 6.2) a permutation of subspaces and clearly preserves inclusions, so the proof is complete.

Proposition 6.46. The perspectivities form a normal subset of the group of collineations, and hence the projective group of $\bar{V}$ is a normal subgroup of the collineation group.

Proof. Let $\bar{f}$ be a perspectivity with axis $\bar{H}$, and let ϕ be a collineation. Then $(\bar{f})\phi^i = \phi^{-1}\bar{f}\phi$ is a collineation leaving exactly one hyperplane, $(\bar{H})\phi$, pointwise fixed. Thus (theorem 6.21), $\phi^{-1}\bar{f}\phi$ is a perspectivity.

The following result is simply a restatement of results we proved in section 6.5; nevertheless, we repeat the proof here for convenience.

Proposition 6.47. If a collineation, ϕ, of $\bar{V}$ leaves a hyperplane, $\bar{H}$, pointwise fixed, then whenever $P \neq (P)\phi$ the line $P + (P)\phi$ is left globally fixed. If, in addition to the points in $\bar{H}$, ϕ leaves two points off $\bar{H}$ fixed, then ϕ is the identity.

Proof. If $P \neq (P)\phi$ and ϕ leaves $\bar{H}$ pointwise fixed, then neither P nor $(P)\phi$ is in $\bar{H}$, so $L = P + (P)\phi$ intersects $\bar{H}$ at a point X. The points $(P)\phi$ and $X = (X)\phi$ are distinct and in both L and $(L)\phi$; hence $L = (L)\phi$.

Now assume that ϕ leaves P_1 and P_2 fixed as well as leaving a hyperplane $\bar{H}$, not containing either P_1 or P_2, pointwise fixed. For each point $Q \notin P_1 + P_2$, the lines $Q + P_1$ and $Q + P_2$ are distinct, and each contains two fixed points of ϕ, P_i and $(Q + P_i) \cap \bar{H}$. Thus, ϕ leaves both lines fixed and hence leaves their intersection, Q, fixed. This implies that every point on $P_1 + P_2$ is also fixed; hence ϕ leaves all of $\bar{V}$ pointwise fixed.

This result is useful in building up the space of fixed points of a collineation by following it with projectivities. We accomplish this in much the same way that we built up the space of fixed points of a projectivity by following it with perspectivities in the proof of theorem 6.36.

Theorem 6.48. A collineation leaving a line L pointwise fixed is a projectivity.

Proof. Assume ϕ is a collineation of $\bar{V}$ leaving $\bar{W}_1$ pointwise fixed and that $1 \leq \dim \bar{W}_1 < \dim \bar{V}$. Let Q be a point not in $\bar{W}_1$. We shall show that there is a projectivity $\bar{f}$ such that $\phi\bar{f}$ leaves $\bar{W}_1 + Q = \bar{W}_2$ pointwise fixed. The reader is then asked to complete the proof as a problem (exercise 6.7, problem 2). We find the necessary projectivity in two steps. First we find $\bar{g}$ such that $\phi\bar{g}$ leaves $\bar{W}_1$ pointwise fixed and also leaves Q fixed; then we find $\bar{h}$ such that $\phi\bar{g}\bar{h}$ leaves $\bar{W}_2$ pointwise fixed. We use several times the fact that a projectivity of any subspace of $\bar{V}$ can be extended to a projectivity of $\bar{V}$ (exercise 6.7, problem 1). Now we find $\bar{g}$.

Case 1. $(Q)\phi \notin \bar{W}_2$. Let R be a third point on $Q + (Q)\phi$. Within $\bar{W}_1 + (Q + (Q)\phi) = \bar{W}_3$, the subspace $\bar{W}_1 + R$ is a hyperplane containing neither Q nor $(Q)\phi$ (Figure 40). Let $\bar{g}$ be a projectivity extending to $\bar{V}$ the elation of $\bar{W}_3$ with axis $\bar{W}_1 + R$, center R, sending $(Q)\phi$ to Q.

Case 2. $(Q)\phi \in \bar{W}_2$, $Q \neq (Q)\phi$. Since $\bar{W}_1$ is a hyperplane in $\bar{W}_2$, the line $Q + (Q)\phi$ meets $\bar{W}_1$ at a point R. Let $\bar{g}$ be a projectivity of $\bar{V}$ extending the elation of $\bar{W}_2$ with axis $\bar{W}_1$, center R, sending $(Q)\phi$ to Q.

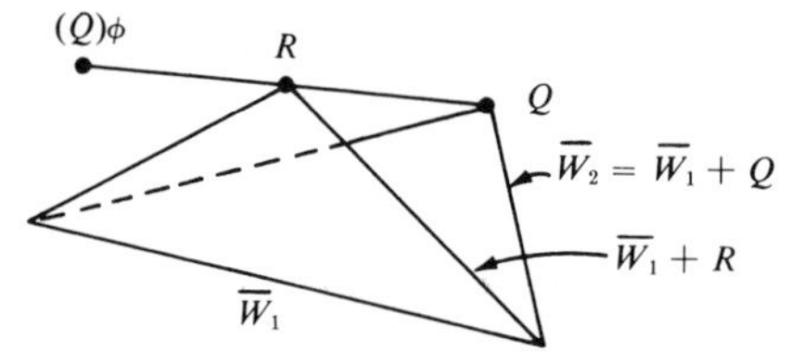

FIG. 40. The subspace $\bar{W}_1 + R$, a hyperplane containing neither Q nor $(Q)\phi$.

Case 3. $Q = (Q)\phi$. Let $\bar{g}$ be the identity.

In all three cases $\bar{g}$ is a projectivity such that $\phi\bar{g}$ leaves Q fixed and leaves $\bar{W}_1$ pointwise fixed. Now choose R in $\bar{W}_2$ but not in $\bar{W}_1$ such that $Q \neq R$. If $(R)\phi\bar{g} = R$, then $\phi\bar{g}$ leaves $\bar{W}_2$ pointwise fixed by proposition 6.47, and $\bar{f} = \bar{g}$ is a suitable projectivity of $\bar{V}$. If $R \neq (R)\phi\bar{g}$, then $(R)\phi\bar{g} \in R + Q$ since $(R + Q) \cap \bar{W}_1$ and Q are fixed points on $R + Q$. Let $\bar{h}$ be a projectivity of $\bar{V}$ extending the homology of $\bar{W}_2$ with axis $\bar{W}_1$, center Q, sending $(R)\phi\bar{g}$ to R. Then, as above, $\phi\bar{g}\bar{h}$ leaves $\bar{W}_2$ pointwise fixed, so $\bar{f} = \bar{g}\bar{h}$ is our desired projectivity.

Let $*:\mathbf{F} \rightarrow \mathbf{F}$ be an automorphism of $\mathbf{F}$, and let $g:\mathbf{V} \rightarrow \mathbf{V}$ be a nonsingular semi-linear mapping with respect to $*$ as defined in section 5.8, i.e., g is a permutation of $\mathbf{V}$ such that $(v + w)g = (v + w)$ and $(\alpha v)g = \alpha^*(v)g$. We call such a mapping a *$*$-symmetry* of $\mathbf{V}$. Note that if $*$ is nontrivial, then $** \neq *$, and hence the $*$-symmetries of $\mathbf{V}$, for a fixed $*$, do *not* form a group. Instead they form a coset of the linear transformations or $\mathbf{1}$-symmetries of $\mathbf{V}$ within the group of all semi-linear permutations (all possible $*$'s) of $\mathbf{V}$. This is the essence of the following result.

Proposition 6.49. If $f:\mathbf{V} \rightarrow \mathbf{V}$ is linear and $g:\mathbf{V} \rightarrow \mathbf{V}$ is semi-linear with respect to the automorphism $*$ of $\mathbf{F}$, then fg and gf are both semi-linear with respect to $*$.

Proof. fg and gf are additive. If $\alpha \in \mathbf{F}$ and $v \in \mathbf{V}$, then

$$(\alpha v)fg = ((\alpha v)f)g = (\alpha(v)f)g = \alpha^*(v)fg$$

and

$$(\alpha v)gf = (\alpha^*(v)g)f = \alpha^*(v)gf.$$

Thus, fg and gf are semi-linear with respect to $*$.

Theorem 6.50. Let g be a $*$-symmetry of $\mathbf{V}$. The induced mapping $\bar{g}:\bar{V} \rightarrow \bar{V}$, defined by $[v]\bar{g} = [(v)g]$, is a collineation of $\bar{V}$. It is a projectivity if and only if $*$ is the identity on $\mathbf{F}$.

Proof. The proof that $\bar{g}$ is well defined and is a collineation is left to the reader (exercise 6.7, problem 3). If $*$ is the trivial automorphism, then $\bar{g}$ is a projectivity by definition, so assume $*$ is nontrivial and choose $\alpha \in \mathbf{F}$ such that $\alpha^* \neq \alpha$. Obviously α is not zero or one. We must show that $\bar{g}$ is not projective. Let v_0 and v_1 be independent vectors in $\mathbf{V}$, and let $w_0 = (v_0)g$, $w_1 = (v_1)g$. The vectors w_0 and w_1 are also independent (exercise 6.7, problem 4). Let f be a nonsingular linear transformation of $\mathbf{V}$ such that $(w_0)f = v_0$ and $(w_1)f = v_1$. The mapping gf is semi-linear with respect to $*$ and leaves v_0 and v_1 fixed. Since $1^* = 1$, gf also leaves $v_0 + v_1$ fixed. Thus, $\bar{g}\bar{f}$ has three distinct fixed points $[v_0]$, $[v_1]$, and $[v_0 + v_1]$ on the line joining $[v_0]$ and $[v_1]$. But consider the point $[v_0 + \alpha v_1]$. Since $\alpha^* \neq \alpha$, $(v_0 + \alpha v_1)gf = (v_0 + \alpha^* v_1)$ is not dependent on $(v_0 + \alpha v_1)$; thus, $[v_0 + \alpha v_1]$ is not left fixed by $\bar{g}\bar{f}$. Thus (theorem 6.34), $\bar{g}\bar{f}$ is not a projectivity. Since $\bar{f}$ is a projectivity, $\bar{g}$ is not in the projective group.

Corollary 6.51. Every automorphism of **F** induces semi-linear permutations of **V**; hence if **F** has a nontrivial automorphism, then there are collineations of $\bar{V}$ that are not projective.

Proof. Let * be an automorphism of **F**, and let $v_0, v_1, \ldots, v_n$ be a basis of **V**. It is easy to verify that the mapping, $g:\mathbf{V} \to \mathbf{V}$, defined by $(\Sigma\alpha_i v_i)g = \Sigma\alpha_i^* v_i$ is a permutation of **V** and is semi-linear with respect to *. If * is nontrivial, then the induced collineation, $\bar{g}$, is not a projectivity.

Example. Let * be an automorphism of **F**. The semi-linear mapping of $\mathbf{V}_{n+1}(\mathbf{F})$ which sends $\langle x_0, x_1, \ldots, x_n\rangle$ to $\langle x_0^*, x_1^*, \ldots, x_n^*\rangle$ induces a collineation of $P_n(\mathbf{F})$ which sends $[x_0, x_1, \ldots, x_n]$ to $[x_0^*, x_1^*, \ldots, x_n^*]$. Note that no collision occurs between * and the homogeneous coordinates, i.e., although $[x_0, x_1, \ldots, x_n] = [\alpha x_0, \ldots, \alpha x_n]$ for all nonzero α, it is also true that $[(\alpha x_0)^*, \ldots, (\alpha x_n)^*] = [x_0^*, \ldots, x_n^*]$ since $(\alpha x_i)^* = \alpha^* x_i^*$ and $\alpha^* \neq 0$ whenever $\alpha \neq 0$. Let us also denote both the semi-linear map and the collineation by *, i.e., $\langle x_0, \ldots, x_n\rangle^* = \langle x_0^*, \ldots, x_n^*\rangle$ and $[x_0, \ldots, x_n]^* = [x_0^*, \ldots, x_n^*]$. If g is any semi-linear mapping of **V** (with respect to *), let A be the matrix whose rows consist of $\langle 1, 0, \ldots, 0\rangle g$, $\langle 0, 1, 0, \ldots, 0\rangle g, \ldots, \langle 0, 0, \ldots, 1\rangle g$, in that order. Then it is easy to verify that $[x_0, x_1, \ldots, x_n]\bar{g} = [\langle x_0, x_1, \ldots, x_n\rangle^* A]$. *Caution!* This need not be the same as $[\langle x_0, \ldots, x_n\rangle A]^*$, i.e., * and A need not commute! Be careful when deciding whether or not the collineation * leaves a point $P = [x_0, x_1, \ldots, x_n]$ fixed. There is always one set of homogeneous coordinates of P left fixed by the semi-linear map *, but not all sets of homogeneous coordinates are left fixed (exercise 6.7, problem 5) if * is nontrivial. For example, if $\mathbf{F} = \mathbf{C}$, $n = 2$, and * is complex conjugation, then the point $[i, -i, i]$ is left fixed even though $\langle i, -i, i\rangle^* = \langle -i, i, -i\rangle \neq \langle i, -i, i\rangle$.

Now we come to the main theorem of this section, another version of the fundamental theorem of projective geometry. There are many proofs of the theorem, some of which do not employ as much group theoretic machinery as the one we have chosen. See, for example, [1], pp. 86–91, [2], pp. 44–50, or [4]. The first two of these proofs include the case in which **F** is noncommutative (see section 6.10), and the third is based on the classical concept of "cross ratio." We have chosen to present the proof below since it emphasizes the analogies between affine and projective geometry and exploits our results about perspectivities.

Theorem 6.52 (fundamental theorem of projective geometry). If $\bar{V}$ is a projective space of dimension at least 2 over **F**, and if ϕ is a collineation of $\bar{V}$, then there is an automorphism, *, of **F** and a *-symmetry, g, of **V** such that $\phi = \bar{g}$. In other words, every symmetry of $\bar{V}$ is induced by a combination of symmetries of **V** and **F.**

Proof. Since the proof is long and complex, we first present an outline.

1. We follow ϕ by a projectivity $\bar{f}$ such that $\phi\bar{f} = \theta$ leaves $n + 2$ independent points fixed.

2. On a line, L, left fixed by θ, we establish affine coordinates using as our origin, unit point, and point at infinity three fixed points, R_0, R_1, and R_∞, of θ.

3. Define the map $*:\mathbf{F} \to \mathbf{F}$ as the coordinate version of the mapping θ restricted to $L - \{R_\infty\}$.

4. To prove that * preserves products we exploit the fact that multiplication of nonzero scalars is mirrored by the composition of stretchings [homologies with center R_0 and axis $\bar{H}$ ($R_\infty \in \bar{H}$) in projective terms].

5. To prove that * preserves sums we use in the same way the fact that addition of scalars is mirrored by the composition of translations along L (elations with axis $\bar{H}$ and center R_∞ in projective terms).

6. Finally we define a *-symmetry, $h:\mathbf{V} \to \mathbf{V}$, such that $\bar{h}$ and θ agree on L and leave the same $n + 2$ independent points fixed. Thus, $\theta\bar{h}^{-1}$ is projective by theorem 6.48 and is, therefore, the identity by theorem 6.34. Hence our original collineation $\phi = \overline{fh}$ is induced by the *-symmetry fh.

Now we carry out the details. We number the parts of the proof according to the outline above.

Part 1. Let $P_0, P_1, \ldots, P_{n+1}$ be $n + 2$ independent points in $\bar{V}$, say, $P_i = [v_i]$, $i = 0, 1, \ldots, n$ and $P_{n+1} = [v_0 + v_1 + \cdots + v_n]$. Let $[v_i]\phi = [w_i]$, $i = 0, 1, \ldots, n$. The vectors w_i, $i = 0, 1, \ldots, n$ form a basis of **V** (exercise 6.7, problem 6); hence there is a linear transformation $f:\mathbf{V} \to \mathbf{V}$ such that $(w_i)f = v_i$, $i = 0, 1, \ldots, n$. Let $\theta = \phi\bar{f}$. Clearly θ leaves $P_0, P_1, \ldots, P_{n+1}$ fixed.

Part 2. Let $L = P_0 + P_1$. Since the P_i are independent, neither P_0 nor P_1 is in the hyperplane spanned by $P_2, P_3, \ldots, P_{n+1}$. Thus, the point of intersection, R_1, of L and this hyperplane is a third point on L left fixed by θ. Relabel the two original fixed points, and let $R_0 = P_0$, $R_\infty = P_1$, i.e., $R_0 = [v_0]$, $R_\infty = [v_1]$. Then $R_1 = [v_0 + v_1]$ since $v_0 + v_1$ is both a linear combination of v_0 and v_1 and also of $v_2, v_3, \ldots, v_n, v_0 + v_1 + v_2 + \cdots + v_n$.

For every point, R, on L, except R_∞, there is a unique scalar λ such that $R = [v_0 + \lambda v_1]$. Thus, we let $R_\lambda = [v_0 + \lambda v_1]$ and note in passing that the correspondence $\lambda \leftrightarrow R_\lambda$ is a one-to-one correspondence between **F** and $L - \{R_\infty\}$.

Part 3. Define $*:\mathbf{F} \to \mathbf{F}$ by $\lambda^* = \mu$ if and only if $(R_\lambda)\theta = R_\mu$. Since θ is a collineation leaving L globally and $\{R_0, R_1, R_\infty\}$ pointwise fixed, θ restricted to $L - \{R_\infty\}$ is a permutation, and hence * is a permutation of $\mathbf{F}$ such that $0^* = 0$, $1^* = 1$.

Part 4. Let $v_0, v_1, \ldots, v_n$ be the basis of $\mathbf{V}$ established in part 1, and let $\bar{H}$ be the hyperplane in $\bar{V}$ spanned by $[v_1], [v_2], \ldots, [v_n]$. Given a nonzero scalar λ, let $f_\lambda: \mathbf{V} \to \mathbf{V}$ be the linear transformation sending v_0 to v_0 and v_i to λv_i, $i = 1, 2, \ldots, n$. The induced projectivity, $\bar{f}_\lambda$, is either the identity (if $\lambda = 1$) or else the homology with center $R_0 = [v_0]$, axis $\bar{H}$, sending R_1 to R_λ (if $\lambda \neq 1$). In either case, $\bar{f}_\lambda$ is the unique collineation leaving $\{R_0\} \cup \bar{H}$ pointwise fixed and sending R_1 to R_λ. Moreover, if $\lambda^* = \mu$, then $\theta^{-1}\bar{f}_\lambda\theta = (\bar{f}_\lambda)\theta^i$ is a collineation leaving $\{R_0\} \cup \bar{H}$ pointwise fixed and sending $(R_1)\theta = R_1$ to $(R_\lambda)\theta = R_\mu$. Thus, $\lambda^* = \mu$ if and only if $(\bar{f}_\lambda)\theta^i = \bar{f}_\mu$. If neither α nor β is zero, then clearly $f_{\alpha\beta} = f_\alpha f_\beta$. But then

$$\bar{f}_{(\alpha\beta)*} = (\bar{f}_{\alpha\beta})\theta^i = (\bar{f}_\alpha\bar{f}_\beta)\theta^i = (\bar{f}_\alpha)\theta^i(\bar{f}_\beta)\theta^i = \bar{f}_{\alpha*}\bar{f}_{\beta*}\,,$$

and thus, $(\alpha\beta)^* = \alpha^*\beta^*$. If either α or β is zero, then $(\alpha\beta)^* = \alpha^*\beta^*$ since $0^* = 0$. Thus, * preserves products.

Part 5. This is left as a problem (exercise 6.7, problem 7) since it is so similar to part 4.

Part 6. Let $h:\mathbf{V} \to \mathbf{V}$ be the *-symmetry such that $h(\Sigma\alpha_i v_i) = \alpha^* v_i$. The two collineations, $\bar{h}$ and θ, agree on L and at $P_2, \ldots, P_{n+1}$, and therefore (see the outline!), $\phi = \bar{f}\bar{h}$ is induced by a *-symmetry. This completes the proof.

Now we use this result to fulfill the promise we gave in section 5.8 and to show that any collineation of an affine space is the combination of a translation and a collineation induced by a semi-linear map of the translations. Recall the method (theorem 5.33) in which a nonsingular semi-linear mapping, g, of $\mathbf{T}$ induces a collineation, γ, of A. We first choose an origin, P, and then define γ by $(Q)\gamma = R$ if and only if $(\overrightarrow{PQ})g = \overrightarrow{PR}$.

Theorem 6.53. Let ϕ be a collineation of an n-dimensional ($n \geq 2$) affine space, A, over a field with at least three elements. Let $\mathbf{T}$ be the vector space of translations of A, and fix a point P in A to serve as our origin. The mapping $g:\mathbf{T} \to \mathbf{T}$ defined by $(\overrightarrow{PQ})g = \overrightarrow{(P)\phi(Q)\phi}$ is a nonsingular semi-linear mapping of $\mathbf{T}$ onto itself; moreover, ϕ is the product of the collineation, γ, induced by g and the translation $t = \overrightarrow{P(P)\phi}$.

Proof. Define g, γ, and t as above. It is easy to verify that g is a one-to-one mapping of $\mathbf{T}$ onto itself. Since

$$\overrightarrow{P(Q)\phi} = \overrightarrow{P(P)\phi} + \overrightarrow{(P)\phi(Q)\phi} = \overrightarrow{(P)\phi(Q)\phi} + \overrightarrow{P(P)\phi} = (\overrightarrow{PQ})g + t\,,$$

it is clear that $\phi = \gamma t$. Thus, we need only show that g is semi-linear to complete the proof.

Choose an n-dimensional projective space $\bar{V}$ and a hyperplane $\bar{H}$ in $\bar{V}$. Imbed A in $\bar{V}$ using the mappings $k:A \to \bar{V}$ and $j:\mathbf{T} \to \mathbf{H}$ as in the proof of theorem 6.20. The affine collineation ϕ then induces a projective collineation $\phi':\bar{V} \to \bar{V}$ in which, for each $v \in \mathbf{V}$, $v \neq 0$,

$$[v]\phi' = \begin{cases} [v]k^{-1}\phi k & \text{if } [v] \notin \bar{H} \\ [(v)j^{-1}gj] & \text{if } [v] \in \bar{H}. \end{cases}$$

We leave to the reader (exercise 6.7, problem 8) the details of verifying that ϕ' is a collineation. Note carefully how the assumption that the field has at least three elements enters the proof at this point. The projective collineation, ϕ', is induced by a semi-linear mapping of $\mathbf{V}$, call it f, that leaves $\mathbf{H}$ invariant. Let h be the restriction of f to $\mathbf{H}$. Since the projective collineations $\bar{h}$ and $\phi' = \bar{f}$ agree on $\bar{H}$, $[v]\bar{h} = [(v)j^{-1}gj]$ for all $[v] \in \bar{H}$. In other words, for each nonzero v in $\mathbf{H}$ there is a nonzero scalar, λ_v, such that $(v)j^{-1}gj = \lambda_v(v)h$. Unfortunately, since we do not know if g is semi-linear, we cannot appeal to previous results and assert directly that $j^{-1}gj = \lambda h$ for some nonzero scalar λ. This is true; however, it requires proof, i.e., we must show that $\lambda_v = \lambda_w$ for all nonzero v and w in $\mathbf{H}$. First we note that since the affine collineation ϕ maps parallelograms to parallelograms, the mapping $g:\mathbf{T} \to \mathbf{T}$ is additive. Since j is linear, this implies that $j^{-1}gj$ is additive. Now we use the standard technique (see case 2 in the proof of theorem 6.14 or problem 2 in either exercise 4.5 or 4.6) to show that if v and w are independent in $\mathbf{H}$, then $\lambda_v = \lambda_w$. If v and w are nonzero but dependent, then (since $n \geq 2$) we can choose x in $\mathbf{H}$ independent from both v and w, and then $\lambda_v = \lambda_x = \lambda_w$. Thus, there is a scalar λ such that $j^{-1}gj = \lambda h$, and hence $g = j\lambda hj^{-1}$ is semi-linear, and the proof is complete.

Exercise 6.7

1. Show that every projectivity of a subspace of $\bar{V}$ onto itself can be extended to a projectivity of $\bar{V}$ onto itself. When is the extension unique? *Caution!* The field with only two elements leads to an unusual situation.

2. Complete the proof of theorem 6.48. Where, in our proof of the "inductive step" in this proof, did we use the assumption that dim $\bar{W}_1 \geq 1$?

3. Let g be a *-symmetry of $\mathbf{V}$. Show that $\bar{g}:\bar{V} \to \bar{V}$ defined by $[v]\bar{g} = [(v)g]$ is well defined and is a collineation.

4. Let $g:\mathbf{V} \to \mathbf{V}$ be semi-linear with respect to the automorphism *. Prove that g is nonsingular, i.e., is a permutation of $\mathbf{V}$, if and only if for all $k \leq n = \dim \bar{V}$, v_i linearly independent and $[v_i]g = [w_i]$, $i = 0, 1, \ldots, k$ implies that $w_0, w_1, \ldots, w_k$ are linearly independent.

5. Let * be a nontrivial automorphism of $\mathbf{F}$, and consider the natural collineation of $P_n(\mathbf{F})$ induced by *, i.e., $[x_0, \ldots, x_n]^* = [x_0^*, \ldots, x_n^*]$. Show that if the point P

is left fixed by this collineation, then there are always two sets of homogeneous coordinates of P such that one set is left fixed by * and the other is not.

6. Let ϕ be a collineation of $\bar{V}$, and let $v_0, \ldots, v_k$ be linearly independent in **V**. Show that $[v_i]\phi = [w_i]$ implies that $w_0, w_1, \ldots, w_k$ are also linearly independent. *Caution!* Since we use this in the proof of theorem 6.52, you may not base your proof on that theorem and problem 4 above.

7. Complete part 5 of the proof of theorem 6.52. *Hint.* For each scalar λ let g_λ be the linear transformation of **V** for which $(v_0)g_\lambda = v_0 + \lambda v_1$ and $(v_i)g_\lambda = v_i$, $i = 1, 2, \ldots, n$.

8. Verify that the mapping ϕ' introduced in the proof of theorem 6.53 is a collineation of the projective space $\bar{V}$. Be sure to point out where you *need* the assumption that there are at least three distinct scalars.

9. We have avoided projective lines in this section since *any* permutation of the points on a line is a collineation. Find all fields **F** such that *all* permutations of the points on a projective line over **F** are induced by semi-linear transformations of the underlying two-dimensional vector space over **F**. Show that in all but one of these fields all of these permutations are, in fact, projectivities in the sense that they are induced by linear transformations.

10. Let **G** be the group of collineations and **P** be the group of projectivities of a projective space $\bar{V}$ over the field **F**. Show that the cosets of **P** in **G** correspond to the distinct automorphisms of **F**.

11. Look up the number of distinct automorphisms of a finite field **F**. Using this information, determine the number of collineations of an n-dimensional projective space over a finite field **F**.

6.8. Dual spaces and the principle of duality

In section 6.3 we discussed the principle of duality for projective planes. To generalize this to higher dimensional projective spaces we first define a "natural" dual space, $\bar{V}^t$, for a given projective space $\bar{V}$ and then establish a "natural" inclusion *reversing* correspondence between the proper subspaces of $\bar{V}$ and $\bar{V}^t$. All of this is based on the concept of the dual space for a vector space. The basic facts appear in the following theorem. We will explain our use of the exponent t when we consider the coordinate version of the dual space.

Theorem 6.54. If **V** is an n-dimensional vector space over **F**, then so is $\mathbf{V}^t$, the space of all linear functions from **V** into **F**. In $\mathbf{V}^t$ addition and multiplication by scalars are defined as for linear transformations, i.e.,

(1) $f + g$ is that function such that $(v)(f + g) = (v)f + (v)g$.

(2) λf is that function such that $(v)(\lambda f) = \lambda(v)f$.

This theorem is a special case of theorem 4.10 in which we stated (and asked the reader to prove) that $L(\mathbf{V}, \mathbf{W})$ is always a vector space and dim

$(L(\mathbf{V}, \mathbf{W})) = (\dim \mathbf{V})(\dim \mathbf{W})$. The basis for $L(\mathbf{V}, \mathbf{W})$ given there specializes to a standard relation between bases of $\mathbf{V}$ and $\mathbf{V}^t$ if we make the natural choice of the scalar 1 as our single basis "vector" in $\mathbf{F}$. This relationship between bases is as follows.

Definition. Let $v_1, v_2, \ldots, v_n$ be a basis of the vector space $\mathbf{V}$. The *dual basis* of $\mathbf{V}^t$ consists of the linear functions $v_1^t, v_2^t, \ldots, v_n^t$ such that

$$(v_i)v_j^t = \begin{cases} 1 & \text{if } i = j \\ 0 & \text{if } i \neq j. \end{cases}$$

The "dual basis" concept provides a simple computational device closely related to the dot product of elementary vector analysis. Given a basis $v_1, v_2, \ldots, v_n$ of $\mathbf{V}$, consider the coordinate map $\phi{:}\mathbf{V} \to \mathbf{V}_n(\mathbf{F})$ relative to this basis, i.e., $(v)\phi = \langle \alpha_1, \ldots, \alpha_n \rangle$ if and only if $v = \alpha_1 v_1 + \cdots + \alpha_n v_n$. It is more convenient to use column vectors to represent linear functions (or *linear functionals* as they are usually called), so we let $\mathbf{V}_n^t(\mathbf{F})$ be the set of all column vectors,

$$\begin{pmatrix} \beta_1 \\ \beta_2 \\ \vdots \\ \beta_n \end{pmatrix},$$

with all n entries in $\mathbf{F}$. Recalling the definition of the transpose, A^t, of a matrix, we can denote this awkward column vector by $\langle \beta_1, \beta_2, \ldots, \beta_n \rangle^t$. Now let $\phi^t{:}\mathbf{V}^t \to \mathbf{V}_n^t(\mathbf{F})$ be the coordinate map associated with the dual basis $v_1^t, v_2^t, \ldots, v_n^t$ of $\mathbf{V}^t$, i.e., $(f)\phi^t = \langle \beta_1, \beta_2, \ldots, \beta_n \rangle^t$ if and only if $f = \beta_1 v_1^t + \cdots + \beta_n v_n^t$. How can we compute $(v)f$ in terms of these two coordinate maps? The result is as follows.

Proposition 6.55. If $\phi{:}\mathbf{V} \to \mathbf{V}_n(\mathbf{F})$ and $\phi^t{:}\mathbf{V}^t \to \mathbf{V}_n^t(\mathbf{F})$ are coordinate maps relative to dual bases of $\mathbf{V}$ and $\mathbf{V}^t$, then for any $v \in \mathbf{V}$ and any $f \in \mathbf{V}^t$, $(v)f = ((v)\phi)((f)\phi^t)$, i.e., to compute $(v)f$ we simply compute the matrix product of their coordinates.

Proof. Assume $(v)\phi = \langle \alpha_1, \alpha_2, \ldots, \alpha_n \rangle$ and $(f)\phi^t = \langle \beta_1, \beta_2, \ldots, \beta_n \rangle^t$. Then $v = \Sigma \alpha_i v_i$, and since v_j^t is linear, $(v)v_j^t = \Sigma \alpha_i (v_i) v_j^t = \alpha_j$. In view of the way $f = \Sigma \beta_j v_j^t$ is defined, $(v)f = (v)\Sigma \beta_j v_j^t = \Sigma \beta_j (v) v_j^t$. Substituting $(v)v_j^t$ into this last equation, we find that $(v)f = \Sigma \beta_j \alpha_j$. But $\Sigma \beta_j \alpha_j = \Sigma \alpha_j \beta_j$ is just the matrix product of $\langle \alpha_1, \ldots, \alpha_n \rangle$ and $\langle \beta_1, \ldots, \beta_n \rangle^t$ so the proof is complete.

If we focus on the special case $\mathbf{V} = \mathbf{V}_n(\mathbf{F})$ and use the natural basis and its associated coordinate map, $\mathbf{1}$, then the coordinate map for the dual basis is the map $\mathbf{1}^t{:}(\mathbf{V}_n(\mathbf{F}))^t \to \mathbf{V}_n^t(\mathbf{F})$. This map provides a simple and attractive (perhaps too attractive) means of identifying the space of linear functionals on

$\mathbf{V}_n(\mathbf{F})$ with the space of column vectors. The temptation to identify these two spaces often leads to confusion and should be resisted!

Now we turn to the projective dual space in which the "points" are classes of linear functions.

Definition. The (projective) *dual space* of a projective space $\bar{V}$ is the projective space obtained by collapsing in the standard way the (linear) dual space $\mathbf{V}^t$. For simplicity in typesetting, we denote the dual space by $\bar{V}^t$ instead of the more accurate notation with the bar extending over the t.

Once again it is convenient to use column vectors (with $n + 1$ entries) as homogeneous coordinates of points in the dual space, for if (using the notation above) $(f)\phi^t = \langle\beta_o, \beta_1, \ldots, \beta_n\rangle^t$ and $f = \lambda g$, then $(g)\phi^t = \langle\lambda\beta_o, \lambda\beta_1, \ldots, \lambda\beta_n\rangle^t$. Thus, $[\beta_o, \beta_1, \ldots, \beta_n]^t$ is the natural notation to use for the homogeneous coordinates of points in $\bar{V}^t$. Once again we caution the reader to remember that these column vectors are only homogeneous *coordinates* of the points in the dual space and that the "real points" are classes of linear functionals on $\mathbf{V}$.

Now we are ready to state the basic duality relation between the subspaces of $\bar{V}$ and those of $\bar{V}^t$. The proof of this result illustrates the advantages of the linear functional approach. We state two versions of duality, first the linear version and then the projective version.

Theorem 6.56 (duality–linear). Let $\mathbf{V}$ be an n-dimensional vector space with dual $\mathbf{V}^t$. For each subspace, $\mathbf{W}$, of $\mathbf{V}$ let $\mathbf{W}^a = \{f \in \mathbf{V}^t | (w)f = 0$ for all $w \in \mathbf{W}\}$. Similarly for each subspace, $\mathbf{W}$, of $\mathbf{V}^t$ let $\mathbf{W}^a = \{v \in \mathbf{V} | (v)f = 0$ for all $f \in \mathbf{W}\}$. The mapping a is then a permutation of the subspaces of $\mathbf{V}$ and $\mathbf{V}^t$ such that:

(1) Inclusions are reversed by a, i.e., if $\mathbf{W}_1$ and $\mathbf{W}_2$ are subspaces, then $\mathbf{W}_1 \subseteq \mathbf{W}_2$ if and only if $\mathbf{W}_1^a \supseteq \mathbf{W}_2^a$.

(2) For any subspace, $\mathbf{W}$, $\mathbf{W}^{aa} = \mathbf{W}$ and $\dim(\mathbf{W}^a) = n - \dim \mathbf{W}$.

Proof. For each $f \in \mathbf{V}^t$ the kernel of f is a subspace of $\mathbf{V}$; hence if X is any *subset* of $\mathbf{V}^t$, then $X^a = \{v \in \mathbf{V} | (v)f = 0$ for all $f \in X\}$ is simply the intersection of the kernels of all f in X and is therefore a *subspace* of $\mathbf{V}$. Moreover, if we throw more linear functionals into X, then we consider the intersection of more kernels to form X^a, and therefore X^a is diminished if it is changed at all. Similarly for each $v \in \mathbf{V}$, the *annihilator* of v, i.e., $\{f \in \mathbf{V}^t | (v)f = 0\}$, is clearly a subspace of $\mathbf{V}^t$; hence if X is any *subset* of $\mathbf{V}$, then X^a is a *subspace* of $\mathbf{V}^t$, and if we insert additional vectors in X, we either leave X^a unchanged or else diminish it. From these remarks, it is clear that if $\mathcal{S}$ is the set of all *subspaces* of $\mathbf{V}$ or $\mathbf{V}^t$, then $a:\mathcal{S} \to \mathcal{S}$ and a (weakly) reverses inclusions.

It is obvious from the definition that for any $\mathbf{W} \in \mathcal{S}$, $\mathbf{W} \subseteq \mathbf{W}^{aa}$. To show that $\mathbf{W}^{aa} \subseteq \mathbf{W}$ it is enough to prove the following two lemmas.

Lemma 1. If $\mathbf{W}$ is a subspace of $\mathbf{V}$ and $v \notin \mathbf{W}$, then there is a linear functional, f, such that $(w)f = 0$ for every $w \in \mathbf{W}$ but $(v)f \neq 0$, i.e., if v is not in $\mathbf{W}$, then v is not in $\mathbf{W}^{aa}$.

Proof. Let $w_1, \ldots, w_k$ be a basis for $\mathbf{W}$. Since $v \notin \mathbf{W}$, $w_1, \ldots, w_k, v$ are linearly independent and hence can be completed to a basis of $\mathbf{V}$. Let $f_1, \ldots, f_n$ be the dual basis of $\mathbf{V}^t$. Then f_{k+1} is a suitable linear functional that is 0 on all of $\mathbf{W}$ but not at v.

Lemma 2. If $\mathbf{W}$ is a subspace of $\mathbf{V}^t$ and $g \notin \mathbf{W}$, then there is a vector v such that $(v)f = 0$ for all $f \in \mathbf{W}$ but $(v)g \neq 0$, i.e., if $g \notin \mathbf{W}$, then $g \notin \mathbf{W}^{aa}$.

Proof. In order to exploit the preceding lemma we first consider $\mathbf{V}^{tt}$, the space of all linear functionals on $\mathbf{V}^t$. It is a curious but convenient fact that for each vector $v \in \mathbf{V}$ there is a linear functional, v^*, on $\mathbf{V}^t$ defined by $(f)v^* = (v)f$ for all $f \in \mathbf{V}^t$. We leave it to the reader (exercise 6.8, problem 7) to verify that v^* is actually linear and that $^*: \mathbf{V} \to \mathbf{V}^{tt}$ is a one-to-one linear transformation. Since $\dim \mathbf{V}^{tt} = \dim \mathbf{V}$, the transformation $*$ must map $\mathbf{V}$ *onto* $\mathbf{V}^{tt}$, i.e., for each linear functional, h, in $\mathbf{V}^{tt}$ there is a vector $v \in \mathbf{V}$ such that $(f)h = (v)f$ for all $f \in \mathbf{V}^t$. Now apply lemma 1 and choose $h \in \mathbf{V}^{tt}$ such that $(f)h = 0$ for all $f \in \mathbf{W}$ but $(g)h \neq 0$. The vector $v \in \mathbf{V}$ such that $v^* = h$ is our desired vector.

The two lemmas above and the remark just preceding them show that aa is the identity on $\mathcal{S}$. Therefore (theorem 1.1), a is a permutation of $\mathcal{S}$. Since it is one to one on *subspaces*, and we know it reverses inclusions weakly for *subsets*, it must reverse inclusions strictly for subspaces.

If $\mathbf{W}$ is a k-dimensional subspace, then include $\mathbf{W}$ in a maximal strictly increasing chain of subspaces, $\{0\} = \mathbf{W}_0 \subseteq \mathbf{W}_1 \subseteq \cdots \subseteq \mathbf{W}_n$ ($\mathbf{W}_n$ is either $\mathbf{V}$ or $\mathbf{V}^t$). In such a chain the dimensions must also strictly increase from 0 to n; hence $\mathbf{W} = \mathbf{W}_k$ and $\dim \mathbf{W}_j = j$. Applying a, we have a strictly decreasing maximal chain of subspaces $\mathbf{W}_0^a \supseteq \mathbf{W}_1^a \supseteq \cdots \supseteq \mathbf{W}_n^a$, and since the dimensions must strictly decrease from n down to 0, $\dim \mathbf{W}_k^a = n - k$. This completes the proof of the linear version of duality.

Now we pass to projective spaces. Since the linear subspace $\{0\}$ does not correspond to a projective subspace (we exclude $\varnothing$ as a projective subspace), we omit both the zero subspaces and their mates under a ($\mathbf{V}$ and $\mathbf{V}^t$) when we consider projective duality. Although this is strictly a matter of preference and not necessary, it is useful as a mnemonic device for the formula given below for $\dim \bar{W}^\alpha$.

Theorem 6.57 (duality–projective). Let $\bar{V}$ be an n-dimensional projective space with dual $\bar{V}^t$. For each *proper* subspace, $\bar{W}$, of $\bar{V}$ let $\bar{W}^\alpha = \{[f] \in \bar{V}^t \mid (w)f = 0 \text{ for all } w \in \mathbf{W}\}$. Similarly, for each *proper* subspace, $\bar{W}$, of $\bar{V}^t$ let

$\bar{W}^{\alpha} = \{[v] \in \bar{V} | (v)f = 0 \text{ for all } f \in \mathbf{W}\}$. The mapping α is a permutation of the proper subspaces of $\bar{V}$ and $\bar{V}^t$ such that:

(1) Inclusions are reversed by α.

(2) For any proper subspace $\bar{W}$, $\bar{W}^{\alpha\alpha} = \bar{W}$ and $\dim \bar{W}^{\alpha} = (n-1) - \dim \bar{W}$. In particular, α interchanges points of $\bar{V}$ (respectively, $\bar{V}^t$) and hyperplanes of $\bar{V}^t$ (respectively, $\bar{V}$).

Proof. $\bar{W}^{\alpha}$ is the projective subspace corresponding to $\mathbf{W}^a$. Hence, with the exception of the result on $\dim \bar{W}^{\alpha}$, all of these properties of α follow trivially from their analogs for a. As to the formula for $\dim \bar{W}^{\alpha}$, we need only remember that the linear dimension of a subspace is always one greater than its projective dimension, for

$$\dim \bar{W}^{\alpha} + \dim \bar{W} = (\dim \mathbf{W}^a - 1) + (\dim \mathbf{W} - 1) = \dim \mathbf{W}^a + \dim \mathbf{W} - 2 = (n+1) - 2 = n - 1 .$$

Caution! The symmetry between the behavior of a on the subspaces of $\mathbf{V}$ and $\mathbf{V}^t$ depends strongly on the fact that $\dim \mathbf{V}$ is finite. For a careful analysis of the infinite dimensional case, see [2], pp. 28–33.

In view of the existence of this inclusion reversing correspondence between the subspaces of $\bar{V}$ and $\bar{V}^t$, the principle of duality for projective planes can be generalized as follows.

Meta-theorem 6.58 (the principle of duality). If, in a statement concerning only "inclusion" relations among subspaces of an n-dimensional projective space, all k-dimensional subspaces are replaced by $((n-1)-k)$-dimensional subspaces and all inclusions are reversed, then the resulting statement is valid in $\bar{V}^t$ if and only if the original is valid in $\bar{V}$. If the original is valid in *all* n-dimensional projective spaces over the field $\mathbf{F}$, then so is its dual.

Note that "statements concerning only inclusion relations" include statements about joins and intersections, for the join of certain subspaces is the smallest subspace containing all of them, and their intersection is the largest subspace contained in all of them. We give two examples of new theorems that can be obtained "by duality" from earlier theorems in this chapter.

Theorem 6.6 (dual). An n-dimensional projective space, $\bar{V}$, contains subspaces of dimensions j and k whose join is all of $\bar{V}$ if and only if $n \leq j + k + 1$.

Theorem 6.24 (dual). A perspectivity is uniquely determined by its center and its action on two hyperplanes not containing the center.

For a deeper study of duality, we recommend section II.3 of [2]. The duality between $\mathbf{V}$ and $\mathbf{V}^t$ can be generalized to the concept of *paired vector spaces*. For an introduction to this topic, see either Appendix II to Chapter II in [2] or section 4 of Chapter 1 in [1].

Exercise 6.9

1. State the duals of theorem 6.2 and corollary 6.5.

2. Discuss problem 4 in exercise 6.4 from the point of view of duality.

3. Let $h:\mathbf{V}_1 \rightarrow \mathbf{V}_2$ be a linear transformation of the vector space $\mathbf{V}_1$ into the vector space $\mathbf{V}_2$. Show that $h^t:\mathbf{V}_2^t \rightarrow \mathbf{V}_1^t$ defined by $(f)h^t = hf$ is a linear transformation. What is the rank of h^t?

4. Let $\mathbf{W}_1$ and $\mathbf{W}_2$ be complementary subspaces of the vector space $\mathbf{V}$, and let $h:\mathbf{V} \rightarrow \mathbf{V}$ be the projection of $\mathbf{V}$ onto $\mathbf{W}_1$ along $\mathbf{W}_2$. Show that $h^t:\mathbf{V}^t \rightarrow \mathbf{V}^t$ (as defined in problem 3) is the projection of $\mathbf{V}^t$ onto $\mathbf{W}_2^a$ along $\mathbf{W}_1^a$.

5. Show that the mapping t sending h to h^t as in problem 3 is an anti-isomorphism of $L(\mathbf{V}_1, \mathbf{V}_2)$ into $L(\mathbf{V}_2^t, \mathbf{V}_1^t)$.

6. Assume that in ϕ-coordinates the linear transformation $h \in L(\mathbf{V}, \mathbf{V})$ is represented by the matrix A. Find the matrix representing $h^t \in L(\mathbf{V}^t, \mathbf{V}^t)$ in the dual coordinates, i.e., with respect to the dual basis.

7. For each vector $v \in \mathbf{V}$, define the function $v^*:\mathbf{V}^t \rightarrow \mathbf{F}$ by $(f)v^* = (v)f$ for all $f \in \mathbf{V}^t$. Verify that v^* is linear and hence in $\mathbf{V}^{tt}$. Thus, $*:\mathbf{V} \rightarrow \mathbf{V}^{tt}$. Verify that $*$ is linear and is one-to-one.

8. In section 6.7 (theorem 6.45), we proved that a collineation of $\bar{V}$ may be viewed as an inclusion preserving permutation of the subspaces of $\bar{V}$. Show that if ϕ is a collineation of $\bar{V}$, then $\alpha\phi\alpha$ is a collineation of $\bar{V}^t$. Describe the collineation $\alpha\phi\alpha$ if the collineation ϕ is (a) a perspectivity with axis $\bar{H}$, center P, and sends Q to R, or (b) a j-involution as defined in exercise 6.6, problem 10.

9. Let $f_1, f_2, \ldots, f_n$ be a basis of $\mathbf{V}^t$. Show that there is a basis $v_1, v_2, \ldots, v_n$ of $\mathbf{V}$ such that $f_1, \ldots, f_n$ is its dual basis.

6.9. *Correlations and semi-bilinear forms*

In the last section we developed the basic inclusion reversing permutation, α, of proper subspaces of a projective space $\bar{V}$ and its dual space $\bar{V}^t$. There are also inclusion *preserving* maps which map proper subspaces of $\bar{V}$ to proper subspaces of $\bar{V}^t$. For example, let $[\phi]:\bar{V} \rightarrow P_n(\mathbf{F})$ and $[\theta]:\bar{V}^t \rightarrow P_n(\mathbf{F})$ be coordinate maps for $\bar{V}$ and $\bar{V}^t$. Then the obvious extension of $[\phi][\theta]^{-1}$ from points to proper subspaces of $\bar{V}$ is an inclusion preserving map. If we follow such an inclusion preserving mapping, β, by α (restricted to proper subspaces of $\bar{V}^t$), then we have an inclusion reversing permutation, $\beta\alpha$, of the proper subspaces of $\bar{V}$. This type of mapping is the object of study in this section. As with collineations, these mappings are uninteresting if $\bar{V}$ is a line, so *we again assume* dim $\bar{V} \geq 2$ *in this section.*

Definition. A *correlation* of the projective space $\bar{V}$ is an inclusion reversing permutation of the proper subspaces of $\bar{V}$.

Theorem 6.59. If ϕ is a correlation of $\bar{V}$, then ϕ interchanges joins and intersections, and for any subspace $\bar{W}$, dim $((\bar{W})\phi) = (n - 1) -$ dim $\bar{W}$. In particular, ϕ interchanges points and hyperplanes and maps collinear points to "concurrent" hyperplanes, i.e., three distinct points, P, Q, and R, are collinear if and only if $(P)\phi \cap (Q)\phi \subseteq (R)\phi$.

Proof. Since ϕ is a permutation of the proper subspaces, it reverses inclusions strictly; hence the argument used to prove the dimension formula for a (theorem 6.56) can be applied to ϕ. The only modification needed is to observe that maximal chains for a terminated at dimension n, whereas those for ϕ terminate at dimension $n - 1$ since we only consider proper subspaces. Since joins and intersections are dual concepts with respect to inclusion, ϕ interchanges them, i.e., $(\bar{W}_1 \cap \bar{W}_2)\phi = (\bar{W}_1)\phi + (\bar{W}_2)\phi$ and $(\bar{W}_1 + \bar{W}_2)\phi = (\bar{W}_1)\phi \cap (\bar{W}_2)\phi$.

We have shown (theorem 6.45) that collineations can be viewed either as permutations of the points of $\bar{V}$ or as permutations of the proper subspaces of $\bar{V}$. Now we prove a similar result for correlations using the result above as our guide.

Theorem 6.60. Let ϕ be a one-to-one function from the set of points of $\bar{V}$ onto the set of hyperplanes of $\bar{V}$ such that three distinct points, P, Q, and R, are collinear if and only if $(P)\phi \cap (Q)\phi \subseteq (R)\phi$. There is a unique way to extend ϕ to a correlation.

Proof. Let θ be a correlation of $\bar{V}$. (We proved that at least one exists at the beginning of this section!). Then $\phi\theta:\bar{V} \rightarrow \bar{V}$ is a collineation (of points) and hence (theorem 6.45) has an extension to a collineation, σ, of proper subspaces. Thus, $\sigma\theta^{-1}$ is one correlation extending ϕ to the proper subspaces of $\bar{V}$. If ϕ' were another, then $\phi'\theta$ would be a collineation (of proper subspaces) agreeing with $\phi\theta$ on points. But then (again by theorem 6.45) $\phi'\theta = \sigma$ so $\phi' = \sigma\theta^{-1}$ also. Thus, the extension of ϕ to subspaces is unique.

The proof above is just one example of the way results about collineations may be used to derive analogous results about correlations.

Theorem 6.61. If we adopt the viewpoint that collineations as well as correlations are permutations of the proper subspaces of $\bar{V}$, then the collineations and correlations form a group **G**. The group of collineations is a subgroup, **H**, of index 2 in **G**, and the correlations form the nontrivial coset of **H** in **G**.

Proof. Let $\mathcal{S}$ be the set of proper subspaces of $\bar{V}$. Clearly the set of permutations of $\mathcal{S}$ that either preserve or reverse inclusions is non-empty, closed under products, and closed under inverses. Thus, it is a subgroup of the group of all permutations of $\mathcal{S}$. Moreover, the mapping sending the elements of this subgroup to the two-element group {"reverse," "preserve"} is a homomorphism whose kernel is the group of collineations.

This group is essentially the group of automorphisms of the projective group on $\bar{V}$, i.e., any automorphism of $PGL_{(n+1)}(\mathbf{F})$ is induced by a unique collineation or correlation of $P_n(\mathbf{F})$. One technique for proving this is very similar to the technique we used to show that the similarities group of euclidean space is complete. Namely, one uses the 1-involutions defined in exercise 6.6, problem 10 to give group theoretic analogs of the geometry of the space and then exploits these to show that each automorphism is induced by a geometric symmetry. We shall not take the time to present the details here and instead refer the reader to [6].

We turn now to the problem of representing correlations algebraically. Using the basic duality mappings, a and α, we derive the following preliminary result.

Theorem 6.62. For each correlation, ϕ, of a projective space $\bar{V}$, there is a semi-linear mapping $g{:}\mathbf{V} \to \mathbf{V}^t$ such that $\phi = \overline{ga} = \bar{g}\alpha$.

Proof. Given ϕ, choose a "basic" nonsingular linear transformation $b{:}\mathbf{V}^t \to \mathbf{V}$ and also call b the unique inclusion preserving mapping induced by b on the set of subspaces of $\mathbf{V}^t$. Then ab is an inclusion reversing mapping of the set of subspaces of $\mathbf{V}$ onto itself. This map induces a correlation, $\overline{ab}$; hence $\phi\overline{ab}$ is a collineation of $\bar{V}$. By the fundamental theorem $\phi\overline{ab} = \bar{f}$, f a semi-linear mapping of $\mathbf{V}$ onto itself. Thus, $\phi = \bar{f}\bar{b}^{-1}\bar{a}$. Since b is linear, $g = fb^{-1}$ is semi-linear. Thus, $\phi = \overline{ga} = \bar{g}\alpha$ since α is the projective version of a.

A semi-linear mapping, $g{:}\mathbf{V} \to \mathbf{V}^t$, sending vectors, w, in $\mathbf{V}$ to linear functionals, $(w)g$, on $\mathbf{V}$ may seem rather elusive at first sight. How can we keep track of the value, $(v)(w)g$, of the linear functional $(w)g$ at a vector v? The key is to examine the way in which $(v)(w)g$ depends on both v and w. Since $(w)g$ is linear, $(v)(w)g$ depends in a linear fashion on v, but since g is semi-linear, it depends in a semi-linear fashion on w. Thus, we have a function mapping ordered pairs of vectors to scalars, which fits the following definition.

Definition. A *semi-bilinear form* on a vector space $\mathbf{V}$ over a field $\mathbf{F}$ is a function $b{:}\mathbf{V} \times \mathbf{V} \to \mathbf{F}$ such that for some automorphism, $*$, of $\mathbf{F}$:

(1) b is additive in both variables, i.e., $(v_1 + v_2, w)b = (v_1, w)b + (v_2, w)b$ and $(v, w_1 + w_2)b = (v, w_1)b + (v, w_2)b$.

(2) b is homogeneous in the first variable and $*$ homogeneous in the second, i.e., $(\lambda v, w)b = \lambda(v, w)b$ and $(v, \lambda w)b = \lambda^*(v, w)b$.

The semi-bilinear forms associated with a given automorphism, *, are called **-forms*. Those associated with the identity are called *bilinear forms*.

With this definition we can rephrase the last theorem and state its converse as follows.

Theorem 6.63. If ϕ is a correlation of the projective space $\bar{V}$, then there is a semi-bilinear form, b, on $\mathbf{V}$ such that for each proper subspace, $\mathbf{W}$, $(\bar{W})\phi = \{[v] | (v, w)b = 0 \text{ for all } w \in \mathbf{W}\}$. Conversely, if b is a semi-bilinear form on $\mathbf{V}$, then the mapping on proper subspaces defined by this equation is a correlation if and only if b satisfies any one of the following conditions:

(1) $(v, w)b = 0$ for all $v \in \mathbf{V}$ implies $w = 0$.

(2) $(v, w)b = 0$ for all $w \in \mathbf{V}$ implies $v = 0$.

(3) Each linear functional f in $\mathbf{V}^t$ is of the form $(\ )f = (\ , w)b$ for some $w \in \mathbf{V}$.

Two semi-bilinear forms induce the same correlation if and only if one is a scalar multiple of the other.

Proof. Let ϕ be a correlation of $\bar{V}$. By theorem 6.62 there is a semi-linear mapping $g: \mathbf{V} \to \mathbf{V}^t$ such that $(\bar{W})\phi = \{[v] | (v)(w)g = 0 \text{ for all } w \in \mathbf{W}\}$. Thus, the semi-bilinear form, b, such that $(v, w)b = (v)(w)g$ induces ϕ in the prescribed manner.

Now assume b is a semi-bilinear form on $\mathbf{V}$. Guided by the discussion above, we define h and θ as follows:

(h) For each $w \in \mathbf{V}$, let $(w)h$ be the function $f: \mathbf{V} \to \mathbf{F}$ such that $(v)f = (v, w)b$ for all v in $\mathbf{V}$.

(θ) For each proper subspace, $\bar{W}$, of $\bar{V}$, let $(\bar{W})\theta = \{[v] | (v, w)b = 0 \text{ for all } w \in \mathbf{W}\}$.

It is easy to verify that h is a semi-linear mapping of $\mathbf{V}$ into $\mathbf{V}^t$ and that θ maps proper subspaces of $\bar{V}$ to subspaces (which might not be proper!) of $\bar{V}$.

Lemma. The semi-linear mapping, h, is nonsingular if and only if θ is a correlation of $\bar{V}$.

Proof. If θ is a correlation, then $w \neq 0$ implies $[w]\theta$ is a hyperplane in $\bar{V}$; hence there is a $[v] \notin [w]\theta$, i.e., there is a nonzero vector v such that $(v, w)b = (v)(w)h \neq 0$. Thus, $w \neq 0$ implies $(w)h \neq 0$, so h is nonsingular.

Conversely, if h is nonsingular, then by an argument analogous to the proof of theorem 6.50 it follows that $\bar{h}$ is a one-to-one, inclusion preserving mapping of the proper subspaces of $\bar{V}$ onto the set of proper subspaces of $\bar{V}^t$. Since (theorem 6.57) α is a one-to-one inclusion reversing map of the proper subspaces of $\bar{V}^t$ onto those of $\bar{V}$, the mapping $\theta = h\alpha$ is a correlation of $\bar{V}$.

Since conditions (1) and (3) in the theorem are simply different ways of asserting that h is nonsingular, the lemma implies that θ is a correlation if and only if b satisfies either of conditions (1) or (3).

As for condition (2), if $v \neq 0$, then there is a linear functional, f, such that $(v)f \neq 0$. Thus, b satisfies condition (2) if and only if h maps $\mathbf{V}$ onto $\mathbf{V}^t$, i.e., conditions (2) and (3) are equivalent.

Finally, assume that the semi-bilinear forms, b and c, induce the same correlation. Let $g:\mathbf{V} \to \mathbf{V}^t$ and $h:\mathbf{V} \to \mathbf{V}^t$ be the semi-linear mappings associated with b and c as in the lemma above. By that lemma both g and h are nonsingular, and thus gh^{-1} is defined and is a semi-linear mapping of $\mathbf{V}$ onto itself. Since $\bar{g}\alpha = \bar{h}\alpha$ and α has an inverse (itself!), $\bar{g} = \bar{h}$. Thus, $[w]g = [w]h$ for any nonzero vector w. But this implies that w is a characteristic vector of gh^{-1} for every nonzero vector w, so $gh^{-1} = \lambda\mathbf{1}$ for some scalar λ. This completes the proof of the theorem.

Definition. A correlation is *projective* if it is induced by a bilinear form.

Example 1. Let $\mathbf{V} = \mathbf{V}_{n+1}(\mathbf{F})$ and define b by

$$(\langle\alpha_0, \ldots, \alpha_n\rangle, \langle\beta_0, \ldots, \beta_n\rangle)b = \langle\alpha_0, \ldots, \alpha_n\rangle \langle\beta_0, \ldots, \beta_n\rangle^t = \Sigma\alpha_i\beta_i .$$

This bilinear form reduces to the ordinary dot product if $\mathbf{F}$ is the field of real numbers. The correlation of $P_n(\mathbf{F})$ induced by b is the "canonical" one in which the point $[\alpha_0, \ldots, \alpha_n]$ is mapped to the kernel of the linear functional with dual coordinates $[\alpha_0, \ldots, \alpha_n]^t$.

Example 2. Let f_1 and f_2 be linear functionals on $\mathbf{V}$. The function $f_1 \otimes f_2:\mathbf{V} \times \mathbf{V} \to \mathbf{F}$, defined by $(v, w)f_1 \otimes f_2 = (v)f_1 \cdot (w)f_2$ is clearly a bilinear form on $\mathbf{V}$. Such a bilinear form never induces a correlation since conditions (1) and (2) of theorem 6.63 obviously do not hold, i.e., $(v, w)f_1 \otimes f_2 = 0$ for all pairs v, w with $v \in \operatorname{kernel} f_1$ or $w \in \operatorname{kernel} f_2$. It is easy to show that every bilinear form on $\mathbf{V}$ is a linear combination of "*tensor products*," i.e., of bilinear forms of the type $f_1 \otimes f_2$. This (for finite dimensional vector spaces) is the quickest approach to the tensor product of two vector spaces, *cf.* [10], p. 40 or [12], pp. 438–441. If $f_0, f_1, \ldots, f_n$ is a basis of $\mathbf{V}^t$, then a linear combination $\Sigma\alpha_{ij}(f_i \otimes f_j)$ is a bilinear form inducing a correlation of $\bar{V}$ if and only if the matrix (α_{ij}) is nonsingular.

Example 3. Let * be an automorphism of $\mathbf{F}$, and define $b:\mathbf{V}_n(\mathbf{F}) \times \mathbf{V}_n(\mathbf{F}) \to \mathbf{F}$ by

$$(\langle\alpha_1, \ldots, \alpha_n\rangle, \langle\beta_1, \ldots, \beta_n\rangle)b = \Sigma\alpha_i\beta_i^* .$$

This is a *-form which reduces to the standard hermitian product if $\mathbf{F}$ is the field of complex numbers and * is complex conjugation. This *-form induces a correlation which is "projective" if and only if * is the identity.

Example 4. Bilinear forms and the *-form of example 3 can be combined as follows. Let B be an $(n + 1)$ by $(n + 1)$ matrix with entries in $\mathbf{F}$, and let * be an automorphism of $\mathbf{F}$. The function b defined by

$$(\langle \alpha_0, \ldots, \alpha_n \rangle, \langle \beta_0, \ldots, \beta_n \rangle)b = (\alpha_0, \ldots, \alpha_n)B(\beta_0^*, \ldots, \beta_n^*)^t$$

is a *-form which induces a correlation of $P_n(\mathbf{F})$ if and only if B is nonsingular. Some tedious computations with coordinate mappings show that this type of *-form is representative of all *-forms.

6.10. *Quadrics and polarities*

We complete our study of projective geometry with an introduction to projective theory of conics and their higher dimensional analogs, quadrics. We first define quadrics analytically, present some elementary results and intuitive examples, and then turn to a more geometric way of defining quadrics —based on the pole, polar relation—and show that the two methods of attack are equivalent. In this section we keep the field of scalars arbitrary except that we avoid characteristic 2, i.e., *in all theorems of this section we assume that the field of scalars does not have characteristic 2.* In the next section we present the sharper results that are possible if one restricts the field of scalars to be the field of real numbers.

Definition. A subset, $\mathcal{Q}$, of an n-dimensional projective space $\bar{V}$ is called a *quadric* (*conic* if $n = 2$) if for some coordinate system, $[\phi]:\bar{V} \to P_n(\mathbf{F})$, there is a matrix B such that $(\mathcal{Q})\phi$ is the solution set of the equation $XBX^t = 0$, i.e., of the equation $\sum_{i,j=0}^{n} b_{ij}x_i x_j$. In other words, a quadric is simply the graph of a homogeneous quadratic equation. We call the matrix B *a representative* of $\mathcal{Q}$ in the ϕ-coordinate system.

First we should liberate this definition from its dependence on a particular coordinate system.

Proposition 6.64. The graph in $\bar{V}$ of a homogeneous quadratic equation in one coordinate system is also the graph of a homogeneous quadratic equation in any other coordinate system of $\bar{V}$. More precisely, if B represents the quadric $\mathcal{Q}$ in the ϕ-coordinate system and P is the matrix converting θ-coordinates to ϕ-coordinates (*Caution!*), then PBP^t represents $\mathcal{Q}$ in the θ-coordinate system.

Proof. Given $[v] \in \bar{V}$, $[v] \in \mathcal{Q}$ if and only if $(v)\phi B((v)\phi)^t = 0$. However, $(v)\phi = (v)\theta P$, so $[v] \in \mathcal{Q}$ if and only if $((v)\theta P)B((v)\theta P)^t = 0$. Since $((v)\theta P)^t = P^t((v)\theta)^t$, $\mathcal{Q}$ is therefore the graph of $XPBP^tX = 0$ in the θ-coordinate system.

Note that we called B a representative, not *the* representative of $\mathcal{Q}$ in the ϕ-coordinate system. Clearly αB is another representative, but worse than that, so is $\alpha B + C$ for any *skew-symmetric* matrix C, i.e., for any matrix C

such that $C^t = -C$. There is a standard result in linear algebra that a quadratic form has only one *symmetric* matrix representative in a particular coordinate system. However, a quadric may have too few points to determine the quadratic form even up to a scalar multiple, so we cannot assert that all symmetric matrices representing a given quadric in a given coordinate system are scalar multiples of each other.

Two trivial examples of quadrics should not be overlooked, $\bar{V}$ and $\emptyset$. $\bar{V}$ is always a quadric (let B be any skew-symmetric matrix), but the empty set may or may not be, depending on the field of scalars (and perhaps the dimension). For example, $\emptyset$ is the graph in $P_n(\mathbf{R})$ of $\sum_{i=1}^{n} x_i^2 = 0$, but it is not a quadric in $P_n(\mathbf{C})$.

Theorem 6.65. If $\mathbb{Q}$ is a quadric, then there is always a coordinate map θ such that in the θ-coordinate system $\mathbb{Q}$ is the graph of $\sum_{i=0}^{k} \gamma_i x_i^2 = 0, k \leq n$ and $\gamma_i \neq 0$.

Proof. Let B be the matrix representing $\mathbb{Q}$ in the ϕ-coordinate system; then $C = (B + B^t)/2$ is a symmetric matrix that also represents $\mathbb{Q}$ in ϕ-coordinates. There are several "standard" methods of determining a nonsingular matrix P such that PCP^t is a diagonal matrix with its nonzero entries appearing first. We leave it as a problem for the reader (exercise 6.10, problem 1) to consult a text on linear algebra and present a proof of this. Be careful that your proof is applicable for any field of scalars not of characteristic 2.

Thus, we have a set of simple equations for quadrics which represent all possible types of quadrics—though, of course, several of these equations may yield the same type of quadric. We now exploit these simple equations to gain some insight into the nature of quadrics, in particular to examine the various types of unusual or pathological behavior.

Example 1. An ellipse, hyperbola, or parabola in $P_2(\mathbf{R})$. The graph of the equation $\alpha^2\beta^2x_0^2 = \alpha^2x_1^2 + \beta^2x_2^2$ (α, β nonzero) contains such points as $[1, \pm\beta, 0]$, $[1, 0, \pm\alpha]$ and $[1, \beta\sin\theta, \alpha\cos\theta]$. If we choose the line at infinity to be the graph of $x_0 = 0$ and convert to affine coordinates $x = x_1/x_0$ and $y = x_2/x_0$, then the equation becomes $1 = (x^2/\beta^2) + (y^2/\alpha^2)$, the familiar equation of an ellipse. However, if we choose as the line at infinity the graph of $x_1 = \beta x_0$ or of $x_1 = 0$ and make similar conversions to affine coordinates, then the affine curve is a parabola or a hyperbola (exercise 6.10, problem 2).

Example 2. A pair of lines in $P_2(\mathbf{Q})$. The graph of the equation $x_1^2 - x_2^2 = 0$ in $P_2(\mathbf{Q})$ consists of all points $[x_0, x_1, x_2]$ with $x_1 = \pm x_2$. This is an example of a degenerate conic. Note that it consists of all points on the lines joining $[1, 0, 0]$ to the two points of the conic on the line $\bar{H}$, with equation $x_0 = 0$. Viewed as a subset of $\bar{H}$, these two points form a quadric.

Example 3. Hyperboloid of one sheet or hyperbolic paraboloid. The graph of $x_0^2 - x_1^2 + x_2^2 - x_3^2 = 0$ contains all points on the line with equations $x_0 = x_1$, $x_2 = x_3$, as well as all points on the line with equations $x_0 = x_1$, $x_2 = -x_3$. Thus, the point $[1, 1, 0, 0]$ lies on two lines entirely contained in the quadric. This is true of any point on the quadric, i.e., the quadric is ruled. If we take the graph of $x_0 = 0$ as our plane at infinity, then the affine form is $1 = x^2 - y^2 + z^2$, the equation of hyperboloid of one sheet (Figure 41). If, however, our plane at infinity has equation $x_0 = x_1$, then the affine form with $x = (x_0 + x_1)/(x_0 - x_1)$, $y = x_2/(x_0 - x_1)$, and $z = x_3/(x_0 - x_1)$ is $x = z^2 - y^2$, the equation of a hyperbolic paraboloid (Figure 41). The familiar string models of these two surfaces illustrate the manner in which they are ruled.

Example 2 above is an example of an "awkward" type of quadric. We now define proper and degenerate quadrics and prove that the study of degenerate quadrics can be reduced to the study of proper quadrics in lower dimensions.

Definition. A quadric $\mathbb{Q}$ is *degenerate* if there is a symmetric matrix, B, representing $\mathbb{Q}$ such that B is singular, i.e., $\det B = 0$. It is *proper* if it has a nonsingular symmetric matrix representative.

Caution! It is not obvious whether or not a quadric could be both degenerate and proper. We suspect not, but since it is not crucial to our treatment of quadrics, we do not attempt to prove it. The reader is invited to investigate on his own.

Once again we have a definition that seems to depend on the coordinate system chosen; however, this is easily remedied.

Proposition 6.66. A quadric is degenerate (respectively proper) in one coordinate system if and only if it is degenerate (respectively proper) in every coordinate system.

Proof. Let B be a symmetric matrix representing $\mathbb{Q}$ in the ϕ-coordinate system, and let P be the matrix converting θ-coordinates to ϕ-coordinates. Then PBP^t is symmetric and represents $\mathbb{Q}$ in the θ-coordinate system, but $\det P \neq 0$; hence $\det(PBP^t) = (\det P)^2(\det B)$ is zero if and only if $\det B$ is zero.

Theorem 6.67. If $\mathbb{Q}$ is a degenerate quadric, then there is a (non-empty!) subspace $\bar{W}_1$ such that either $\mathbb{Q} = \bar{W}_1$ or else there is a complementary subspace $\bar{W}_2$ and a proper quadric, $\mathbb{Q}' = \mathbb{Q} \cap \bar{W}_2$, in $\bar{W}_2$ such that $\mathbb{Q} = \{P | P \in Q + R, Q \in \bar{W}_1, R \in \mathbb{Q}'\}$.

the empty set

$x_0^2 + x_1^2 + x_2^2 + x_3^3 = 0$, Index $= -1$

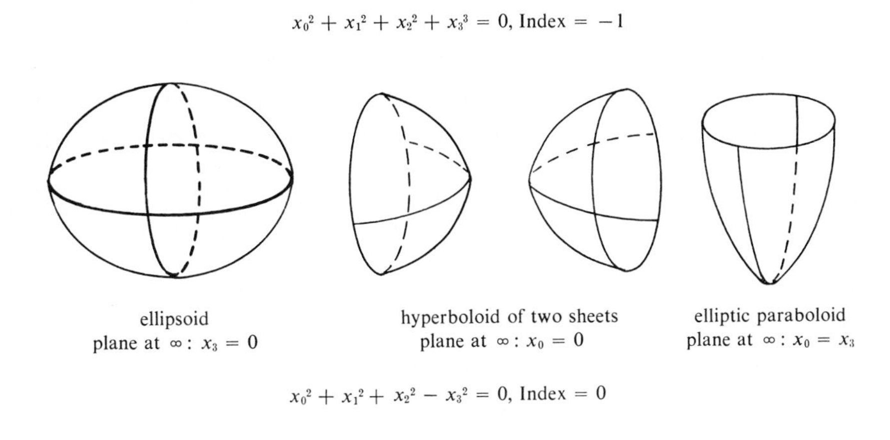

ellipsoid
plane at $\infty : x_3 = 0$

hyperboloid of two sheets
plane at $\infty : x_0 = 0$

elliptic paraboloid
plane at $\infty : x_0 = x_3$

$x_0^2 + x_1^2 + x_2^2 - x_3^2 = 0$, Index $= 0$

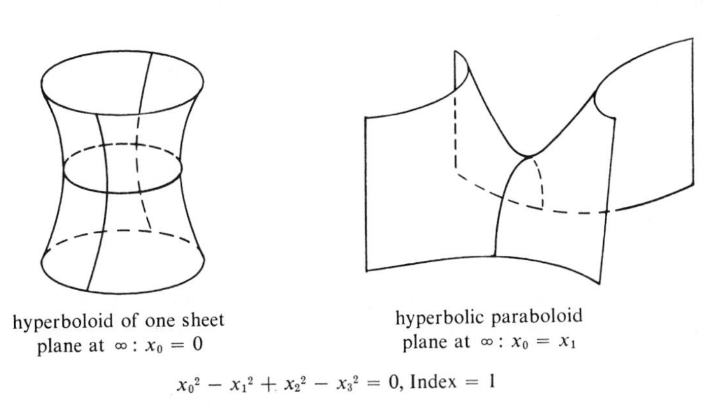

hyperboloid of one sheet
plane at $\infty : x_0 = 0$

hyperbolic paraboloid
plane at $\infty : x_0 = x_1$

$x_0^2 - x_1^2 + x_2^2 - x_3^2 = 0$, Index $= 1$

FIG. 41. Proper quadrics in real affine space.

Proof. Let B be a symmetric singular matrix representing $\mathcal{Q}$. Diagonalize B, i.e., choose a coordinate system θ such the $\mathcal{Q}$ is the θ-graph of the equation $\gamma_0 x_0^2 + \cdots + \gamma_k x_k^2 = 0$ with $\gamma_i \neq 0$, and $PBP^t = \text{diag}(\gamma_0, \ldots, \gamma_k, 0, 0, \ldots, 0)$. Since B is singular, so is the diagonal matrix; hence it must have some 0's on its diagonal, i.e., $k < n$. Let $\mathbf{W}_1$ be the (linear) subspace spanned by the last $n - k$ basis vectors in the θ-system. $\overline{W}_1 \neq \varnothing$ and $\overline{W}_1 \subseteq \mathcal{Q}$. Either $\overline{W}_1 = \overline{V}$ or else we can let $\mathbf{W}_2$ be the complementary (linear) subspace spanned by the first $k + 1$ θ-basis vectors. The quadric $\mathcal{Q} \cap \overline{W}_2$ is a proper quadric in $\overline{W}_2$ since its matrix representative in the θ-coordinates (restricted to $\mathbf{W}_2$) is just $\text{diag}(\gamma_0, \ldots, \gamma_k)$ with determinant $\gamma_0 \gamma_1 \cdots \gamma_k \neq 0$. Thus, $\mathcal{Q}$ is the set of points on lines joining points in $\overline{W}_1$ to points in $\mathcal{Q} \cap \overline{W}_2$.

Looking back at our examples, we see that the conic in example 1 is proper, as is the quadric in example 3. The conic in example 2 is degenerate and consists of all points on lines joining [1, 0, 0] (the subspace $\bar{W}_1$) to the two-point graph of $x_1^2 = x_2^2$ in the hyperplane, $\bar{H}$, with equation $x_0 = 0$. Since x_1, x_2 are acceptable coordinates in $\bar{H}$ and $\det \begin{bmatrix} 1 & 0 \\ 0 & -1 \end{bmatrix} \neq 0$, these two points constitute a proper quadric in $\bar{H}$. The proper quadrics in low dimensions are considered further in exercise 6.10, problem 3. These determine the degenerate quadrics whose *subspace of degeneracy*, $\bar{W}_1$ in the result above, is of dimension $n - 1$ or $n - 2$ (exercise 6.10, problem 4).

From now on we restrict ourselves to proper quadrics. If $\mathbb{Q}$ is a proper quadric represented, say, by the nonsingular symmetric matrix B in ϕ-coordinates, then the bilinear form, $(v, w)b = (v)\phi B((w)\phi)^t$, induced on $\mathbf{V}$ satisfies the conditions of theorem 6.63 and hence induces a correlation, σ. Since B is symmetric, the bilinear form b is symmetric, $(v, w)b = (w, v)b$, and hence (as we prove later) the correlation σ is of order 2. Moreover, $[v] \in \mathbb{Q}$ if and only if $[v] \in [v]\sigma$. Since we are not sure that σ is the only correlation induced by all matrix representatives of $\mathbb{Q}$, we are in a rather awkward spot. Perhaps the best manner of recovery is to use these comments as a guide and exploit correlations of order 2 to define quadrics in an intrinsic, i.e., coordinate-free, way. We now turn to this, show that not all correlations of order 2 define proper quadrics, and formulate the restrictions that one must impose. *Caution!* Since we use the results of the last section on correlations, we assume that $\dim \bar{V} \geq 2$ as well as characteristic $\mathbf{F} \neq 2$.

Proposition 6.68. Let ϕ be a correlation of $\bar{V}$ represented by the semi-bilinear form b, i.e., $(\bar{W})\phi = \{[v] | (v, w)b = 0 \text{ for all } w \in \mathbf{W}\}$. The order of ϕ is 2 if and only if $(v, w)b = 0$ implies $(w, v)b = 0$.

Proof. If ϕ is of order 2 and $(v, w)b = 0$, then $[v] \in [w]\phi$. Since ϕ reverses inclusions, this implies $[w] = [w]\phi\phi \in [v]\phi$. But this means that $(w, v)b = 0$, so the condition on b is necessary.

Now assume $(v, w)b = 0$ implies $(w, v)b = 0$. If $\bar{W}$ is a proper subspace of $\bar{V}$, then for every $[w] \in \bar{W}$, $(v, w)b = (w, v)b = 0$ for every $[v] \in (\bar{W})\phi$. However, $(w, v)b = 0$ for every $[v] \in (\bar{W})\phi$ means that $[w] \in (\bar{W})\phi\phi$; hence $\bar{W} \subseteq (\bar{W})\phi\phi$. By theorem 6.59 $\dim \bar{W} = \dim (\bar{W})\phi\phi$; hence $\bar{W} = (\bar{W})\phi\phi$ and therefore $\phi\phi = \mathbf{1}$.

We cannot study the fixed points of a correlation, ϕ, except in the rather unnatural case of subspaces of dimension $(n - 1)/2$, but we can study those points $[w]$ such that $[w] \in [w]\phi$. Thus, we are led to the following definition.

Definition. Let ϕ be a correlation of $\bar{V}$. We say that ϕ is *symplectic on a subspace* $\bar{W}$ if $[w] \in [w]\phi$ for every $[w] \in \bar{W}$. If ϕ is symplectic on $\bar{V}$ itself, then we simply say that ϕ is *symplectic*.

Proposition 6.69. Let ϕ be a correlation of $\bar{V}$ represented by the semi-bilinear form b. ϕ is symplectic on a subspace $\bar{W}$ if and only if $(w, w)b = 0$ for all $w \in \mathbf{W}$. Moreover, if ϕ is symplectic on $\bar{W}$, then $(w_1, w_2)b = -(w_2, w_1)b$ for all $w_1, w_2 \in \mathbf{W}$.

Before proving this, we comment that we have stated the proposition in this awkward fashion so that it is valid for characteristic 2. If we insist that the characteristic is not 2, then we could simply say that ϕ is symplectic on $\bar{W}$ if and only if b is skew-symmetric on $\mathbf{W}$. In characteristic 2, $(w, w)b = -(w, w)b$ does not imply $(w, w)b = 0$.

Proof. If ϕ is represented by b, then $[w] \in [w]\phi$ if and only if $(w, w)b = 0$; hence ϕ is symplectic on $\bar{W}$ if and only if $(w, w)b = 0$ for all $w \in \mathbf{W}$. If $(w, w)b = 0$ for all $w \in \mathbf{W}$, then for all $w_1, w_2 \in \mathbf{W}$,

$$0 = (w_1 + w_2, w_1 + w_2)b = (w_1, w_1)b + (w_1, w_2)b + (w_2, w_1)b + (w_2, w_2)b = (w_1, w_2)b + (w_2, w_1)b \,,$$

which proves the "moreover" clause.

A semi-bilinear form, b, on $\mathbf{V}$ such that $(v, v)b = 0$ for all $v \in \mathbf{V}$ and yet b induces a correlation (theorem 6.63) is called a *null system* on $\mathbf{V}$. The proof of the following theorem is modeled on the proof appearing in [2], pp. 106–109. For another proof depending more on coordinates, see [4].

Theorem 6.70 (characterization of symplectic correlations). If the correlation ϕ is symplectic and b is the semi-bilinear form representing ϕ, then:

(1) b is a null system on $\mathbf{V}$, i.e., $(v, v)b = 0$ for all $v \in \mathbf{V}$.
(2) b is skew-symmetric and bilinear.
(3) ϕ is of order 2.
(4) $\dim \bar{V} = n$ is odd, and there is a basis $w_0, w_1, \ldots, w_n$ of $\mathbf{V}$ such that $(w_{2i}, w_{2i+1})b = 1$, but otherwise $(w_i, w_j)b = 0$.

Proof. If ϕ is a symplectic correlation represented by b, then it follows immediately from the proposition above that b is a null system and is skew-symmetric. By theorem 6.63 we can find vectors v_1, v_2 in $\mathbf{V}$ such that $(v_1, v_2)b \neq 0$. Adjust v_1 if necessary such that $(v_1, v_2)b = 1$. If $*$ is the automorphism of $\mathbf{F}$ belonging to b, then for any $\alpha \in \mathbf{F}$ we have:

$$\alpha = (\alpha v_1, v_2)b = -(v_2, \alpha v_1)b = -\alpha^*(v_2, v_1)b = \alpha^*(v_1, v_2)b = \alpha^* \,,$$

and thus, b is bilinear. This completes the proof of (1) and (2).

Since b is skew-symmetric, $\phi\phi = \mathbf{1}$ by proposition 6.68, so (3) is proved.

We construct the basis for (4) inductively. Choose any $w_0 \neq 0$ as the first basis vector. Since (theorem 6.63) $(w_0, v)b \neq 0$ for some v, we can choose a second basis vector, w_1, such that $(w_0, w_1)b = 1$. Let $\mathbf{W}_1$ be the subspace of $\mathbf{V}$

spanned by w_0 and w_1. If $\bar{W}_1 = \bar{V}$, then stop, but if not, let w_2 be any vector representing a point in $(\bar{W}_1)\phi$. $\bar{W}_1 \cap (\bar{W}_1)\phi = \varnothing$ since $(\alpha w_0 + \beta w_1, w)b \neq 0$ for at least one of the choices $w = w_0$ or w_1 if $\alpha w_0 + \beta w_1 \neq 0$. But then $\bar{W}_1 + (\bar{W}_1)\phi$ cannot be contained in any hyperplane (dimension count!). In particular, $(\bar{W}_1)\phi \nsubseteq [w_2]\phi$ since $[w_2] \in (\bar{W}_1)\phi$ implies $(\bar{W}_1) \subseteq [w_2]\phi$. Therefore, there is a point $[v_3]$ in $(\bar{W}_1)\phi$ but not in $[w_2]\phi$, i.e., $(w_i, v_3)b = 0$ for $i = 0$ or 1, but $(w_2, v_3)b = \alpha \neq 0$. Let $w_3 = (1/\alpha)v_3$, and let $\mathbf{W}_2$ be the subspace spanned by w_0, w_1, w_2, w_3.

Once again $(\Sigma\alpha_i w_i, w_i)b = \alpha_i$, so $\bar{W}_2 \cap (\bar{W}_2)\phi = \varnothing$. Thus, if $\bar{W}_2 \neq \bar{V}$, we can repeat this process to find w_4 and w_5. An obvious formalization by induction is possible, and thus (4) is proved.

Corollary 6.71. If ϕ is a symplectic correlation, then there are k lines, $L_1, L_2, \ldots, L_k$, with $k = (\dim \bar{V} + 1)/2$ such that $\bar{V}$ is the join of $L_1, L_2, \ldots, L_k$, and $(L_j)\phi$ is the join of the other $k - 1$ lines, $j = 1, 2, \ldots, k$.

Proof. Let $L_{i+1} = [w_{2i}] + [w_{2i+1}]$, $i = 0, 1, \ldots, k - 1$, with w_i chosen as in part (4) of the theorem.

Corollary 6.72. If ϕ_1 and ϕ_2 are both symplectic correlations of $\bar{V}$, then there is a *projective* transformation, θ, such that $\phi_2 = \theta^{-1}\phi_1\theta$.

Proof. Choose bases $w_0, \ldots, w_n$ and $w'_0, \ldots, w'_n$ for $\mathbf{V}$ to conform to part (4) of the theorem. Let $f{:}\mathbf{V} \to \mathbf{V}$ be the linear transformation sending w_j to w'_j for $j = 0, 1, \ldots, n$. Then $\bar{f}$ is the desired projective transformation.

We did not attempt to use the k skew lines of corollary 6.71 in this proof since the action of ϕ on these lines is not quite enough to determine ϕ. We need to know what happens to one more point.

The symplectic correlations are therefore one type of correlation we must avoid as we attempt to define a proper quadric in a coordinate-free manner. Moreover, we do not have to be concerned about them if the dimension is even; in particular, symplectic correlations do not exist in projective planes over a field (even of characteristic 2 since we carefully included that case in proposition 6.69 and in theorem 6.70). The following theorem identifies the only other type of correlation we must avoid.

Theorem 6.73. If ϕ is a correlation of order 2, then ϕ is either symplectic or else ϕ can be represented by a *-symmetric, *-form in which the automorphism, *, is either **1** or of order 2. Conversely, any symmetric bilinear or *-symmetric *-form (with * an involution) which meets the conditions of theorem 6.63 represents a correlation of order 2 that is not symplectic.

Proof. If ϕ is symplectic, we are finished. If not, let b_1 be any *-form representing ϕ. By theorem 6.63 we can find w such that $(w, w)b = \alpha \neq 0$. The

*-form $b = (1/\alpha)b_1$ also represents ϕ; moreover, $(w, w)b = 1$. Let $\bar{H} = [w]\phi$.

We first show that b is *-symmetric on the (linear) hyperplane **H**. Given $x_1, x_2 \in \mathbf{H}$, let $(x_1, x_2)b = \gamma$. Then $(\gamma w + x_1, w - x_2)b = \gamma - \gamma = 0$. Since ϕ is of order 2, it follows (proposition 6.68) that $0 = (w - x_2, \gamma w + x_1)b = \gamma^* - (x_2, x_1)b$. Thus, $((x_1, x_2)b)^* = (x_2, x_1)b$, i.e., b is *-symmetric.

Next we show that $** = \mathbf{1}$, i.e., that * is either **1** or an involution on the field of scalars. Clearly, $\bar{H} \nsubseteq (\bar{H})\phi$ since $(\bar{H})\phi = [w]$, and so we can choose x_1, x_2 in **H** such that $(x_1, x_2)b = 1$. Given $\beta \in \mathbf{F}$,

$$\begin{aligned}\beta = \beta(x_1, x_2)b = (\beta x_1, x_2)b &= ((x_2, \beta x_1)b)^* = (\beta^*(x_2, x_1)b)^* \\ &= \beta^{**}((x_2, x_1)b)^* = \beta^{**}((x_1, x_2)b)^{**} = \beta^{**}1^{**} = \beta^{**}.\end{aligned}$$

Thus, $** = \mathbf{1}$ on **F**.

Finally we show that b is *-symmetric on all of **V**. Given v_1, v_2 in **V**, since $\bar{V} = [w] + \bar{H}$, there are scalars α_1, α_2 and vectors x_1, x_2 in **H** such that $v_i = \alpha_i w + x_i$, $i = 1, 2$. On the one hand, $(v_1, v_2)b = \alpha_1\alpha_2^* + (x_1, x_2)b$; on the other hand, $(v_2, v_1)b = \alpha_2\alpha_1^* + (x_2, x_1)b$. Since $\alpha_i^{**} = \alpha_i$ and $(x_2, x_1)b = ((x_1, x_2)b)^*$, it clearly follows that b is *-symmetric.

The proof of the converse is left to the reader (exercise 6.10, problem 5).

We can paraphrase the theorem above as follows. Nonsymplectic correlations of order 2 are either represented by symmetric bilinear forms—in analogy with the ordinary dot product—or else by *-symmetric forms analogous to the standard hermitian product in complex vector spaces.

Definition. A *polarity* of $\bar{V}$ is a projective, nonsymplectic correlation, ϕ, of order 2. If P is a point and $\bar{H}$ a hyperplane, then we call $\bar{H}$ the *polar* of P and P the *pole* of $\bar{H}$ if and only if $(P)\phi = \bar{H}$ or, equivalently, $(\bar{H})\phi = P$. Two points P and Q such that $P \in (Q)\phi$ [and hence $Q \in (P)\phi$] are said to be *conjugate points* with respect to ϕ. The set of all *self-conjugate* points, i.e., all points P such that $P \in (P)\phi$, is called the *quadric induced by* ϕ.

Theorem 6.74. The quadric induced by a polarity is a bona fide, proper quadric. Conversely, every proper quadric is induced by at least one polarity.

Proof. If σ is a polarity of $\bar{V}$, then let b be the *-symmetric form representing σ. Since σ is projective, $* = \mathbf{1}$; hence b is a bilinear form. Let ϕ be any coordinate map of **V**. The matrix, B, representing b in ϕ-coordinates is clearly symmetric and is nonsingular by theorem 6.63. Thus, $\mathcal{Q}$, the set of self-conjugate points of σ, is a proper quadric represented by B.

The converse follows easily from the comments preceding proposition 6.68.

Even though a polarity, σ, is never symplectic, it may very well be symplectic on a subspace $\bar{W}$. Recall the hyperboloid of one sheet! By definition, σ is symplectic on $\bar{W}$ if and only if $\bar{W}$ is entirely contained in the quadric induced by σ. As our final result, we show that this is equivalent to $\bar{W} \subseteq (\bar{W})\sigma$ so that, in particular, a polarity is never symplectic on a subspace of dimension greater than $[(n - 1)/2]$.

Definition. Let ϕ be a polarity. A subspace, $\bar{W}$, is *isotropic* with respect to ϕ if $\bar{W} \subseteq (\bar{W})\phi$.

The word isotropic (same in all directions) is used since $\bar{W} \subseteq (\bar{W})\phi$ implies that a vector in **W** is "perpendicular" to all vectors in **W**, regardless of their "direction."

Theorem 6.75. If $\mathbb{Q}$ is the quadric induced by a polarity ϕ, then $\bar{W} \subseteq \mathbb{Q}$ if and only if $\bar{W} \subseteq (\bar{W})\phi$.

Proof. Clearly $\bar{W} \subseteq (\bar{W})\phi$ implies $\bar{W} \subseteq \mathbb{Q}$. If $\bar{W} \subseteq \mathbb{Q}$, then ϕ is symplectic on $\bar{W}$, and hence by proposition 6.69, $(w_1, w_2)b = -(w_2, w_1)b$ for all vectors w_1, w_2 in **W** and *any* semi-bilinear form representing ϕ. But ϕ is a polarity; hence there is a symmetric bilinear form representing ϕ. For this bilinear form we also have $(w_1, w_2)b = (w_2, w_1)b$, and since characteristic $\mathbf{F} \neq 2$, this means that $(w_1, w_2)b = 0$ for all $w_1, w_2 \in \mathbf{W}$, i.e., $\bar{W} \subseteq (\bar{W})\phi$.

Definition. The *index* of a proper quadric $\mathbb{Q}$ is the dimension of the largest subspace contained in $\mathbb{Q}$ (and is -1 if $\mathbb{Q} = \varnothing$). In view of the result above, the index can also be defined as max $\{\dim \bar{W} | \bar{W} \subseteq (\bar{W})\phi\}$ for *any* polarity ϕ that induces the quadric $\mathbb{Q}$.

For further results on quadrics, we refer the reader to either [1] or [2]. Unfortunately, the terminology in each reference differs slightly from the way it is used in this book. One of the interesting results just out of our reach is that each point on a quadric lies on an isotropic subspace whose dimension is the index of $\mathbb{Q}$, *cf.* theorem 3.10 in [1] or corollary 4, p. 121, of [2].

As we mentioned in paraphrasing theorem 6.73, polarities are the "reasonable" generalization of the ordinary dot product which we use to define distance in euclidean space. Thus, we have come full circle to meet again the topic that we studied in Chapter 2, because the study of the symmetries of euclidean space generalizes to the study of the linear transformations leaving a particular symmetric bilinear form invariant, or, in projective terms, to the study of projectivities commuting with a given correlation. This study of "classical groups" is beyond our scope so we must leave it to other texts and other courses.

Exercise 6.10

1. Find and write up (with references) a proof that any quadratic form can be diagonalized if the characteristic is not 2. In other words, prove that for any symmetric matrix B there is a nonsingular matrix P such that PBP^t is diagonal.

2. Show that the conic in $P_2(\mathbf{R})$ represented (in natural coordinates) by

$$\begin{bmatrix} -\alpha^2\beta^2 & 0 & 0 \\ 0 & \alpha^2 & 0 \\ 0 & 0 & \beta^2 \end{bmatrix}$$

with α, β nonzero has as its affine equation:

(1) $\alpha^2 x = \beta^2 y^2$ if we use the graph of $x_1 = \beta x_0$ as the line at infinity and let $x = (\beta x_0 + x_1)/(\beta x_0 - x_1)$ and $y = x_2/(\beta x_0 - x_1)$.

(2) the equation of a hyperbola if $x_1 = 0$ is the equation of the line at infinity.

3. Determine all proper quadrics in zero- or one-dimensional spaces.

4. Using the result above, show that the only degenerate quadrics whose subspace of degeneracy is of dimension $n - 1$ or $n - 2$ consist of all points on either one or two hyperplanes.

5. Let b be a *-symmetric *-form on $\mathbf{V}$ such that $** = \mathbf{1}$. Assume, in addition, that $(v, w)b = 0$ for all $w \in \mathbf{V}$ implies $v = 0$. Show that b induces a correlation of order 2 that is not symplectic.

6. Let $\mathcal{Q}$ be the proper quadric induced by a correlation, ϕ, of a projective space. Let L be a line such that $L \cap \mathcal{Q} = \{P, Q\}$, $P \neq Q$. Show that if $S \in L$ but $S \notin \mathcal{Q}$, then $R = L \cap (S)\phi$ if and only if R is the harmonic conjugate of S with respect to P and Q.

6.11. Real projective spaces

The field of real numbers has a much more rigid, yet at the same time, richer structure than most fields. It is rigid in the sense that it has no nontrivial automorphisms. It is richer in that any polynomial can be factored into irreducible factors of degree at most 2, and then, of course, it also has the very powerful completeness property. Thus, when we specialize from projective spaces over more or less arbitrary fields to real projective spaces, we are able to state much more powerful results. Perhaps the most powerful is the specialization of the fundamental theorem to real projective spaces.

Theorem 6.76. Every collineation or correlation of an n-dimensional real projective space ($n \geq 2$) is a projectivity or a projective correlation. Thus, a collineation is completely determined by its action on $n + 2$ points, provided every $n + 1$ of them is independent.

Since every correlation is projective, the only correlations of order 2 which we must avoid when considering polarities are the symplectic correlations, and if the dimension is even, we need not worry about them. Thus, it is easy to see the natural simplicity of the pole-polar definition of conics in the real projective plane. This approach to conics was mapped out by Von Staudt.

We can state better results concerning the index of a quadric in real projective spaces.

Theorem 6.77. If $\mathcal{Q}$ is a proper quadric in an n-dimensional real projective space, there is a coordinate map, ϕ, such that $\mathcal{Q}$ is the ϕ-graph of the equation

$$x_0^2 + x_1^2 + \cdots + x_k^2 = x_{k+1}^2 + \cdots + x_n^2 .$$

In other words, the equation of $\mathcal{Q}$ in at least one coordinate system is in the form: the sum of the squares of $k + 1$ coordinates equals the sum of the squares of the other $j + 1$ coordinates, $j + k = n - 1$. The smaller of the two integers j and k is the index of $\mathcal{Q}$. The larger is the greatest integer, i, such that for some subspace $\overline{W}$, dim $\overline{W} = i$ and $\overline{W} \cap \mathcal{Q} = \varnothing$. Thus, two proper quadrics are projectively related if and only if they have the same index.

Proof. By theorem 6.65 there is a coordinate system in which the equation for $\mathcal{Q}$ is a mixed sum and difference of perfect squares. By permuting the basis vectors, we can arrive at a coordinate system in which the equation for $\mathcal{Q}$ is $\gamma_0^2 x_0^2 + \cdots + \gamma_k^2 x_k^2 - \gamma_{k+1}^2 x_{k+1}^2 - \cdots - \gamma_n^2 x_n^2 = 0$. By multiplying the ith basis vector by $1/\gamma_i$, $i = 0, 1, \ldots, n$, we arrive at a coordinate system, ϕ, in which the equation for $\mathcal{Q}$ is of the desired form, let us say with $k \leq n - k - 1 = j$.

The ϕ-graph of the $k + 1$ equations $x_0 = x_1 = \cdots = x_k = 0$ is a subspace, $\overline{W}_1$, of dimension j (theorem 6.59) which clearly contains no points of $\mathcal{Q}$. On the other hand, the ϕ-graph, $\overline{W}_2$, of the $j + 1$ equations $x_0 = x_{k+1}$, $x_1 = x_{k+2}, \ldots, x_k = x_{2k+1}, x_{2k+2} = 0, \ldots, x_n = 0$ is a subspace of dimension k which is entirely contained in $\mathcal{Q}$. Since $j + k = n - 1$, theorem 6.4 yields the following two results:

(1) Every subspace of dimension larger than k has points in common with the j-dimensional subspace $\overline{W}_1$.

(2) Every subspace of dimension larger than j has points in common with the k-dimensional subspace $\overline{W}_2$.

Since $\overline{W}_1 \cap \mathcal{Q} = \varnothing$, it is clear from 1 that the index of $\mathcal{Q}$ is at most k, but $\overline{W}_2 \subseteq \mathcal{Q}$, so the index must be k. Similarly, since $\overline{W}_2 \subseteq \mathcal{Q}$, it is clear from 2 that no subspace of dimension greater than j could be disjoint from $\mathcal{Q}$. However, $\overline{W}_1 \cap \mathcal{Q} = \varnothing$, so j is the maximum dimension of subspaces disjoint from $\mathcal{Q}$.

Finally, in view of the geometric nature of the index, it is clear that if ϕ is any collineation (automatically projective in *real* projective spaces) of $\overline{V}$ sending the proper quadric $\mathcal{Q}$ to the proper quadric $\mathcal{Q}'$, then $\mathcal{Q}$ and $\mathcal{Q}'$ must have the same index. Conversely, if $\mathcal{Q}$ and $\mathcal{Q}'$ have the same index, then we can find two (linear) coordinate systems such that the equation of $\mathcal{Q}$ in one is the same as the equation of $\mathcal{Q}'$ in the other. The linear transformation converting the first coordinate system to the second induces a projectivity carrying $\mathcal{Q}$ to $\mathcal{Q}'$. In other words, the algebraic nature of the index guarantees that quadrics with the same index are projectively equivalent.

With this fundamental result, we can classify all proper quadrics in real projective spaces. For low dimensions, the results are as follows.

Dimension	*Index*	*Type of proper quadric*
0	−1	∅
1	−1	∅
1	0	two distinct points
2	−1	∅
2	0	ellipse
3	−1	∅
3	0	ellipsoid
3	1	ruled quadric— hyperboloid of one sheet

The last two examples are interesting in terms of the largest subspaces disjoint from the quadric. The largest subspace not intersecting an ellipsoid is of dimension $(3 - 1) - 0 = 2$, whereas for a hyperboloid of one sheet it is a line since $(3 - 1) - 1 = 1$.

If we combine the results above with theorem 6.67, then we can also classify the degenerate quadrics. For low dimensions the degenerate quadrics *not of the form* $\bar{W}_1$, i.e., $\mathcal{Q}$ = the set of all points on lines joining points in $\bar{W}_1$ to points in a non-empty *proper* quadric, $\mathcal{Q}'$, in a complementary subspace $\bar{W}_2$, are as follows:

dim $\bar{V}$	$\bar{W}_1$	$\bar{W}_2$	$\mathcal{Q}' \neq \varnothing$	*The degenerate quadric* $\mathcal{Q}$
0				
1	a point	a point		
2	a point	a line	two points	two lines
2	a line	a point		
3	a point	a plane	ellipse	cone
3	a line	a line	two points	two planes
3	a plane	a point		

Thus, the "interesting" degenerate quadrics in projective 3-space are the cones and the "two planes." Remember that any single subspace, $\bar{W}$, of dimension $k = 0, 1, \ldots, n$ is also a quadric in an n-dimensional real projective space (exercise 6.11, problem 4).

Some final comments on the connections between euclidean, affine, and projective spaces are in order here since they involve quadrics and polarities in real projective spaces. The group of collineations of a real affine n-space is isomorphic to the group of collineations of a real projective n-space leaving a hyperplane, $\bar{H}$, globally invariant; moreover, the most natural subgroup of the affine group, the group of dilatations, corresponds to the group of projective collineations leaving $\bar{H}$ pointwise invariant. The connection between the symmetries (similarities) of euclidean n-space and those of projective

n-space is a little more complicated. To see it we embed euclidean n-space in $P_n(\mathbf{R})$ using coordinates such that the hyperplane at infinity, $\bar{H}$, is the graph of $x_0 = 0$. Each euclidean line, m, say with direction $[v] = [0, \alpha_1, \ldots, \alpha_n]$, determines a unique point, $[v]$, in $\bar{H}$. The similarities are those affine collineations that preserve perpendicularity. To put this in projective terms we first observe that if the direction of m is $[0, \alpha_1, \ldots, \alpha_n]$ and the direction of n is $[0, \beta_1, \beta_2, \ldots, \beta_n]$ then $m \perp n$ if and only if $\Sigma\alpha_i\beta_i = 0$. The bilinear form, b, defined by $(\langle\alpha_1, \ldots, \alpha_n\rangle\langle\beta_1, \ldots, \beta_n\rangle)b = \Sigma\alpha_i\beta_i$ is the canonical symmetric bilinear form and therefore induces a polarity, θ, on $\bar{H}$ which has no self-conjugate points. A projective collineation, ϕ, leaving $\bar{H}$ globally invariant corresponds to a similarity if and only if its restriction to $\bar{H}$ commutes with θ. Thus the similarities group is isomorphic to the group of projective collineations commuting with a fixed polarity of the hyperplane at infinity without isotropic points.

Exercise 6.11

1. Show that if a line contains at least three points on a proper quadric, then it is entirely contained in the quadric.

2. Let $\mathcal{Q}$ be a proper quadric induced by the polarity ϕ of a real n-dimensional projective space. Show that if $P \in \mathcal{Q}$, then $(P)\phi$ is the tangent hyperplane to $\mathcal{Q}$ at the point P. To make this precise you are asked to show that $(P)\phi = \{Q | P + Q$ is a tangent line to $\mathcal{Q}\}$. A *tangent line* is a line, L, such that $L \cap \mathcal{Q}$ is either a *single* point or else L itself.

3. Let $\mathcal{Q}$ be an ellipse in the euclidean plane. Show that the midpoints of parallel chords of $\mathcal{Q}$ lie on a line passing through the center of $\mathcal{Q}$ and through the points of tangency of the two tangents parallel to these chords. *Hint.* Recall that the midpoint of a segment is the harmonic conjugate of the point at infinity on the line with respect to the two endpoints of the segment. Also, review exercise 6.10, problem 6.

4. We know that a subspace, $\bar{W}$, of $P_n(\mathbf{R})$ is a quadric in $P_n(\mathbf{R})$ since the empty set is a proper quadric in any real projective space and can therefore serve as the proper quadric $\mathcal{Q}'$ in theorem 6.67. Prove this result in a different way by showing that $\bar{W}$ is the graph of a quadratic equation. Be careful to show that the graph of your quadratic equation is not too large.

5. Let $\mathcal{Q}$ be an ellipse in the real projective plane. Show that the points not on the ellipse fall into two disjoint sets: the *interior*, consisting of all points through which no tangents pass, and the *exterior*, consisting of all points through which two tangents pass. Show that the polar of an exterior point, P, is the line joining the points of tangency of the two tangents through P.

6. Find two distinct polarities of $P_2(\mathbf{R})$, both of which induce the quadric $\mathcal{Q} = \varnothing$. Using problem 5, show that there is only one polarity inducing a given ellipse.

7. Is each proper, non-empty quadric in $P_3(\mathbf{R})$ induced by a unique polarity?

6.12. Projective spaces over noncommutative fields

In this final section we comment briefly on the changes that must be made to extend the results of this chapter to projective spaces over noncommutative fields, i.e., over algebraic systems having all the properties of a field (section 4.1), except commutativity of multiplication. The simplest example of such a system is the field of quaternions (exercise 6.12, problem 1). *Caution!* We use the term "noncommutative" to mean definitely not commutative in this section. Other authors often use it to mean not necessarily commutative.

A celebrated theorem of Wedderburn asserts that there are no finite noncommutative fields (see [1], pp. 35–37, or [8], p. 375, for a proof). Thus, the comments of this section are of no importance for projective spaces with only a finite number of points.

One basic change in the noncommutative case is that one must distinguish between a *left* vector space and a *right* vector space over a noncommutative field, for the associative law (see part 3 of axiom M, section 4.2) has two different forms. We must decide, in multiplying the vector v by the scalar $\alpha\beta$, whether the product should be: first, β times v, then α times the result, or, first, α times v, then β times the result. If $\alpha\beta \neq \beta\alpha$, then this makes a difference! There is an easy way to keep track of our decision. If we choose the first alternative above, then it is natural to put the scalar on the left when multiplying a vector by a scalar, whereas in the second alternative, it is more natural to put it on the right. Thus, the adjectives "left" and "right" in the two types of vector spaces over noncommutative fields refer to the position of scalars in products of scalars and vectors. The definition of vector space in section 4.2 is also the definition of a *left vector space* over a noncommutative field. Simply reverse *all* 12 products (including $\alpha\beta$) for the definition of a *right vector space.* A *two-sided vector space* over the field of scalars **F** is an abelian group **V** whose elements may be multiplied on the left or right by scalars in such a way that **V** is both a left and a right vector space. The simplest example of a two-sided vector space over **F** is **F** itself.

For either left or right vector spaces, one can form the associated projective spaces in exactly the same way as for vector spaces over commutative fields. The theory of projective subspaces given in section 6.2 is all valid for projective spaces over noncommutative fields.

When we consider homogeneous coordinates and the "canonical" projective space of dimension n over a noncommutative field, then the difference between left and right appears again. The set of $(n + 1)$-tuples out of **F** is, with the natural addition and multiplications by scalars, a two-sided vector space, $\mathbf{V}_n(\mathbf{F})$, over **F**. Thus, there is no problem with left or right here. However, when we collapse to a projective space, we must decide whether $[\alpha_0, \alpha_1, \ldots, \alpha_n] = [\beta_0, \ldots, \beta_n]$ means that there is a scalar $\lambda \neq 0$ such that $\lambda\alpha_i = \beta_i$

for all i or a scalar $\mu \neq 0$ such that $\alpha_i\mu = \beta_i$ for all i. This makes a difference! For example, if i, j, and k are the quaternions in exercise 6.12, problem 1 below, then $\langle 1, i, j\rangle$ and $\langle k, j, -i\rangle$ are proportional on the left but not on the right (exercise 6.12, problem 2). For this reason there are **two** standard n-dimensional projective spaces over a noncommutative field **F**, one for left vector spaces and one for right vector spaces. Similarly when one considers homogeneous equations, one must distinguish between left homogeneous of degree k and right homogeneous of degree k according as $f(tx_0, tx_1, \ldots, tx_n) = t^kf(x_0, \ldots, x_n)$ or $f(x_0t, \ldots, x_nt) = f(x_0, \ldots, x_n)t^k$. In the standard *left* n-dimensional projective space over **F**, the *left* homogeneous equations are the equations whose graphs are truly projective, i.e., are independent of the particular homogeneous coordinates used to represent a point. The *left* homogeneous linear equations are of the form $\Sigma x_i\alpha_i = 0$, i.e., the coefficients must be on the *right!*

The results of section 6.5 on perspectivities are valid for projective spaces over noncommutative fields. In particular, the proof of theorem 6.29 is valid for any projective space derived from a left vector space over a noncommutative field. Minor modifications are required for right projective spaces. The existence of perspectivities shows that Desargues theorem is valid in projective spaces (especially planes!) over noncommutative fields. In a projective space over a "not necessarily commutative" field, the theorem of Pappus (exercise 6.4, problem 7) is the geometric equivalent of the commutativity of multiplication.

In passing to projective transformations, we first note that since we write linear transformations on the right, it is more convenient to have the scalars on the left, for then the homogeneity of a linear transformation appears as an associative property and meshes nicely with products. Thus, we focus our attention primarily on left vector spaces from now on. If f is a nonsingular linear transformation of a left vector space **V**, then, as in section 6.6, it induces a collineation, $\bar{f}$, of $\bar{V}$, and the mapping sending f to $\bar{f}$ is a homomorphism. In the noncommutative case, not all scalar transformations are linear transformations! Although it is true that $\bar{f} = \mathbf{1}$ if and only if there is a scalar λ such that $(v)f = \lambda v$ for all vectors v, the mapping f, defined by $(v)f = \lambda v$ for all v, is linear on a left vector space if and only if λ commutes with every scalar α, i.e., if and only if λ is in the center of **F.** If λ is not in the center, then the mapping f is only *-linear with * the inner automorphism induced by λ^{-1} (exercise 6.12, problem 3). The proof (as presented!) of theorem 6.34 is valid in the noncommutative case so the fundamental theorem for projectivities (theorem 6.35) is still true in the noncommutative case. Projectivities of order 2 are a little more awkward in the noncommutative case (see [6], pp. 8–10 and 13) since a scalar may be a perfect square in **F** but not in the center of **F,** e.g., -1 is a perfect square in the quaternions but not in the reals.

A collineation of a projective space over a noncommutative field (dimension at least 2) is induced by a semi-linear map as in section 6.7; however, the automorphism of **F** is only determined up to an inner automorphism in agreement with the comments above about the identity collineation. For a thorough discussion of the various groups and homomorphisms between groups involved in the study of collineations and the associated automorphisms of **F**, we highly recommend a careful study of the commutative diagram (with exact rows and columns) on page 93 of [1].

The unexpected interchange of left and right as we pass from homogeneous coordinates to homogeneous linear equations is simply the coordinate manifestation of the same interchange as we pass from **V** to its dual space. If **V** is a left vector space over a noncommutative field, then $\mathbf{V}^t$, the set of linear functionals, is in a very natural way a right vector space but never, at least not in a natural way, a left vector space. For once again, if f is a linear functional, then the transformation, αf, defined by $(v)\alpha f = \alpha(v)f$ is *-linear with * the inner automorphism induced by α^{-1}. Since **F** is also a right vector space, we can define, for a linear functional f, the map $f\alpha{:}\mathbf{V} \rightarrow \mathbf{F}$ by $(v)f\alpha = ((v)f)\alpha$; moreover, $f\alpha$ is still linear. With this interchange of left and right, "duality" is recovered, except that the dual of a left projective space is always a right projective space, see [2], pp. 27–33. This switch also causes trouble with semi-bilinear forms, and the only natural way to define a *-form on a left vector space is with respect to an anti-automorphism, *, of **F**, i.e., $(\beta\alpha)^* = \alpha^*\beta^*$. The definition of a semi-bilinear form in the noncommutative case is as follows.

Definition. A *semi-bilinear form* on a vector space over a noncommutative field **F** is a function $b{:}\mathbf{V} \times \mathbf{V} \rightarrow \mathbf{F}$ such that b is additive in both variables and for some anti-automorphism, *, of **F**, $(\lambda v, w)b = \lambda(v, w)b$ and $(v, \lambda w)b = (v, w)b\lambda^*$ for all $\lambda \in \mathbf{F}$ and all $v, w \in \mathbf{V}$.

If the dimension is at least 2, then every correlation of $\bar{V}$ is represented by a semi-bilinear form; in particular, the field **F** must possess at least one anti-automorphism in order for any correlations of $\bar{V}$ to exist, [2], pp. 95–106.

Exercise 6.12

1. *The field of quaternions.* Let $\mathbf{F} = \{a + bi + cj + dk \mid a, b, c, d \text{ real numbers}\}$. Define addition in **F** in the natural way, i.e., $(a + bi + cj + dk) + (a' + b'i + c'j + d'k) = (a + a') + (b + b')i + (c + c')j + (d + d')k$. Define multiplication in **F** so that it is compatible with the usual multiplication for real numbers, the associative and distributive laws, and the special products $i^2 = j^2 = k^2 = -1$, $ij = k, jk = i$, and $ki = j$.

(1) Show that $ji = -k$, $kj = -i$, and $ik = -j$.

(2) Find two concrete models of the quaternions, one using 4×4 matrices

with real entries, and the other using 2×2 matrices with complex entries. This proves that the requirements above are not contradictory!

(3) Show that $*:\mathbf{F} \to \mathbf{F}$, defined by $(a + bi + cj + dk)^* = a - bi - cj - dk$, is an anti-automorphism of **F** and that $** = \mathbf{1}$.

(4) Find the center of **F** and show that xx^* is always in the center of **F**.

(5) Using the two preceding results, find a simple method to compute x^{-1} for any quaternion $x \neq 0$.

(6) Show that the anti-automorphism, *, above is not an inner automorphism of **F**.

2. Let $\mathbf{V}_n(\mathbf{F})$ be the two-sided vector space of n-tuples over a noncommutative field **F**. Show that $n > 1$ if and only if there are vectors in $\mathbf{V}_n(\mathbf{F})$ that are left proportional but not right proportional.

3. Let $f:\mathbf{V} \to \mathbf{W}$ be a linear transformation of a left vector space **V** into a two-sided vector space **W**, both over the same noncommutative field **F**.

(1) Given $\lambda \in \mathbf{F}$, show that the mapping $\lambda f:\mathbf{V} \to \mathbf{W}$ defined by $(v)\lambda f = \lambda(v)f$ is *-linear if * is the inner automorphism of **F** induced by λ^{-1}. When is λf linear?

(2) Given $\alpha \in \mathbf{F}$, show that $f\alpha:\mathbf{V} \to \mathbf{W}$ defined by $(v)f\alpha = ((v)f)\alpha$ is linear.

Bibliography and suggestions for further reading

[1] Artin, E., *Geometric Algebra*, Interscience Tracts in Pure and Applied Math., No. 3, Wiley, New York, 1957.

[2] Baer, Reinhold, *Linear Algebra and Projective Geometry*, Pure and Applied Math. Series, Vol. 2, Academic Press, New York, 1952.

[3] Baer, Reinhold, "Projectivities with fixed points on every line of the plane," *Bull. Am. Math. Soc.* **52,** 273–286 (1946).

[4] Brauer, Richard, "A characterization of null systems in projective spaces," *Bull. Am. Math. Soc.* **42,** 247–254 (1936).

[5] Coxeter, H. S. M., *Projective Geometry*, Blaisdell, Boston, 1964.

[6] Dieudonné, Jean, "On the automorphisms of the classical groups," *Memoirs Am. Math. Soc.*, No. 2, (1951). Reprinted by University Microfilms, Ann Arbor, Mich., 1964.

[7] Gleason, Andrew M., "The definition of a quadratic form," *The Am. Math. Monthly* **73,** 1049–1056 (1966).

[8] Hall, Marshall, *The Theory of Groups*, Macmillan, New York, 1959.

[9] Hall, M., Swift, J. D., and Killgrove, R., "On projective planes of order nine," *Math. Tables and Other Aids to Computation* **13,** 233–246 (1959).

[10] Halmos, Paul R., *Finite-Dimensional Vector Spaces*, 2nd. ed., Van Nostrand, Princeton, N.J., 1958.

[11] Killgrove, R. B., "Quadrangles in projective planes," *Can. J. Math.* **17,** 155–165 (1965).

[12] Mostow, George D., J. H. Sampson, and Jean-Pierre Meyer, *Fundamental Structures of Algebra*, McGraw-Hill, New York, 1963.

[13] Pickert, Gunter, *Projektive Ebenen*, Vol. 80 of *Die Grundlehren der Mathematischen Wissenschaften*, Springer, Berlin, 1955.

[14] Schreier, Otto, and Emanuel Sperner, *Projective Geometry of n Dimensions*, Vol. 2 of *Introduction to Modern Algebra and Matrix Theory*, Chelsea, New York, 1961.

[15] Seidenberg, A., *Lectures in Projective Geometry*, Van Nostrand, Princeton, N.J., 1962.

[16] Veblen, Oswald, and J. W. Young, *Projective Geometry* Vols. 1 and 2, Ginn, New York, 1910. Also published recently by Blaisdell as a paperback.

[17] Wesson, J. R., "An elementary construction of a finite nonarguesian projective plane," *The Am. Math. Monthly* **73,** 183–186 (1966).

The two basic references for this chapter are the books by Artin and Baer, [1] and [2]. Artin's terminology in chapter 3 of [1] is designed to fit the study of bilinear and quadratic forms rather than quadrics in projective spaces; however, his results are stated in a "geometric" fashion, and their applications to projective spaces are easily perceived. Baer's account of projective spaces is authoritative and detailed. It is designed to cover both the commutative and noncommutative cases. Be careful about changing back and forth between Baer's terminology and that used here. A glossary of some of the more troublesome terms is as follows:

	Baer	*This text*
	rank	linear dimension
	dimension (if rank is finite)	projective dimension
	point	vector or linear point
	line	projective point
	plane	projective line
	adjoint space	dual space (linear)
Caution!	projectivity	collineation
	collineation	projective transformation
	auto-duality	correlation
	polarity	correlation of order 2
	$\mathbf{W}$ is an N-subspace of ϕ	ϕ is symplectic on $\overline{W}$
	strictly isotropic	isotropic

Brauer's derivation in [4] of the fundamental results concerning collineations, correlations, and null systems uses coordinates rather heavily; however, it is easy to understand and complements our treatment very well.

It would be appropriate to list all of Coxeter's excellent books on geometry. We chose the most recent, [5], to mention here and commend in particular the results in chapters 7, 8, and 9 on conics and polarities and chapter 10 on the finite plane of order 5.

In the highly condensed, technical monograph, [6], Dieudonné covers in the first few sections all of our results concerning collineations and correlations. He then proceeds to study not only the classical groups, i.e., the groups of linear transformations leaving a certain bilinear or quadratic form invariant, but also the automorphisms of these groups. Thus, in a certain sense, this book is a very tame introduction to his monograph.

For lack of space, we only touched on bilinear forms and quadratic forms when we discussed quadrics. Both [10] and [12] contain useful information on these important topics. For an elegant discussion of intrinsic ways to define quadratic forms without leaning on bilinear forms, we recommend [7].

The subject of projective planes has a vast literature all its own. The encyclopedic work, [13], by Pickert contains an extensive bibliography. Among the finite projective planes, those of order 9 are especially interesting, *cf.* [9], [11], and [17]. The last chapter of Hall's book on group theory, [8], is devoted to those aspects of algebra useful in studying projective planes, especially finite projective planes. Seidenberg's text contains a readable (and teachable!) account of the way Desargues theorem, the theorem of Pappus, the fundamental theorem, and the introduction of coordinates in projective planes are related to each other.

By restricting themselves to $P_n(\mathbf{R})$ or $P_n(\mathbf{C})$ in [14], Schreier and Sperner present collineations and correlations in an easy to understand manner. The affine classification of quadrics (called hypersurfaces of the second order) is carried out in detail in their book.

No bibliography on projective geometry can omit the classic, [16], by Veblen and Young. These two volumes should be in the library of (and read by) every serious student of projective geometry.

Term-paper topics

There are many excellent topics for term papers buried in the suggestions for reading above. Among our favorites are the following:

(1) Read Artin's proof of Witt's theorem on extending "isometries" in Chapter 3 of [1], and prove that every point on a quadric lies on a subspace entirely in the quadric and whose dimension is the index of the quadric.

(2) Investigate projective planes of order 9 and discuss their collineations and correlations. For a *much* simpler topic, replace 9 by 4.

(3) Discuss the field of quaternions. There are expository articles available, one in the MAA *Studies on Algebra* and another by Coxeter in a back issue of the *Monthly*. We are purposely vague with our references here. Include a proof that the only automorphisms of the quaternions are inner automorphisms and the relevance of this for projective spaces over the quaternions.

(4) Discuss the affine classification of quadrics.

(5) Discuss conics in the real projective plane.

(6) Discuss noncommutative fields of characteristic p. (*Monthly* April, 1962)

Index

Abelian group, 4
Affine form of a linear transformation, 181
Affine space, 159
Affine subspace, 161
Affine transformation, 175
Aleksandrov, A. D., 199
Algebraic element over a field, 121
Allendoerfer, Carl B., 83
Alternating group of degree n, 23
Alternating $2q$-prism, 97
Annihilator, 254
Anti-representation, 23
Artin, Emil, 83, 210, 278
Associative product, 4
Automorphism, of a field, 187
  of a group, 24
Axis, of a half-turn, 52
  of a perspectivity, 224
  of a rotation, 57
  of a screw displacement, 58

Bachmann, Friedrich, 83
Baer, Reinhold, 278
Basic parallelepiped, 101
Basis of a vector space, 126
Beaumont, R. A., 83
Beckenbach, Edwin F., 43
Bilinear forms, 260
Birkhoff, Garrett, 43, 118, 157, 199
Body-centered lattice, 113
Bounded set, in an affine space, 191
  in euclidean space, 86
Box, 189
Brauer, Richard, 278
Bravais lattice, 113
Buckley, J. T., 199
Burckhardt, J. J., 118
Burnside, William, 37, 43

Canonical basis, 129
Cardinality of a set, 2
Cassels, J. W. S., 118
Cauchy sequence, 88
Cayley representation theorem, 22
Cayley-Hamilton theorem, 153
Center, of a dilatation, 68
  of a group, 28
  of an inversion, 52
  of a pencil of lines, 212
  of a perspectivity between lines, 212
  of a perspectivity of space, 228
Central collineation, 225
Central dilatation, affine, 172
  euclidean, 68
Centralizer, 36
Characteristic, of an affine space, 176
  of a field, 122
  polynomial, 149
  subset of a group, 28
  subspace, 149
  value, 149
  vector, 149
Choice principle, 39
Class equation, 33
Codomain, 2

Collineation, affine, 184
euclidean, 52
projective, 226
Commutative diagram, 142
Commutative product, 4
Commutator, 28
Complementary subspaces, affine, 177
projective, 241
vector, 128
Complete group, 27
Conic, 262
Conjugacy class in a group, 13
Conjugate points in a polarity, 269
Conjugate subgroups, 15
Contraction mapping, affine, 197
general, 50
Convex set, 188
Coordinate mapping, affine, 170
projective, 215
vector, 138
Correlation, 258
Countable set, 87
Coxeter, H. S. M., 43, 83, 118, 199, 278
Crystal class, 106, 114
Crystal system, 109
Crystallographic group, 114
Crystallographic point group, 103
Crystallographic restriction, 104
Crystallographic space group, 114
Cubic crystal system, 112
Cubic lattice, 102, 103
Curie, P., 91, 118
Cycle notation for permutations, 10
Cyclic group, 5

Danzer, Ludwig, 199
De Bruijn, N. G., 43
Degenerate quadric, 264
Degree, of an element algebraic over a field, 121
of a representation of a group, 21
of a symmetric group, 9
Desargues' theorem, 205–207
Determinant, of a lattice, 101
of a matrix, 149
Dieudonné, Jean, 157, 278
Dihedral group, 93
Dilatation, affine, 171
euclidean, 68
Dilatation ratio, 174
Dimension, of an affine space, 164
of a lattice, 99, 194
of a projective space, 202, 203
of a vector space, 127
Direct similarity, 60
Directing space of an affine space, 161
Direction, of an affine reflection, 178
of a translation, 57
Discrete group, affine, 192
euclidean, 87
Displacement, 62
Domain, 2
Dual basis, 253
Dual figures, 93
Dual space of a projective space, 254
Dual theorems in projective geometry, 213
Duality, linear, 254
principle of, 212
projective, 255

Ebey, Sherwood, 188
Echelon form, 129
Eigenvalues, 149
Elation, 228
Elliptic projectivity, 243
Enantiomorphic space groups, 115
End-centered lattice, 113
Escher, M. C., 83
Euclidean group, 46
Eulerian angles, 84
Exactly transitive, 159

Face of a parallelepiped, 191
Face-centered lattice, 102, 113
Faithful representation, 20
Fiber group, 117
Field, 121
Field extension, 121
Fixed, globally, 11
hyper-, 227
linewise, 227
pointwise, 11
Fomin, S. V., 83
Full symmetric group, 9
Function, 2
Fundamental domain, 101
Fundamental theorem of geometry, affine version, 171
classical projective version, 236
version for projectivities, 236
version for collineations, 249

Gardner, Martin, 199
General linear group, 154
General position, in an affine subspace, 164
with respect to a group of isometries, 77
Generators, of an affine subspace, 163
of a projective subspace, 209
of a subgroup, 5
of a vector subspace, 126
Gleason, A. M., 278
Glide reflection, 58

Globally invariant, 11
Golomb, S. W., 43
Graph, of a homogeneous system in a projective space, 221
of a linear system in an affine space, 170
Grassmann space, 244
Group, 4
of the cube, 12
of the square, 16
Grünbaum, Branko, 199
Guggenheimer, Heinrich, 83, 118

Half-turn, 52
Hall, Marshall, 43, 278
Halmos, Paul R., 278
Harmonic conjugate, 239, 241
Harmonic homology, 242
Hemihedry, 112
Henry, N. F. M., 118
Herstein, I. N., 43, 157
Hexagonal crystal system, 112
Hilbert, David, 118
Hirsch, K. A., 43
Holohedry, 109
Homogeneous coordinates, 214
Homogeneous equations, 220
Homogeneous mappings, 48
Homology, 228
Homomorphism, of a group, 18
of a ring, 122
Homothetic transformation, 68
Hyperbolic projectivity, 243
Hyper-fixed, 227
Hyperplane, affine, 164
at infinity, 202
projective, 202

Icosahedral group, 93
Identity element in a group, 4
Identity function, 3
Improper congruence, 62
Independent points, affine, 164
euclidean, 46
projective, 216
Independent vectors, 126
Index, of a quadric, 270
of a subgroup, 34
Initial vertex of a parallelepiped, 189
Inner automorphism, 25
Integral basis of a lattice group, 193
Integral matrix, 195
International Tables of X-Ray Crystallography, 116, 118
Invariant, globally, 11
pointwise, 11
subset of a group, 28
Inverse element in a group, 4
Inverse function, 3
Inversion with center X, 52
Involution, affine, 176–178
euclidean, 52
linear, 135–137
projective, 238–243
Isometry, 46
Isomorphism, 20
Isotropic, 270

Join, of two projective subspaces, 209
of two subgroups, 7

Kernel, of a homomorphism, 18
of a linear transformation, 132
Killgrove, R., 278
Klee, Victor, 199
Kolmogorov, A. M., 83, 199
Kurosh, A. G., 43

Lagrange's theorem, 33
Lang, Serge, 199
Latent values, 149
Lattice, 99, 192
body-centered, 113
body-centered cubic, 102, 103
end-centered, 113
face-centered, 113
face-centered cubic, 102
primitive, 112
primitive cubic, 102
Lattice group, 99, 192
Lavrent'ev, M. A., 199
Le Corbeiller, Phillipe, 118
Length of a translation, 57
Levine, Jack, 157
Lightstone, A. H., 157
Limit point, 87
Linear combination, 125
Linear functional, 253
Linear transformation, 132
Linearly dependent, 126
Linewise fixed, 227
Lomont, J. S., 119
Lonsdale, K., 118

MacGillavry, Caroline H., 83
MacLane, Saunders, 43, 157, 199
Mapping, 2
Matrix of a linear transformation, 140
Matrix representation, 21
Meyer, Jean-Pierre, 199, 278
Midpoint, 177
Miller, D. W., 43
Minimal polynomial, 121

Mirror, of an affine reflection, 178
of a euclidean reflection, 52
of a perspectivity, 225
Mitchell, Barry, 157
Modenov, P. S., 83
Monoclinic crystal system, 112
Moser, W. O. J., 43
Mostow, George D., 199, 278
Motions, 62

Nahikian, H. M., 157
Natural basis, 126
Natural matrix, 139
Net group, 117
Nomizu, K., 199
Noncommutative field, 275
Nonsingular linear transformation, 154
Nonsingular semi-linear mapping, 187
Nontrivial permutation, 8
Normal subgroup, 20
Normal subset, 28
Normalizer, 33, 36
Null system, 267
Nullity of a linear transformation, 221

Octahedral group, 93
One-to-one correspondence, 2
One-to-one function, 2
Onto function, 2
Opposite similarity, 60
Orbit, 34
Order, of an element in a group, 6
of a group, 7
of a projective plane, 213
Ordered basis, 126
Oriented space group, 115
Origin, for affine coordinates, 169
in $A_n(\mathbf{F})$, 160
Orthogonal complements, 129
Orthogonal group, 79
Orthorhombic crystal system, 112
Outer automorphism, 25

Parker, E. T., 43
Parkhomenko, A. S., 83
Paired vector spaces, 256
Pappus, theorem of, 223
Parabolic projectivity, 243
Parallel axiom, the, 167
Parallel subspaces, 167
Parallelepiped, 189
Parallelotope, 189
Permutation, 8
Permutation representation, 20

Pencil, of lines, 212
of planes, 62
Perspectivity, from m_1 to m_2, 212
of a projective space, 224
Pickert, Gunter, 278
Point group, crystallographic, 103
of a crystallographic space group, 114
of a group of isometries, 77
Points, affine, 159, 201
ideal, 201
projective, 201
Pointwise fixed or invariant, 11
Polar, 269
Polarity, 269
Pole, of a group of isometries, 91
with respect to a quadric, 269
Polya, G., 37, 43
Polya-Burnside theorem, 37
Position vector of a point in $A_n(\mathbf{F})$, 160
Prime edges of a parallelepiped, 189
Prime field, 122
Primitive cell of a lattice, 101, 194
Primitive lattice, 112
Principle of contractive mappings, 50
Principle of duality, 212, 256
Prism, 95
Projection, 135
of $\mathbf{V}$ onto $\mathbf{W}_1$ along $\mathbf{W}_2$, 135
Projective correlation, 261
Projective group on $\overline{V}$, 235
Projective plane, 210
over a field, 210
Projective transformation, 234
Projectivity, 234
classical, 237
induced by a linear mapping, 235
Proper motion, 62
Proper quadric, 264
Proper root, 149
Pyramid, 95

Quadric, 262
degenerate, 264
induced by a polarity, 269
proper, 264

Range of a function, 2
of a linear transformation, 132
Rank of a linear transformation, 132
Reflection, affine, 178
euclidean, 52
Regular 2-gon, 95
Representation of a group, 20
right regular, 23
Reversal, 62
Rhombohedral system, 116

Right coset, 31
Rigid motion, 62
Rotation, 57
Rotatory inversion, 58
Rotatory reflection, 58
Rotman, J. J., 43

Sampson, J. H., 199, 278
Satz von den drei Spiegelungen, 64
Scalars, 125
Schiff, L. I., 119
Schoenflies symbols for space groups, 116
Schreier, J., 43
Schreier, Otto, 279
Screw displacement, 58
Segal, I. E., 43
Seidenberg, A., 279
Self-conjugate point with respect to a polarity, 269
Semi-bilinear form, 259
  noncommutative case, 277
Semi-direct product, 114
Semi-linear mapping, 187
Shaved prism, 96
Shear, 180
Similar groups, 90
Similar matrices, 147
Similarities group, 46
Similarity, 46
Simple group, 78
Skew subspaces, 168
Snapper, Ernst, 199
Space group, 114
Spectral values, 149
Sperner, Emanuel, 279
Spiral similarity, 71
Standard n-space, affine, 160
  projective, 214
  vector, 126
Stretching factor of a similarity, 49
Strip group, 117
Subfield, 121
Subgroup, 4
Subspace, affine, 161
  projective, 202
  vector, 125
Subspace of degeneracy for a quadric, 266
Sum of two vector subspaces, 128
Swift, J. D., 278
Symmetric group of degree n, 9
Symmorphic, 114
Symplectic, 266

Tangent line to a quadric, 274
Tensor product, 261
Tetartohedry, 112
Tetragonal crystal system, 112
Tetrahedral group, 93
Thomsen, Gerhard, 83
Trace of a matrix, 149
Transcendental element over a field, 121
Transformation, 2
Transitive constituents or sets, 34
Transitive group, 34
Transitivity axiom for an affine space, 159
Translation, 57, 159
Translation group of an affine space, 159
Transpose of a matrix, 156
Transvection, 180
Triclinic crystal system, 112
Trigonal crystal system, 112
Trivial cases of parallelism, 167
Trivial permutation, 8
Twisted q-prism, 96

Ulam, S., 43
Unimodular affine group, 180
Unimodular matrix, 195
Unit cell of a lattice, 112, 194
Unit point, of an affine coordinate system, 170
  of a projective coordinate system, 217
Unit vectors of an affine coordinate system, 170

Valentine, F. A., 199
Veblen, Oswald, 279
Vector axiom for an affine space, 159
Vector from P to Q, 159
Vector space over **F**, **F** commutative, 125
  F noncommutative, 275
Vertices of a projective coordinate system, 217
Volume, of a box, 190
  of a lattice, 101, 194
Von Staudt theory of conics, 271

Wedderburn's theorem on finite fields, 275
Welch, L. R., 43
Wesson, J. R., 279
Weyl, Hermann, 43, 83, 119, 159, 199
Wielandt, Helmut, 43

Young, J. W., 279

Zuckerman, H. S., 83

Index of Notation

Bracket conventions

$\{a_1, a_2, \ldots, a_n\}$	set containing $a_1, a_2, \ldots, a_n$, 1
$(a_1, a_2, \ldots, a_n)$	affine coordinates, 160, 169
	permutation cycle of length n, 10
	cyclic subgroup generated by a_1, if $n = 1$, 5
	ordered n-tuple, 2
$\langle a_1, a_2, \ldots, a_n\rangle$	vector in $\mathbf{V}_n(\mathbf{F})$, 126, 160
$[a_0, a_1, a_2, \ldots, a_n]$	projective point in $P_n(\mathbf{F})$, 214

Functions and sets

$\#(X)$	cardinality of the set X, 2
$(x)f$ or x^f or x_f	image of x under the function f, 2
$(Z)f$ or Z^f or Z_f	set of images under f of elements in the set Z, 2
fg	composite of the functions f and g, first apply f, then g, 3
$\mathbf{1}$ or $\mathbf{1}_X$	identity function on the set X, 3
Y^X	set of all functions with domain X and codomain Y, 7
$f\colon X \to Y$	f is a function with domain X and codomain Y, 8

Groups, subgroups, and automorphisms

G, $\mathbf{G}$	$\mathbf{G}$ is a group consisting of the set G and a product on G, 4
g^i	inner automorphism, $(x)g^i = g^{-1}xg$, induced by g, 25
$\mathbf{H}g$	right coset of $\mathbf{H}$ containing g, 31

$[\mathbf{G}:\mathbf{H}]$	index of the subgroup **H** in **G**, 34
$\mathbf{A}(\mathbf{G})$	automorphism group of **G**, 24
(x)	cyclic subgroup generated by x, 5
$\mathbf{N}_g$ and $\mathbf{N}_S$	normalizer of g and S in a group, 33, 36
$\mathbf{Z}_S$	centralizer of the set S in a group, 36
$\mathbf{Z}_G$	center of the group **G**, 28
$\mathbf{G}'$	commutator subgroup of **G**, 28

Groups operating on a set

$\mathbf{S}_X$ or $\mathbf{S}_n$	group of all permutations of X or of $\{1, 2, \ldots, n\}$, 8, 9
$\mathbf{A}_n$	alternating group of degree n, 23
$[\mathbf{G}, Y; pi]$	subgroup of **G** leaving Y pointwise invariant, 11
$[\mathbf{G}, Y; gi]$	subgroup of **G** leaving Y globally invariant, 11
$[\mathbf{G}, x]$	subgroup of **G** leaving x invariant, 30
$(x)\mathbf{G}$	**G**-orbit of x, 34, 87
$X/\mathbf{G}$	set of **G**-orbits in X, 34

Euclidean symmetries

$\mathbf{S}$	similarities group, 46
$\mathbf{E}$	group of isometries, i.e., the euclidean group, 46
$\mathbf{G}^+$	group of direct similarities in **G**, **G** a subgroup of **S**, 61
x_α	central dilatation with center x and ratio α, 68
$\mathbf{D}$	group of dilatations, 68
$\mathbf{T}$	group of translations, 72
$\mathbf{C}_q, \mathbf{D}_q, \mathbf{T}, \mathbf{O}, \mathbf{I}$	finite groups of direct isometries, 94
$\mathbf{G}^+[\mathbf{H}$	a group derived from $\mathbf{G}^+$ and **H**, $\mathbf{G}^+$ of index two in **H**, 90, 95, 98

Vector spaces, subspaces, and symmetries

$\mathbf{V}_n(\mathbf{F})$	standard vector space of n-tuples over **F**, 126
$\mathbf{V}_S(\mathbf{F})$	vector space of functions with domain S and codomain **F**, 127
$\dim(\mathbf{W})$ or $\dim_F(\mathbf{W})$	dimension over **F** of the vector space **W**, 127
$\mathbf{W}_1 + \mathbf{W}_2$	sum of two vector subspaces, 128
$L(\mathbf{V}, \mathbf{W})$	space of linear mappings with domain **V** and codomain **W**, 134
$M_n(\mathbf{F})$	ring of all n by n matrices over **F**, 140

$GL_n(\mathbf{F})$ — general linear group of n by n matrices ($\det \neq 0$) over $\mathbf{F}$, 154

$SL_n(\mathbf{F})$ — special or unimodular ($\det = 1$) subgroup of $GL_n(\mathbf{F})$, 195

$\mathbf{V}^t$ — space of linear functionals on $\mathbf{V}$, 252

ϕ^t — coordinate map dual to the coordinate map ϕ, 253

$\mathbf{W}^a$ — annihilator of $\mathbf{W}$, 254

Affine spaces, subspaces, and symmetries

$A_n(\mathbf{F})$ — standard affine space of n-tuples over $\mathbf{F}$, 160

$(P)\mathbf{W}$ — affine subspace through P in the direction of $\mathbf{W}$, i.e., the $\mathbf{W}$-orbit of P, 162

$[B]$ — affine subspace generated by B, 163

$P + Q$ — affine line through P and Q, 163

ϕ^* — affine coordinate map, 169

p_μ — central dilatation with center p and ratio μ, 172

$\mathbf{D}$ or $\mathbf{D}_A$ — group of dilatations of the affine space A, 171

Projective spaces, subspaces, and symmetries

PA — projective space constructed by extending the affine space A, 201

$\bar{V}$ — projective space constructed by collapsing the vector space $\mathbf{V}$, 203

$[v]$ — projective point represented by the vector v, $v \neq 0$, 208

$\bar{U} + \bar{V}$ — join of two projective subspaces, 209

$[v] + [w]$ — projective line joining $[v]$ and $[w]$, 208

$P_n(\mathbf{F})$ — standard projective n-space over $\mathbf{F}$, 214

$[\phi]$ — projective coordinate map of $\bar{V}$ onto $P_n(\mathbf{F})$ induced by ϕ, ϕ a linear coordinate map of $\mathbf{V}$ onto $\mathbf{V}_{n+1}(\mathbf{F})$, 215

$\mathbf{D}_H$ — group of perspectivities with axis H, 231

$\mathbf{T}_H$ — group of elations with axis H, 231

g — projectivity induced by g, 235

$PGL_n(\mathbf{F})$ — projective general linear group, 235